Organic Light-Emitting Diodes

Luiz Pereira

Organic Light-Emitting Diodes

The Use of Rare-Earth and Transition Metals

PAN STANFORD PUBLISHING

Published by

Pan Stanford Publishing Pte. Ltd.
Penthouse Level, Suntec Tower 3
8 Temasek Boulevard
Singapore 038988

Email: editorial@panstanford.com
Web: www.panstanford.com

British Library Cataloguing-in-Publication Data
A catalogue record for this book is available from the British Library.

Organic Light-Emitting Diodes: The Use of Rare-Earth and Transition Metals

ISBN 978-981-4267-29-8 (Hardcover)
ISBN 978-981-4267-95-3 (eBook)

Printed in the USA

To my mother and my father…

…because without their stimulus,
encouragement, and persistence,
I would have never gone so far

Contents

Preface

The world of the organic light-emitting diodes (OLEDs) is currently a fascinating area spreading far beyond the simple laboratory research. In fact, the commercially available displays have projected this new technology to the "top ten" in materials science. Although for most people, displays are the most appellative application of OLEDs, there are many other areas of attraction, perhaps even more surprising, being discovered nowadays. These include energy-efficient lighting and the trendy, sought-after light decoration. Simultaneously, the concept of a truly flexible light emitter opens application in the fields never imagined before. In the early market technological products, polymeric materials were extensively used for almost all purposes, but currently the new production routes are looking for different organic electroluminescent materials, simultaneously able to deliver high color purity across the color spectrum, brightness, easy synthesis with good chemical and electrical stability, low cost, and simple processing. In this new strategy, complexes based on rare-earth (RE — lanthanides) and transition metals (TM) gain more importance because of their unique and special properties.

This book started to be written in the end of 2008, precisely when a few important companies were launching OLED displays of large emitting areas. Still, none of them employed the RE or TM complexes. The commercial application of these complexes remains open, and the scientific exploration of these materials is viewed as the next major break/opportunity. So, the specificity and timelines of this book are self-evident. The initial approach is to briefly describe how to fabricate an OLED, based on a physical point of view. The common perception is that this question is often overlooked by most people starting their work in this field, although the basic notions are simple to understand and will clearly contribute to the success. Some fabrication processes are effortlessly explained, showing the main advantages and disadvantages; next, the most common electrical and optical characterization techniques of an OLED are described in a clear and straightforward way. These discussions are clearly intended to help new starters, in particular undergraduate students, understand the basics of OLEDs.

The main part of the book is devoted to the application of RE and TM complexes in the fabric of a new class of organic light emitters. The reasoning starts by relating simple physics concepts with the molecular structure of the materials and with their luminescence properties, then proceeding to their application in OLED active layers, explaining the choice of each device structure, on the basis of previous discussions. These are illustrated by many practical examples. The book ends with a concise description of the basic synthetic routes for RE and TM complexes, interfacing with the field of chemistry, but completing the book strategy.

With this focus, I am convinced that the information and explanations presented in this book will be immensely useful not only to students but also to scientists and researchers looking for new metallo-organic complexes and their specific applications, always with help of the comprehensive schemes and figures for a better understanding. Also, this book is expected to help open further pathways in both fundamental science and applied research.

Luiz Pereira

Acknowledgments

Many people have contributed, even if indirectly, to this book, especially colleagues and students, with laboratory work and many fruitful discussions. First of all, I wish to acknowledge Dr. Susana Braga (CICECO, University of Aveiro, Portugal), my researcher colleague, for several important things: the help with the manuscript English revision, chemical synthesis of some coordination complexes used, and, of course, for her contribution in Chapter 6 with a concise and simple description of the methods for the basic chemical synthesis of the rare earth and transition metal complexes; my former (unofficial!) PhD student and great friend, Dr. Gerson Santos (University of S. Paulo, Brazil) for exceptional laboratory work; my colleagues Dr. Fernando Fonseca and Dr. Adnei Andrade (from the University of S. Paulo) for interesting discussions and friendship; my colleagues (and their students) from various chemistry departments across Brazil and Portugal for the synthesis of materials , in particular, Dr. Neyde Ilha (University of S. Paulo), Prof. Leni Akcelrud (Federal University of Paraná, Brazil), and Dr. Isabel Gonçalves (University of Aveiro); my colleague, Dr. Luís Rino, for discussions on photo-physics; naturally, all of my former and present students who work or have worked in the laboratory, developing new devices structures useful for organic based devices and testing new materials: Wilson Simões, António Trindade, Tiago Simões, Licínio Ferreira, Mauro Gonçalo, Ricardo Pinho, Daniel Duarte, Carina Ferreira, João Corte-Real, João Costa, and Cláudia Martins; and new students starting now, who will bring further developments. Without their "network" of contributions to the Laboratory of Organic Semiconductors, however small and indirect, I believe we would have never achieved this fast our current level of knowledge in this field. To all of you, many thanks.

Chapter 1

Introduction

The world of organic light-emitting diodes (OLEDs) represents one of the most interesting areas of scientific research in the field of optoelectronic devices. The strong efforts made in recent years and the current OLEDs scientific state of the art have led to a fast transition from the academic research to the market, although there are still some problems to overcome, such as the low life span of these products compared with the traditional ones. Meanwhile, the scientific community still considers that the organic electroluminescent materials and devices will be the next generation of light emitters. The applications include all of the known products in the market, ranging from displays to general solid-state lighting, passing to the artistic and architectural environment utilization. Such wide-span use is only possible due the exceptional specifications of the organic electroluminescence materials that allow us to have a really thin (nanometer-scale) emitter, suitable for incorporation virtually anywhere. Additionally, nowadays new organic materials are presently being synthesized, opening even further possibilities for the field.

1.1 OLEDs: Background and Basic Facts

Since the discovery of the electroluminescence properties of anthracene in the 1960s [1], followed by the first low-driving-voltage

Organic Light-Emitting Diodes: The Use of Rare-Earth and Transition Metals
Luiz Pereira

www.panstanford.com

OLED in the 1980s [2] and the first OLEDs based on lanthanides elements (europium and terbium) [3] and very bright OLEDs based on transition metals phosphors [4] in the 1990s [3], much work has been done, as market applications shows. In fact, the research work on such devices has increased exponentially over the past 30 years [5–9].

An OLED is a light emitter with a thickness of about 100–200 nm, i.e., on average, a thousand times thinner than a human hair. An OLED device has a total thickness of about some millimeters where practically 100% comprises the support (substrate) and the encapsulation structure. It is self-emitter, i.e., it doesn't need a background light. For the same brightness as that of the best inorganic light emitter in the market, it consumes quite less energy; it is also more efficient and has a viewing angle nearing 170°. So, what else is needed to convince the research centers, both academic and industrial, to take their stake in such technology? Quite honestly, nothing else. Evidence is clear, and the current investment proves it. Obviously, there is the other side of the coin: their long-time stability is still unsatisfactory. On the other side, a "quasi-infinite" set of organic electroluminescence complexes has been discovered, allowing for new developments.

This book focuses on the two major areas of OLEDs that cover the use of rare-earth and transition metals complexes with organic ligands as part of active layers for the electroluminescent device. These kinds of complexes are very interesting for device application as they permit either a "quasi-monochromatic" emission or a strong emission, two distinct properties that are both very useful for specific applications. Moreover, a wide range of visible colors is accessible.

These exceptional qualities of the materials have been extensively studied under photoluminescence, but the material incorporation into the active devices, although much desired, poses, in many situations, a huge problem. The active layers, when used alone, are not suitable for light emission due to the poor carrier confinement; in addition, the choice of correct hole and/or electron transport layers for "sandwich" structures is sometimes complex because some of the electroluminescent and/or the corresponding energy levels are not compatible with those of metallo–organic complexes; finally some of these complexes, specially the transition metal ones, can be processed by only a wet process such as spin coating, and sometimes need to be blended in a blend with a polymer (that may bring an additional intrinsic electroluminescence). All of these issues pose serious constrains to the achievements of an efficient final device and represent a challenge to the researchers desiring to achieve the best and, simultaneously, simplest way to produce a viable device.

Given the high interest in the specific applications of OLEDs, the scientific community still works and concentrates many efforts in the fundamental research, searching for new strategies, materials, device structures and their physical, optical and electrical characterization. Concerning materials processing, new approaches are currently under development for the optimizations of the thin-film deposition routes.

These devices also offer several advantages over conventional light emitters, namely low power consumption, high efficiency, large areas of display and the very attractive possibility of flexible devices (impossible for any other light-emitting materials).

The main purpose of this book is to give an idea about the luminescent materials properties, the techniques of device fabrication, the optical and electrical characterization and how to get feedback for device improvement, and finally, the potential applications. As a more physical approach, some insights into chemistry (for better understanding the materials properties and how they are processed) and electronics (the final applications) are shared, leading to a very wide "picture" that, we hope, will be useful for any scientist and researcher working on this field.

1.2 State of the Art

The wide range of OLEDs applications, from displays to solid-state lighting, back-lighting and others, are clearly attractive for the industry. Over the past 5 years, a tremendous progress has been made in all aspects of OLED production, including materials, device architectures and manufacturing processes.

The recent developments in OLED market already include all of the known types, i.e., the OLED itself, polymer light-emitting diode (PLED), white OLED (WOLED), and light electrochemical cell (LEC). Yet the strongest insight in the market was undoubtedly the introduction of an OLED television for the household consumer by Sony at the beginning of 2009. It is also noteworthy that at least since 2005, small displays have been used and commercialized in cell phones, mp3 players, etc. The main constraint for the OLED commercialization (the lifetime) was overcome by the industry with the development of an efficient encapsulation. Unfortunately, a similarly effective encapsulation technique was not yet achieved with market compliance for the flexible OLEDs, although some interesting approaches are currently under development and perhaps a solution will arise in a short time.

Figures 1.1 and 1.2 present examples of specific applications of OLEDs in gadgets and household lighting, respectively.

Figure 1.1 Example of possible applications of OLEDs, with a flexible OLED prototype display by Universal Display. It is a full-color display. Courtesy of Universal Display.

Presently, the market looks for new, cheap, and efficient luminescent organic materials, and also for simple and efficient devices. This causes some of the known scientific results, although quite interesting, to be unviable for industrial-scale production. For example, the WOLED produced by the European consortium OLLA[a] is very appealing but comprises dozens of layers in a

[a] Reports and information at http://www.hitech-projects.com/euprojects/olla/.

complex and very expensive structure. Thus, simpler solutions must be sought; for example, the metallo–organic complexes with "quasi-monochromatic" emissions in the primary spectrum colors can be obtained and joined or highly bright emissions can be obtained. In addition, these complexes are equally interesting to cover specific applications.

Figure 1.2 OLEDs in general lighting. Monochromatic (white or very near) OLEDs.

Solid-state lighting, for instance is one of the most attractive areas either for scientific and research groups or in the industrial field. In fact, this industry is constantly seeking new lighting solutions that are not only cheap and efficient but also aesthetically attractive and easy to incorporate into architectural structures. OLEDs, after their successful application in displays, are currently under consideration. There is a consensus in the industry that OLEDs have become economically viable and adequate for niche lighting applications. In the near future, as their core capabilities improve, they will ultimately become contenders for mainstream applications that require a wide area of emission rather than a single emission point (as on information displays). *NanoMarkets* estimated OLEDs to be the main light sources in the near future. The report predicts that the market share of OLED lamps will virtually grow from zero in 2007 to about $443 million of the overall $619 million lighting market until 2010. Although we are not quite there yet, the ongoing increase of OLED expression in the market is undeniable. In fact, a more recent study conducted by DisplaySearch forecasted the total OLED display market to grow up to $7.1 billion by 2016, from $0.6 billion in 2008. The same source reports that the worldwide active matrix (AM) OLED revenue surpassed the passive matrix (PM) OLED for the first time. Figure 1.3 shows the DisplaySearch OLED Display Revenue Forecast expectations.

Potential applications for OLEDs, besides the displays, include back-lighting, general-purpose illumination, architectural and specialized industrial lighting, vehicular lighting, and signage. And in these specific applications, the RE and TM complexes may bring a remarkable improvement to the traditional PLEDs and can be used in the AMOLED structure. In fact, the RE-based OLEDs combine very low power consumption with a high brightness and a "quasi-monochromatic" emission, whereas the TM-based OLEDs exhibit a higher power requirement but with a very strong brightness in relatively large spectral emission bans. This means that it's actually not difficult to find the correct material for a specific application.

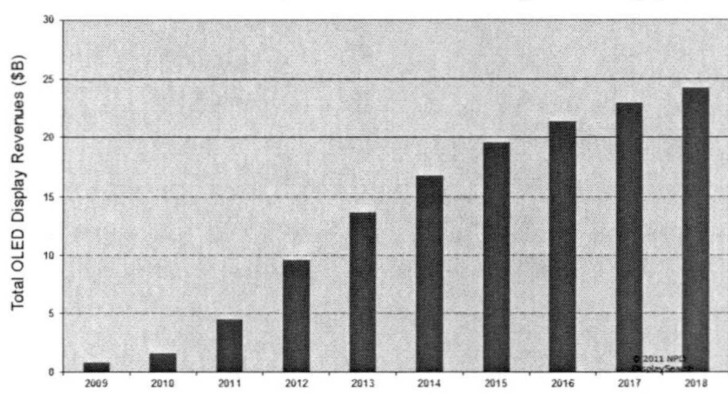

Figure 1.3 DisplaySearch OLED Display Revenue Forecast (Source: DisplaySearch Q2'09 Quarterly OLED Shipment and Forecast Report)

However, there are still several scientific issues to be improved before these devices are suitable for market launching. The most relevant are (i) the lifetime, as not all of the current available encapsulation technologies are suitable for these materials, (ii) the device architecture, and (iii) the optimization of the active layers, either with new materials or by improving the injection/ transport layers.

From the previous years' literature work on RE and TM organic complexes, we see that in RE complexes there are two elements of choice: europium and terbium; while in TM complexes, iridium is the most studied, followed by some ruthenium and rhenium complexes. Others RE and TM elements are also studied in order to be incorporated in OLEDs active layers. In the first group, samarium, thulium and dysprosium are the most interesting; in the second group the focus is on the platinum. Each of them exhibits proper and interesting characteristics.

Others have also been tested, but the main focus is on the aforementioned. Coordination complexes of europium and terbium ions are relatively easy to synthesize, feature nearly pure red and green emissions, respectively, and can be used with several organic ligands, which allows several device configurations. The very high internal efficiency of energy transfer must also be taken into account. On the other hand, iridium, ruthenium, and rhenium complexes show strong emissions covering all the spectrum colors, green color, and red color, respectively, and —like Eu and Tb — may bind to a plethora of organic ligands in a relatively easy chemical synthesis.

Different approaches are often used in the RE- and TM-based OLEDs, either in organic ligands types or in specific device structure application. There is a lot of published work on RE and TM organic complexes as the basis of the OLED emitting layers. Two main routes have been followed: first, obtain all the principal spectral colors with the maximum purity possible, especially, the RGB (red, green, blue) matrix (due to its application for displays) and second, maximize the device efficiency, either by the chemical molecular synthesis improvement or by OLED structure design. In such approach and due to the enormous efforts in the fundamental research, a very big jump has been witnessed since the first RE or TM organic complex was used in OLED development two decades ago.

In a more scientific point of view, those organic complexes have very specific physical mechanisms for light emission. The understanding of such mechanism is the fundamental departing point for further developments for the researchers in chemistry, materials science, and electronic engineering and physicists. With this knowledge, it will be possible to develop new OLED structures for maximizing the electrical carrier confinement and efficiency (with high-brightness output), concomitantly with fabrication methods and parameters.

This book addresses those aspects.

1.3 The Book: Contents and Organization

This book focuses on the two major areas of OLEDs that are becoming more and more important day by day. Naturally, not all of those organic complexes are present and discussed in detailed not only due to the book limit extension but also (and more important) due to the fact that some of them, in spite of different electroluminescence results (e.g. color, efficiency and the rest of the *figures of merit*), belong to specific materials groups with same optical and electrical properties. This not only strongly simplifies the discussion but also is a extraordinary surplus for further comprehension and facility to understanding the physical process and mechanisms involved. So, it is not surprising that many explanations that are given focused into a general model that further will be extended to specific experimental cases with references for the similar ones.

In particular Chapter 1 presents an overview about this book and the present state of the art in OLEDs development with some general insights about the RE- and TM-based devices. Chapter 2 gives a basic description of organic LEDs, including the typical structure and the fundamentals of the energy levels in organic semiconductors, from the simple carbon atoms conjugation. The notion of OLED architecture or design is introduced, briefly explained how to address issues related to the optimization of organic layers sequences. An overview of the main methods of OLED fabrication ends this chapter. Chapter 3 goes more in depth into the physics of OLEDs explaining the principal electrical carrier injection and transport in semiconductors and the discussion about their applicability in OLEDs electrical characterization, specifically in the adaption of the models developed to inorganic semiconductors the new devices. A bridge between the electrical and optical properties is summarized. It is followed by a relatively detailed explanation about the energy levels formation and the respective influence on the energy transitions on the pure organic semiconductor. Finally, the OLED optical properties and characterization is explained, focusing on the OLED light measurements and efficiency calculation.

Chapters 4 and 5 are dedicated to RE- and TM-based OLEDs, respectively, although both have similar structures. Both chapters begin with the explanation of the specific properties of RE or TM organic complexes for active OLED layers, the physical nature of each kind of complex and the relationship with the optical properties that works as the basis of the device structure and development; in particular, the specific energy level formation is presented with its correlation with the molecular basic structure and the description of how those molecules perform under optical and in more detail, electrical excitation. The specific molecular energy levels for RE or TM complexes are discussed, highlighting the importance of the chemical design in obtaining useful molecules for the emitting layers of OLEDs. The different mechanisms of molecular energy absorption and transference are described, with a detailed explanation about the working pathways for a further device study. In addition to this knowledge, a first (and brief) explanation of the luminescence behavior is given to introduce the last (and principal) part of the chapters, that is, the OLEDs based on RE or TM complexes. In that part, several OLED developments, accompanied by experimental data with specific discussion, allow a final overview and comprehension of those kinds of OLEDs. In particular, the triangle formed by the chemical–physical molecule properties/optical and electrical behavior/device structure is the main topic of discussion. The goal is to provide the necessary (and explained) information about the actual work done and, fundamentally, to open further possibilities of developments in this field.

Chapter 6 focuses, in a brief description, the main chemical routes to synthesize a rare earth and a transition metal complex, showing the principal ideas and trying to make a bridge with the most physical discussion presented in Chapters 4 and 5.

The book concludes with some prospects and outlook for the future of RE and TM complexes-based OLEDs. In a condensed way, some scientific improvements and suggestion are presented.

References

1. Pope, M. (1962) Electroluminescence in organic crystals, *J. Chem. Phys.*, **38**, 2042–2043.
2. Tang, C. W. (1987) Organic electroluminescent diodes, *Appl. Phys. Lett.*, **51**, 913–915.
3. Kido, J. (1990) Electroluminescence in a terbium complex, *Chem. Lett.*, **19**, 657–660.
4. Baldo, M. A. (1998) Highly efficient phosphorescent emission from organic electroluminescent devices, *Nature*, **395**, 151–154.
5. Kalinowski, J. (2004). *Organic Light-Emitting Diodes: Principles, Characterization and Process*, CRC Press, USA.
6. Müllen, K. and Scherf, U. (eds) (2006) *Organic Light Emitting Devices — Synthesis, Properties and Applications*, Wiley-VCH, Germany.
7. Shinar, J. (ed) (2004). *Organic Light-Emitting Devices*, Springer, Germany.
8. Li, Z. and Meng, H. (eds) (2007). *Organic Light-Emitting Materials and Devices*, Taylor & Francis, USA.
9. Kafafi, Z. H. (ed) (2005). *Organic Electroluminescence*, Taylor & Francis, USA.

Chapter 2

Organic Light-Emitting Devices: The Basics

In this chapter the basics of an OLED device are explained followed by description of the main constrains to their wide-spread use. The specific necessities for special applications are focused.

2.1 The Organic-LED Basics

Organic LEDs have a more complex structure than inorganic ones for two main reasons. Firstly, the low-carrier mobility and the electrical conduction process impose a new device structure; secondly, the much different carrier mobility between electron transport material ("n-type") and hole transport material ("p-type"), the last being several orders of magnitude higher. In turn, the benefits of using an organic-based self-emissive layer that allows very thin film devices (in some cases below 100 nm), the much sought after flexibility result in devices with unique properties that move the interest of the scientific community.

The traditional LED from inorganic semiconductors is typically a p-n junction, were a hole injected from anode and an electron injected from cathode diffuses in the respective semiconductor type to recombine at the interface resulting in light emission. In this kind of semiconductors, the carrier mobilities are relatively

Organic Light-Emitting Diodes: The Use of Rare-Earth and Transition Metals
Luiz Pereira

www.panstanford.com

similar, the device structure can be kept simple, and one needs to only worry about the electrode contacts and the p-n interface. The III-V semiconductors (due to the bandgap in the visible region) are the best and cheapest choice as basic materials for inorganic LED. Figure 2.1 shows a simple scheme.

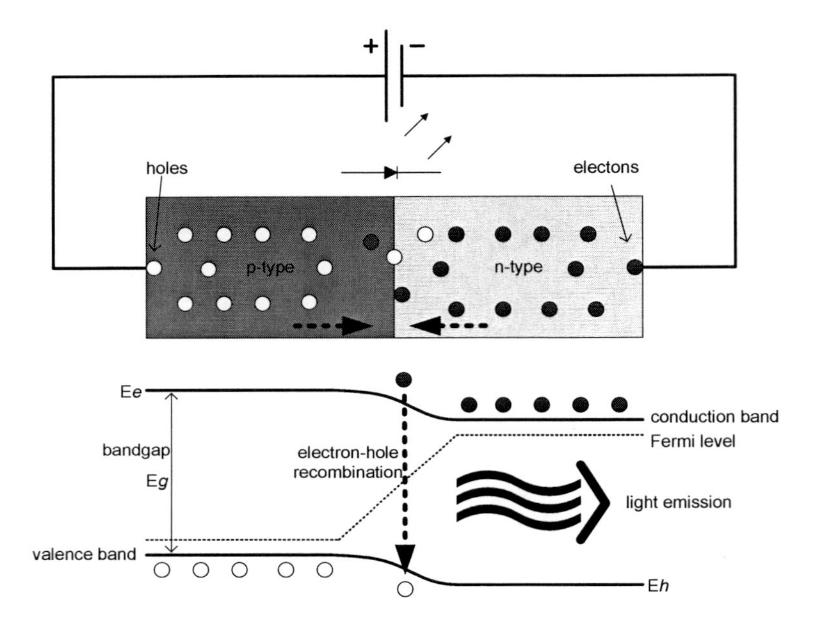

Figure 2.1 Simple LED device structure from inorganic p-n semiconductors junction. Reprinted from *Journal of Non-Crystalline Solids*, 352 (50–51), V.M. Silva, L. Pereira, The nature of the electrical conduction and light emitting efficiency in organic semiconductors layers: The case of [m-MTDATA] – [NPB] – Alq3 OLED, 5429–5436, Copyright (2006), with permission from Elsevier.

The emitted photon energy can be roughly estimated, by the simple difference between the conduction band energy, E_e, and the valence band energy E_h:

$$h\nu = E_e - E_h \cong E_g \tag{2.1}$$

where h is the Planck constant, ν the emitted photon frequency, and E_g the bandgap energy.

Organic semiconductors, though sharing a similar idea and conception basis, present different structure and working process. The basics of carrier transport in organic semiconductors are the carbon atoms ligations and its respective molecular orbital. As a way to understand the whole system, let us analyze the case of the benzene ring, shown in Fig. 2.2. The question is simple: how does charge begins throughout the molecule?

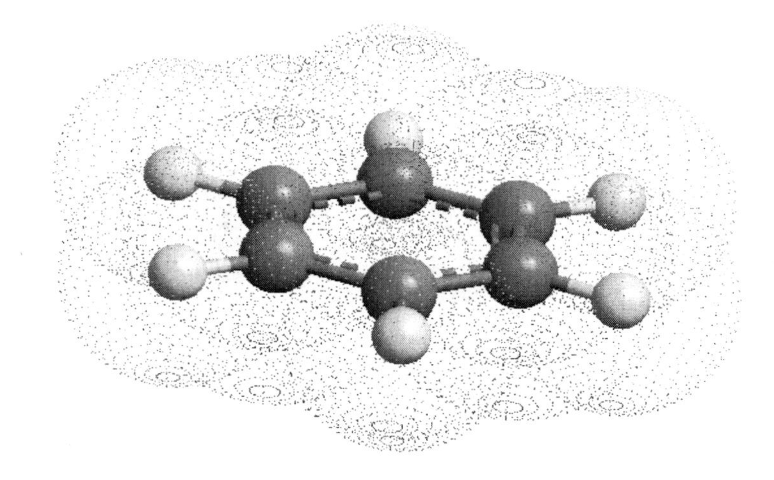

Figure 2.2 The benzene molecule.

Consider one carbon atom, with an atomic orbital configuration of $1s^2 2s^2 2p^2$. Once the $2s$ orbital is full, the carbon will form two bonds with one neighbor by the unmatched $2p^2$ electrons. But we know that four ligations exist. The answer is the hybridization [2]: the *s* and *p* orbitals mix to give hybrid orbitals (sp^1, sp^2, and sp^3—depending on the number of orbitals that are combined), which gives place to simple, double, or triple ligations. In general, we have three *sp* hybrid orbitals, leaving one non-hybridized *p* orbital. Two of those *sp* orbital that are in each carbon atom, make covalent ligations with the neighbor carbon atoms (called π bonds) and the third usually make a covalent bond with hydrogen or lateral group: σ bonds that are a cylindrical bond around the inter-nuclear axis. These bonds are the basic formation of the *Highest Occupied Molecular Orbital* (HOMO) and the *Lowest Unoccupied Molecular Orbital* (LUMO) that have correspondence (respectively) to the valence and conduction band in inorganic semiconductors. The difference between HOMO and LUMO levels is the bandgap in organic semiconductors.

Getting back to the benzene molecule, we can now see what happens when the different orbitals are located near-neighboring.

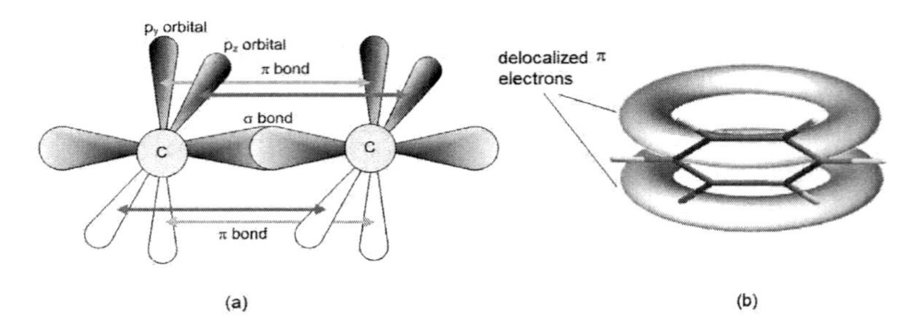

Figure 2.3 (a) σ and π bonds in two neighboring carbon atoms; (b) the delocalized system in the benzene molecule.

When the π orbitals from neighbor molecules overlap, an intermolecular electronic change can occur, inside the HOMO and LUMO levels. However, due to the weakness of the Van der Waals ligations, this change is very inefficient and the electrons stay more fixed into individual molecules. As the covalent ligations are strong, we do not have free electrons like in inorganic semiconductors. The intermolecular interactions in organic materials with π conjugated systems induce an unfolding in the HOMO and LUMO levels, originating narrow bands that consist in energetic steps called bandgap (of organic semiconductors that act like the similar bandgap of inorganic ones) that usually range from 1 eV to 4 eV, corresponding to the minimum energy to generate an excited state. This bandgap usually decreases with the increase of the delocalization, because the π electronic system increases. The delocalized π orbitals account for the organic molecules' essential optoelectronic properties, from light absorption and emission to carrier generation and transport. By locking at the very simple example of the benzene ring, it is very easy to see that electrical conduction occurs when a carbon chain presents a system of the type –C=C–C=C–C= that is an alternate double/single bond set.

This system is called a conjugated system and is, as mentioned, the basic of the electronic properties of most organic semiconductors. Figure 2.4 shows the comparison of conjugation in benzene with that of complex polymer [*poly(9,9'-n-dihexil-2,7-fluorenodiilvinylene-alt-2,5thiophene)*] [3] to illustrate the generalization of this concept.

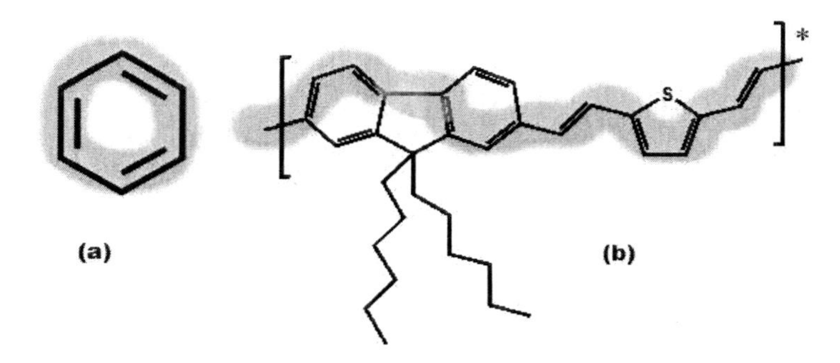

Figure 2.4 (a) The simple scheme of a benzene ring showing the conjugation (–C=C–C=C–C=C–) on the basis of the electrical conduction; and (b) a more general system with a polymer chain. The shadows represent the electrical pathway due to the conjugation.

Although not the object of detailed studies in this book, polymers chains help understand the electrical properties of organic semiconductors. The small molecules based on rare earth or transition metals complexes use carbon–carbon conjugation to "deliver" the electrical charge to the coordinated ions. These two kind of small structures, in which there is an inorganic (rare earth or transition metal) core coordinated with organic ligands, exhibit electroluminescent characteristics that are intimately associated with the electrical properties of those ligands (that can be viewed as organic counterparts).

In a similar way to that we found for inorganic semiconductors, the organic LEDs performance depends on the injection/transport and recombination of electrical charge. For this, the general structure is usually comprised by a hole transport material (also called hole transport layer—HTL) and an electron transport material (electron transport layer—ETL), needed to deliver the charge to an active material between them (in a "sandwich" structure), usually called emissive layer—EL. The choice of HTL and ETL is clearly dependent on the physical and electronic properties of the EL, in particular the location of the HOMO and LUMO levels. In fact, we expect that the interface between HTL and EL will also act as an electron blocking region, and, of course, that the interface between ETL and EL can act as hole blocking region. This strategy is not surprising because increasing the carrier concentration inside the EL helps the electron–hole recombination and therefore the efficiency. This simple system addresses the so-called carrier confinement and is one of the most important configuration issues in organic-LEDs developments. Figure 2.5 shows a schematic of this general architecture.

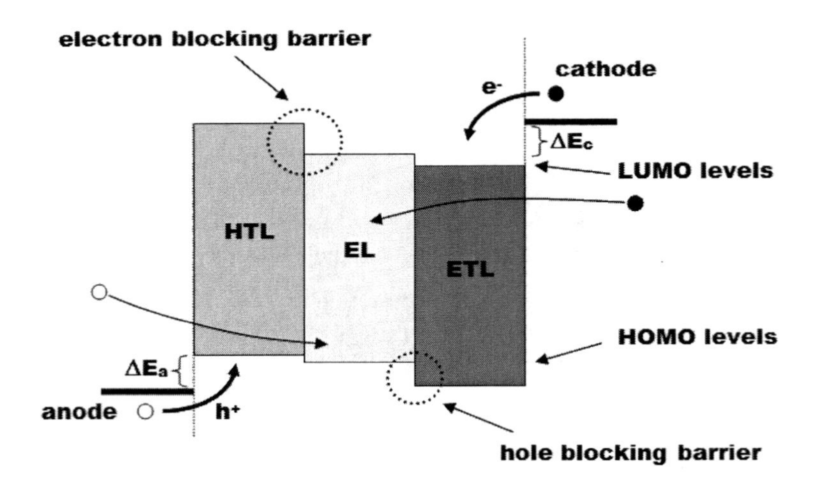

Figure 2.5 Typical OLED structure with the hole transport layer (for hole injection/transport), electron transport layer (for electron injection/transport), and emissive layer. The ideal cathode and anode levels, in respect to the materials LUMO and HOMO levels, are shown. The interfaces at HTL/EL and EL/ETL act as electron and hole barriers, respectively, as desired for a correct carrier confinement inside the EL.

If the HTL and/or ETL cannot efficiently acts as electron/hole blocking barrier, respectively, then, specific organic materials may be employed to specifically perform that functions. In this case, the OLED structure becomes more complex, and, may comprise two more layers as shown in Fig. 2.6.

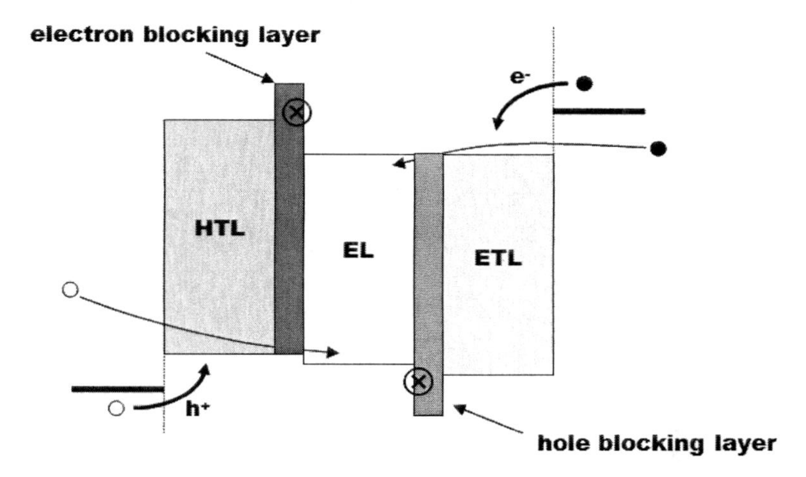

Figure 2.6 OLED structure, with the hole and electron blocking layers, derived from Fig. 2.5. The specific use of such layers must prevent the carrier leakage into HTL and ETL with further helping on carrier confinement inside EL.

In Figs. 2.5 and 2.6, schematics of the anode and cathode levels (or more precisely the respective work-functions) are ideally represented; that is, the anode lies below the HTL HOMO level and the cathode above the ETL LUMO level. This aims at helping the respective carrier injection as the energetic differences (ΔE_a and ΔE_c) to eliminate any electrical potential barrier, providing those interfaces with ohmic electrical characteristics (as described later).

Unfortunately, in most real-case scenarios this is not simple to attain. The main problem resides at the anode that must be transparent for light emission. Usually it is used as a thin layer of a transparent conductive oxide (TCO) with high work-function value (up to 5–5.5 eV, dependent on the material). The most widely employed (and known) is ITO (indium tin oxide) that has a work-function of about 4.7 eV and helps match the anode/HTL barrier for one side, with the low resistivity/high morphological uniformity for the other. This point is very important. As in fact, the TCO morphological uniformity (low roughness, very thin layer, and large area) plays a fundamental role in the device performance. In parallel, it works as substrate for the device fabrication. Some other TCOs commonly used and/or studied include FTO (fluorine doped tin oxide with a work-function about 4.4 eV) and ZnO (zinc oxide with a work-function about 5.2 eV). Other oxides have been tested but remain largely unexplored, e.g. AZO—aluminum doped zinc oxide, and RuO—ruthenium oxide. As one of the most important features of any TCO, and besides the morphological issues, the optical transmittance is absolutely fundamental. Transmittances over 80–85% in the device spectral range emission are needed to avoid light reflection and loss of device emission. Another important point is the layer resistivity, which must be below 20–30 Ω/sq. to allow the best hole injection and reduce the electrode resistance. Thus the TCO of choice for each device structure features the best adjusted conductivity to the HOMO level of the hole transport material; evaluation of this relation is one of the driving forces for research in the field. Finally, a combination of a very low thickness with a strong microscopically adhesion is required for flexible substrates. Regardless of all these points, ITO remains the most acceptable and most employed TCO available.

The cathode commonly employs a high pure metal. Aluminum is mostly used because of its environmental stability, fine grade availability, and relative low cost; generally, a very pure metal form is commonly used. Its work-function is not among the best (4.2 eV) of the widely used electron transport material such as Alq3 (*tris-(8-hydroxyquinolate)-aluminum*) and butyl-PBD (*2-(4-tert-Butylphenyl)-5-(4-biphenyl)-1,3,4-oxadiazole*) but the good adhesion adding to the other cited features make it the right choice for a value-to-performance compromise. Other attractive metals include magnesium and calcium, in particular this latest, due to its low work-function (around 2.8 eV); however, it is difficult to obtain perfect films from magnesium evaporation, and calcium films are too sensitive to environmental conditions, being rapidly oxidized. Some solutions are currently under test to overcome these constrains: a mixture of magnesium and aluminum (still with some problems due to the different melting points) or an evaporation of a "protective layer" of aluminum over the calcium one (a good solution with the drawback of increased device fabrication complexity). Following different pathways another solution is the deposition of a very thin layer of lithium fluoride (LiF) of less than 0.1 nm between the organic electron transport layer and the aluminum. This "dielectric" layer can reduce the potential barrier at the cathode interface, leading to an ohmic electrical behavior. Naturally the resulting benefits depend on the LUMO level of the ETL and the increased fabrication complexity. All these points account for aluminum as metal of choice for making stable and relatively low work-function cathode. Also, its very high-reflection coefficient (that in some cases resembles a "mirror") helps the device light output, acting as a perfect back reflector for the light emitted by the emissive layer, and guiding the photons to the transparent electrode (anode). Independently of the metal used for the cathode, some care must be taken into account to avoid the metal diffusion into organic layer as this phenomenon is known to reduce significantly the efficiency and lifetime of the device.

To summarize, the traditional OLED structure comprises a rigid (glass) or flexible (usually PET—polyester) substrate in which a TCO is previously deposited (traditionally by sputtering) working as anode; on top of this substrate, the organic layers are deposited by several techniques (to be briefly described in next sections) and finally a metal layer (working as cathode) is thermally evaporated. Naturally, the number of the organic layers can be any, depending on the desired emission and organic emissive complex physical properties.

One of the most problematic features regarding devices based on organic semiconductor is their lifetime. In fact, these materials are extremely reactive to the oxygen and because of this elements break the double carbon bounds in organic chains and occupy that position. This leads to a break in the conjugation system that translates into a strong degradation of the organic semiconductor electrical properties (as stated before). Moreover, some organic materials also react to humidity and/or different environmental gases. Lifetime is thus considered as the main technological issue some years ago, but presently seems to be overcome. The solution was the efficient development of encapsulation methods that protect the organic device. Although some doubts about the very long-time protection efficiency remains (in order to allow the organic devices to compete with the inorganic ones), the display market offers several consumer products with some level of confidence. The most common encapsulation technique consists in covering all the device (from the back part) with a capsule that is glued around it with a special epoxy resin, absolutely impermeable to the most common gases. Moreover, the inner surface of the capsule carries a material that captures the major of the problematic gases, getting an additional protection and guarantying that a small amount of reacting gases that might have remained inside the device during the fabrication process are captured by that getter and do not interfere with the normal device operation. Several techniques are used for this process (not be focused in this book), all of them performed under high vacuum inert atmosphere. In spite of the advances in the encapsulation/protection methods, all of these solutions are based on rigid capsules, and therefore not applicable to flexible devices. This is the reason why the flexible OLEDs are not (at the moment) available in the market. The encapsulating techniques for these are being developed based on flexible layers of some special polymers applied to cover all of the device surface followed by several treatment techniques. Usually two main problems arise from this solution: some polymers can react chemically with the organic device materials and/or the protective polymer layers are not really long time efficient for some gases and/or humidity. Much work is still under progress to achieve a solution for flexible device encapsulation, which allows in their future commercialization.

In Fig. 2.7, an OLED device schematic structure is shown.

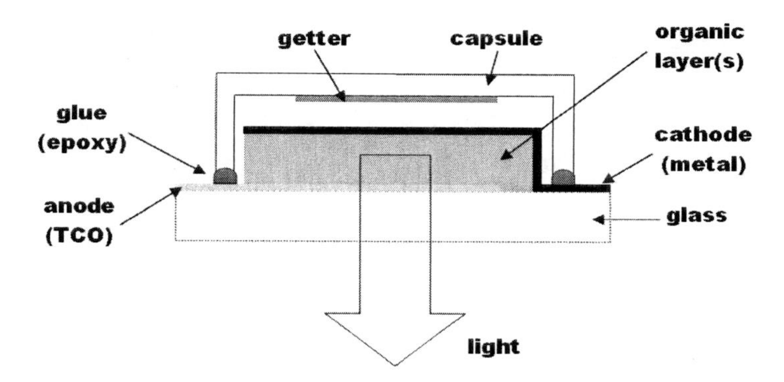

Figure 2.7 Typical OLED structure with the schematic representation of the most important parts.

Apart from chemical degradation, organic devices can also be subject to an electrical degradation, when part of the molecular structure is in an irreversible way, modified by the applied electrical field. Such kind of degradation is very difficult to overcome and dictates, in an expressive number of cases, the abandon of promising emissive materials. One of the most interesting cases was the study of the electrical AC behavior of a single-layer device based on Alq3, with ITO and aluminum as respective anode and cathode [4]. The authors found that, when successive voltage in DC conditions is applied, different electrical behaviors are obtained. Between two successive DC characteristics, the AC data are taken and their modeling to the traditional equivalent circuit shows a pronounced change in the circuit parallel resistance, indicating changes in the semiconductor bulk. These changes, probably due to a restructuration in active film of Alq3, are completely irreversible. Naturally, this situation is a handicap for device viability because the performance shows a noticeable degradation.

Some solutions have been tested to bypass this question but with no clear results. In some cases, different methods for organic layers deposition attempt to make a film in a solid conformation that, in principle, would not allow much electric-induced restructuration; also, using the same deposition technique, different parameters were tested, in a similar strategy; finally, when this kind of degradation is accelerated by the evidence of a simultaneously chemical degradation, the encapsulating solution is proposed. In fact, the most acceptable way to solve this problem is to keep the device in a relatively low-applied voltage regime although this implies a low-driving voltage and a relatively short dynamic operation range. This model can only work in a multi-layer structure, and the choice of the different layers (apart from the emissive one) follows a rigid criterion in order to allow quasi-ohmic barriers between them. Simultaneously, high-carrier confinement in the emissive layer must be achieved for the electrical operative constrains imposed. All these issues are normally addressed to the device layers architecture

2.2 Approaches for OLED Organic Layers Structure

It is well known that OLEDs structure can comprise a different number of organic layers. As mentioned before, this number is closely related to several questions, namely the physical properties of the emitting material (in particular, the localization of the HOMO and LUMO levels), the carrier mobility, the need for special carrier confinement features, etc.

The simplest OLED that can be made comprises of only one organic layer. This "ideal" configuration, due to its simplicity, easy of fabrication, and with a very high probability of good reproducibility, is only available in very special situations and does not correspond to the majority of systems. The most important case is a particular group of OLEDs, called *Light Electrochemical Cells* (LECs) that are explained in Chapter 5 describing in particular Ruthenium-based devices. But this kind of OLEDs follows a clearly different electrical behavior that from of the most usual ones and cannot be considered as a general rule. Two considerations must be taken into account in order to fabricate an OLED with a single layer. First, the organic semiconductor, besides its emissive nature, must have a clear bi-polar characteristic with electron and hole mobilities relatively similar; and, second, its HOMO and LUMO levels must "match" the required localization for an ohmic contact with both electrodes. It is not difficult to conclude that such kind of organic semiconductor is very difficult to find. In many cases, the problem arises from a very poor carrier confinement and, in spite of an efficient emitting material with good bi-polar characteristics, no useful device is obtained. A typical example is the Alq3-based device. The Alq3 is a small molecule that emits strongly in the green spectral region. A device employing only the Alq3 layer can be easily made by thermal evaporation, as Alq3 sublimes in a simple controllable way. Such kind of device, if ITO and aluminum are used as anode and cathode respectively, has an energy level as shown in Fig. 2.8.

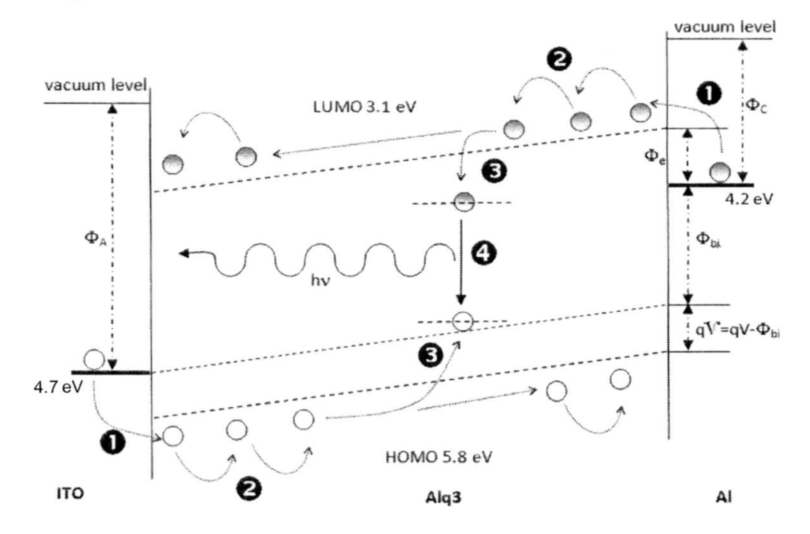

Figure 2.8 Schematic representation of a single layer OLED of Alq3 under forward bias.

The figure has some important sequences to be noted. First, (1) we have the carrier injection inserted directly into active layer followed by the carrier transport throughput the bulk of that layer; (2) the carrier capture; (3) the radiative recombination; and (4) end of the process. For a better understanding of the physical nature of that configuration, some important data are shown: Φ_a (anode work-function), Φ_c (cathode work-function), Φ_h (potential barrier for hole injection), Φ_e (potential barrier for electron injection), Φ_{bi} (built-in potential), V (applied potential), V (effective potential), and q the elementary charge. The light output is given by $h\nu$ (h the Plank constant and ν the frequency) and depends on the energy difference between the carrier-captured levels in the semiconductor bulk.

In general, and for a simple approximation, we can consider that this difference is approximately equal to the LUMO–HOMO level, e.g. the semiconductor bandgap. The process is clearly dominated by the potential barriers for hole and electron injection that in this particular case are about 1.1 eV in both situations. These values are clearly high, so a less-efficient carrier injection is expected, although the difference between both barriers are quite similar a help a balanced carrier injection. But the high barriers combined with the different electron and hole mobilities, must introduce some instability in the carrier injection/-transport. The effective applied potential (built-in potential) is, in a first approximation, estimated by the difference between Φ_a and Φ_c, in this case rounding 0.5 eV, a relatively low value (a more complete explanation is given in next chapter). As expected, from our work, low efficiencies and a relative high-device driving voltage are obtained. Considering that the Alq3 usually gives a strong electroluminescence, a different device structure is clearly needed.

A reasonable solution for hole injection can be obtained by employing a "p-type" organic semiconductor between the anode and the Alq3 layer. However, the physical properties (energy levels) must unfailingly allow a better energy match between ITO and Alq3 (that plays a double function–emissive layer and electron transport layer) allowing a balanced charge injection and transport. The instability in the carrier injection can be solved by a careful choice of the "p-type" organic semiconductor to be located between anode and "n-type" one, in such a way that the barrier for hole injection will be similar to the barrier for electron injection (between cathode and "n-type" material). The instability in carrier transport inside the organic layers (bulk) can be solved with the discerning choice for hole transport material, that in turn depend on the electron transport material properties. It must be noted that even a correct choice of materials is made, the device could not be as efficient as expected. We must also assure that the thickness of the organic layers is correct, in account to carrier mobility, electrical field, global device resistance, etc.

This is the first step to improve OLED performance. The material choice for the "p-type" organic semiconductor is the m-MTDATA (*4,4',4''-tris(3-methyl-phenylphenylamino) triphenylamine*) with HOMO and LUMO levels located at 5.1 eV and 1.9 eV, respectively. The hole barrier is about 0.4 eV and the electron injection barrier remain unchanged: 1.1 eV. The barrier between the organic interface m-MTDATA/Alq3 is 0.7 eV for hole injection from m-MTDATA to Alq3. These low values for hole injection increase the hole density in the emissive layer. The final expected result will be a decrease in the driving voltage that is the device electroluminescence starts at lower applied voltage and an efficiency increase. In fact, the driving voltage decreases about 20% and the efficiency increases about 250 times!

Naturally if a third layer "n-type" is employed (between Alq3 and aluminum) to reduce the electron injection barrier, better results are expected. Unfortunately, the "n-type" organic semiconductors are not so easy to obtain as the "p-type" ones and in the most cases the barrier at the cathode can only be reduced by using a metal with a work-function lower than aluminum. In this particular case, magnesium and calcium will be the correct choice though all the problems cited before.

In unavailable cases (by absence of correct n-type organic semiconductor and/or fabrication difficulty to use low work-function cathodes), a different solution can be implemented: increase the hole transport facility, creating an electron blocking layer between emissive layer and p-type layer and/or creating a very thin "dielectric" layer between cathode and n-type semiconductor (as mentioned before). When the emissive layer is also the electron transport layer, the solution of increasing the hole transport facility is usually the simplest. And, due to the existence of innumerous p-type organic semiconductors, it is relatively simple to choose one for the desired function.

In the sequence of the experimental case used as model, the second step to improve the Alq3 emission is to introduce a thin layer of NPB (*N,N '-bis- (1-naphthyl)-N,N'- diphenyl-1,1'- biphenyl-*

4,4''-diamine) between m-MTDATA and Alq3. NPB is a p-type organic semiconductor with the HOMO and LUMO levels located at 5.7 eV and 2.6 eV, respectively. The hole injection barrier from m-MTDATA to NPB is now about 0.6 eV but the correspondent barrier form NPB to Alq3 is only 0.1 eV. In this situation it must be noted that the electron injection barrier from Alq3 to p-type organic semiconductor is now 0.5 eV (Alq3 to NPB) and not 1.2 eV (Alq3 to m-MTDATA). This difference can bring additional problems in the carrier confinement, which can be only solved by introducing an electron blocking layer between Alq3 and NPB. Consequently, the device complexity increases and the reproducibility may easily decrease. The best way to keep the device structure simple is to achieve an optimal configuration, in which the "problems" are overcome (or at least balanced) by the "benefits." Figure 2.9 shows the schematic band diagram for ITO/m-MTDATA/NPB/Alq3/Al device.

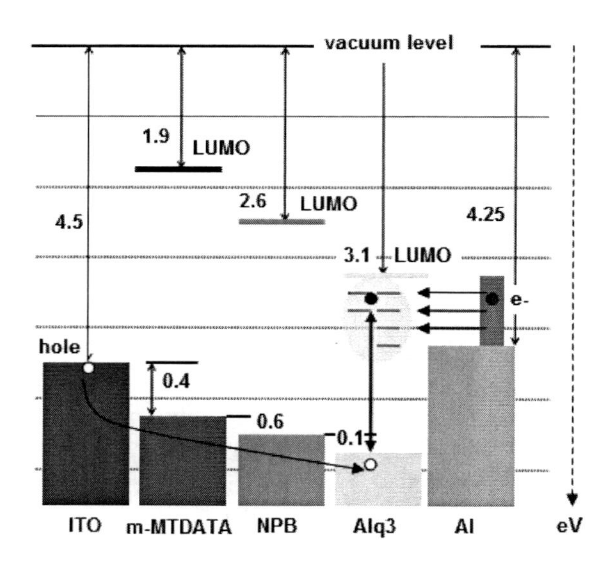

Figure 2.9 Schematic band diagram for ITO/m-MTDATA/NPB/Alq3/Al device. Reprinted from *Journal of Non-Crystalline Solids*, 352 (50–51), V.M. Silva, L. Pereira, The nature of the electrical conduction and light emitting efficiency in organic semiconductors layers: The case of [m-MTDATA] – [NPB] – Alq3 OLED, 5429–5436, Copyright (2006), with permission from Elsevier.

In this new structure, designed to increase the hole density inside Alq3 and to simultaneously facilitate the hole transport in p-type layers, and comparatively to the two layer structure, the driving voltage is reduced by about 15% and an efficiency increases nearly by four times. It is clear that main device figures of merit are improved, but the presence of NPB and the traditional energy states at the Alq3 LUMO level (represented in the figure near the cathode) leads to a small blue shift of the well-known green band that arises from the Alq3 electroluminescence [5]. The next figures show some typical data obtained to fabricate, by this strategy, an Alq3-based efficient device. They must be interpreted as a "picture" concerning the evolution described above. Figure 2.10 shows the changes in the device optical power vs. applied voltage.

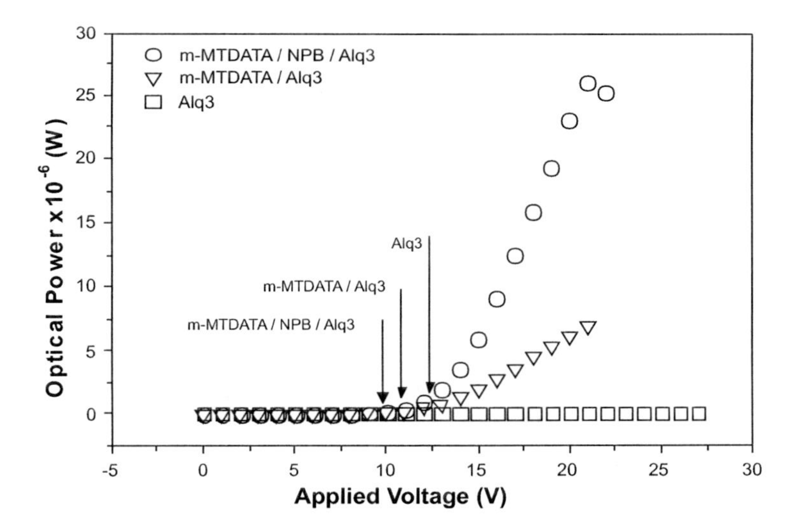

Figure 2.10 Optical power (at single wavelength—the electroluminescent band) of the different devices fabricated with Alq3. The arrows show the driving voltage. The more optical power, the more efficient it will be, as the electrical current does not increase so fast, for the same applied voltage.

As the main emission remains unchanged and is fully dominant, this "side effect" is not a problem. However, it is important to retain that when designing an OLED structure, the physical properties of the organic semiconductors have a much stronger influence than the inorganic ones. The appearance of several "side effects" is very common (we will lock some of them) and in many cases they are so problematic that the entire device structure must be abandoned.

Returning to the general OLED conception, another common problem is the intrinsic luminescence of several organic materials used as n-type or (the major part) p-type semiconductors. Its simultaneously interesting and sometimes irritating behavior implies that we must pay additional attention to the device structure. Moreover, when a new emitting material is tested for the first time, several preliminary tests must be made to get as much precise information as possible about its light absorption and emission as well its energy levels. Only then, we can assess the most compatible structure. Figure 2.11 shows the changes in the electroluminescent band.

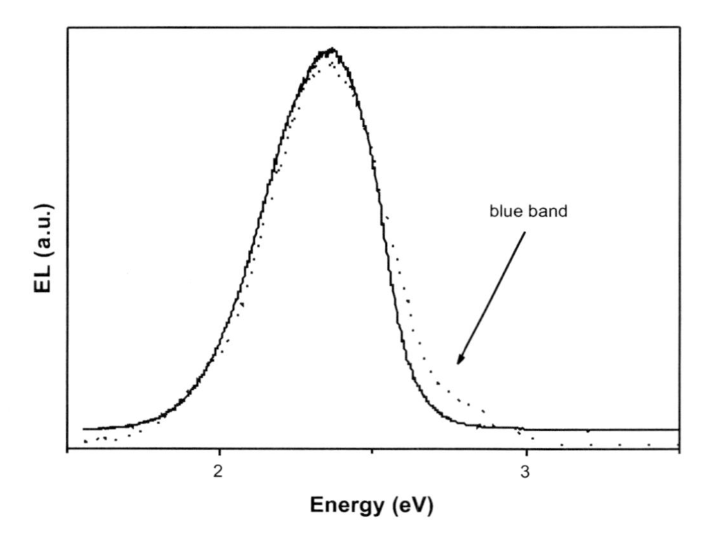

Figure 2.11 Changes in the electroluminescent band of an Alq3 device (full line). One and two layer device and (dot line) three layer device. The small blue band is an attempt ascribed to some emission from NPB [5].

Considering the initial idea of the three layered structure (HTL/EL/ETL), an initial general conception and diagram can help the researchers on the OLED development. Following the explanation of several points in the experimental (and widely known) case of the Alq3-based OLED, the "whole picture" can now be explained. With the Alq3 device case that covers all steps from a single layer to three layer structure, it starts from the conventional "sandwich" due to the double nature of the Alq3 organic semiconductor (ETL and EL simultaneously); but several topics can easy be transported to the main discussion. The ideal OLED structure must consider the following points:

(1) Carrier confinement must be completely optimized inside the EL bulk;

(2) A low (as much as possible) injection barrier is required for holes (anode/HTL interface) and for electrons (cathode/ETL). This can be done by using the appropriate electrode and/or organic material;

(3) If possible, the interface HTL/EL and ETL/EL must act as electron- and hole-blocking interfaces, respectively. See the HOMO and LUMO levels of the organic semiconductors;

(4) Never block a carrier, hole, or electron, inside the HTL and ETL respectively; if necessary, to balance the carrier injection and transport, attention must be paid to the successive HOMO and LUMO levels, respectively;

(5) To increase the carrier confinement, a device structure can be made using an electron blocking layer (between the EL/HTL) and/or an hole-blocking layer (between the EM/ETL). Attention must be paid to the HOMO and LUMO levels;

(6) Pay attention to the device layers thickness: a thin layer is difficult to obtain with good uniformity, whereas a thick layer can increase significantly the series resistance. In some cases, the ideal thickness is also dependent on the carrier mobility;

(7) Care must be taken with organic semiconductors, that at first sight may seem the best choice for HTL or ETL but that are also electroluminescence.

As an additional question about OLEDs structure, we need to bear in mind that their reproducibility is fundamental for technological applications. Sometimes, and due to the huge range of available organic semiconductors, a very complex structure is build in order to achieve the marvelous goal. These complex structures employ, as far as we know, up to dozens of layers, distributed by ETL, HTL, EL, and hole-/electron-blocking layers. But the final result (for the best of our knowledge) is very difficult to reproduce and (obviously) completely unrealizable, from the technological market point of view. So, the plan is always the same: try to keep the device structure as simple as possible even if spurious "side effects" occurs. If such effects do not alter significantly the desired result, it is much more important to guarantee the reproducibility than the original stipulated "perfect device." The exceptional properties of OLEDs deserve an intelligent way to extract the best results with the lowest structure complexity.

In this simple general three layer device we expect the electro-optical behavior to follow the scheme shown in Fig. 2.12.

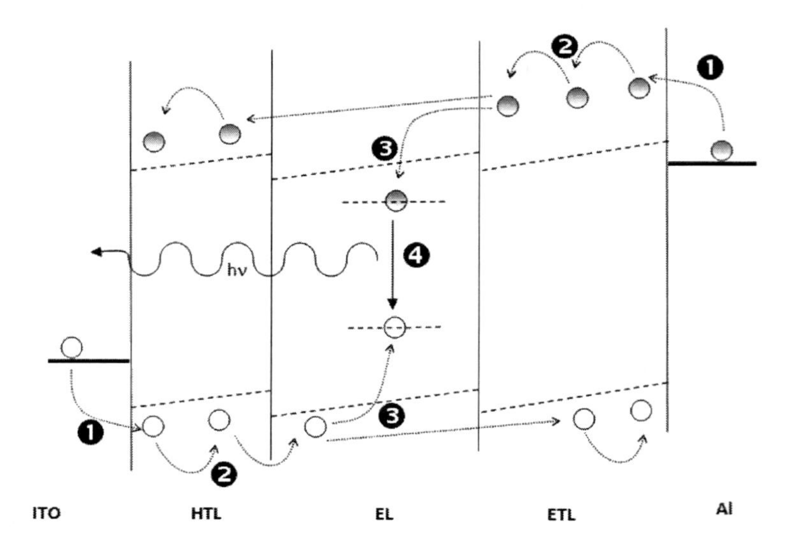

Figure 2.12 General OLED structure based on the "sandwich" idea. The emissive layer (EL) is confined by an electron transport layer (ETL) and a hole transport layer (HTL).

Considering the symbol definitions given for Fig. 2.8, an obviously and needed situation can be derived from the figure: at the cathode, the injection barrier for electrons decrease as the electrons flow through the device; the same is valid for holes at the anode. This simple and easy to understand process is the key for the device structure, as it a not only facilitate the carriers to flows throughput organic bulk, but it also helps on the carrier confinement inside EL (as desired) and makes the HTL/EL an electron-blocking interface and the ETL/EL a hole-blocking interface. It is obvious that this configuration is an ideal structure not exactly obtained in the real world. But it represents the target of any OLED structure. This approach will be followed latter. Results obtained with electron and hole transport organic semiconductors have confirmed its efficacy and practical utility. For more general information about OLEDs structure and device working process, see. references 5–9 of Chapter 1. The specific electrical behavior and respective physical models are focused in the next chapter. For now, the following step is how organic semiconductors can be processed.

2.3 Methods for Processing Organic Semiconductors

All the known organic semiconductors are available, physically speaking, under a form of micro-sized particles powders. This is the result of the chemical synthesis after all the specific synthesis routes and material related purification. Some polymers can have "large" micro dimensions due to the large chains, which are based on a repetitive monomer. Although not explored in this book, it is important to note that these differences in size must be not forgotten when the rare-earth or transition-metals complexes are embedded into organic polymeric matrix. Following the explanations given in last section about OLED structures, it is now important to give a compact and simple, but elucidative overview of the processing techniques. In our specific case (RE and TM materials) two techniques are usually employed: controlled thermal evaporation and spin coating.

2.3.1 Controlled Thermal Evaporation

This is the most used technique for processing the all organic small molecules. The technique is based on the fact that, under a strong heating the small molecules reach the melting point and after that, evaporation can occur. Though simple in theory, practical implementation requires more or less sophisticated equipment that must operate under relative high vacuum and specific sensors to determine the organic layer thickness deposited.

In Fig. 2.13 an evaporation system is shown, indicating the main systems.

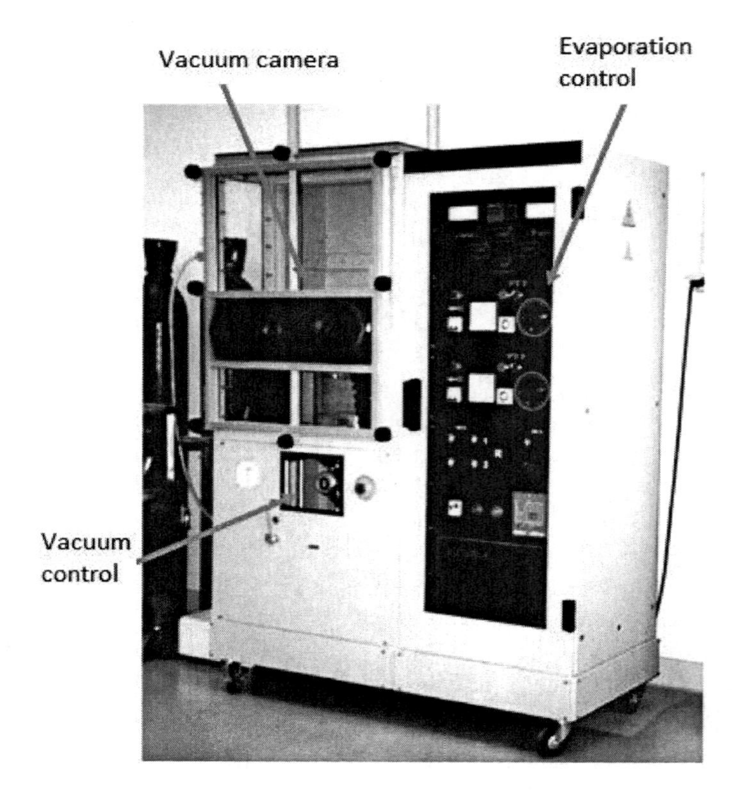

Figure 2.13 Controlled Thermal Evaporation equipment. Laboratory of Organic Semiconductors at the Department of Physics, University of Aveiro.

The main part of the system is empathized in Fig. 2.14. By application of an electric potential, a high electrical current flows through an evaporation boat holding the organic powder. The high temperature obtained allows the melting (or sublimation) of that material, with subsequent evaporation until deposition on the substrate. In order to control the thickness (both total thickness and growth rate) a piezoelectric sensor is located near the substrate and reacts accordingly to the amount of organic material that is deposited.

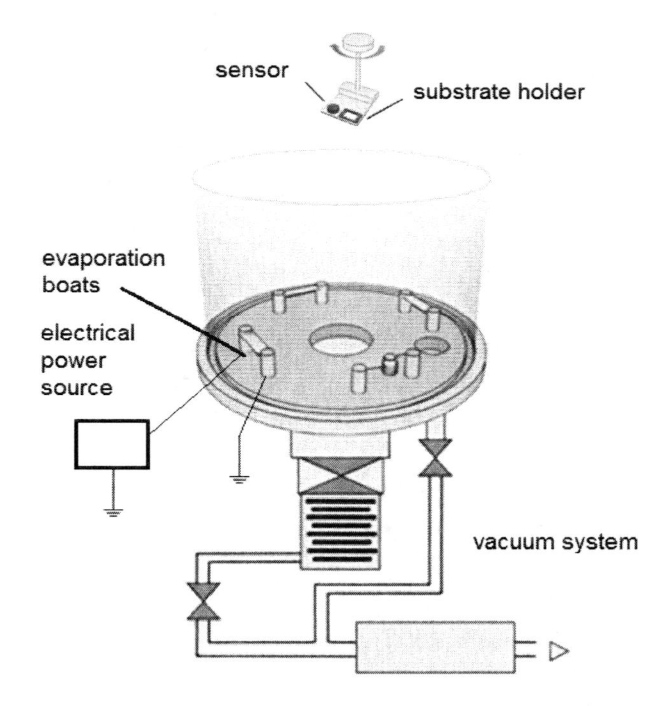

Figure 2.14 Scheme of a typical evaporation chamber with the most common components.

A mask can be used to make any desired film contour. The number of independent boats determines the total number of layers that can be deposited without vacuum break. As advantageous, this controlled thermal evaporation yields a very uniform film with thickness as low as a few dozen angstroms, and growth rates as low as 0.2–0.4 angstroms/second (the precision depends on the correct calibration parameters). The disadvantages include the need for high vacuum (up to 10^{-5} to 10^{-6} Torr) and the exclusion of organic materials that do not support high temperature, undergoing molecular degradation (e.g., semiconducting polymers). Essentially, only small molecules can be easily and safely processed with this system.

2.3.2 Spin-Coating

The spin-coating technique is probably the most widely used for the processing of organic thin films. Its simplicity allied to the relatively low-cost equipment, and the good results that can be obtained, turns it to the first choice for researchers in many fields of activity. Concerning the organic semiconducting materials deposition is very simple to obtain a relatively uniform film (with similar areas obtained by thermal evaporation).

The basic principle of the system is to use the rotation of the substrate to first spread a solution on the top of that substrate and, after this spreading, obtain a uniform thin film by keeping the substrate in rotation for needed time. After this process, the solvent is removed by evaporation in a furnace, at controlled time and temperature. Naturally, the metallic cathode must be thermally evaporated as described before.

In order to use this system for OLEDs fabrication, the organic semiconductor material is previously dissolved in a specific solvent and then as a solution. It is not very difficult to understand that the entire process has many issues that are not simple to solve, if we want a good final deposition. First, the organic material must be able to dissolve in a solvent that does not irreversibly modify its molecular structure and that can simultaneously be used in the spin-coating process; secondly, we must adjust the solution concentration in order to (a) give the desired film morphology and uniformity and (b) keep the solution viscosity within the useful values for spin-coating processing; thirdly, we must adjust the spin-coating parameters (rotation and time) in order to achieve the desired result, in combination with the solution's physical and chemical properties. All these questions are necessarily a hard challenge to address. However, it must be noted that for some large organic molecules and perhaps all semiconducting polymers, this method is a viable alternative to the impossibility of using the thermal evaporation. Of course, there are other methods to process these materials in solution, such as the ink-jet printing, but its costs are very high. Figure 2.15 shows the picture of a typical spin-coating system.

In some cases, the equipment is located inside a glove box, as a part of a processing line that can also have a small furnace and a thermal evaporation system. In these much more complete systems, "hybrid" devices can be fabricated having some layers deposited by spin-coating and others (the last ones) by thermal evaporation. The advantage is that in such a complete system all the steps of organic materials processing are performed under controlled atmospheric environment which brings huge advantages.

Figure 2.15 Spin-coating system at the Department of Physics, University of Aveiro.

Preparation of the solutions for spin-coating involves several points: first it requires finding a solvent that fully dissolves the organic material and that can be evaporated at mild temperatures, not to destroy or irreversibly modifies the organic material; also when using more than one layer deposited by spin-coating, we must use different solvents (one for each material that will be used in one layer) that are exclusive, cannot dissolve to cut more than one material, otherwise destroying the already deposited film.

In Fig. 2.16 the procedure for a thin film processing is schematically represented.

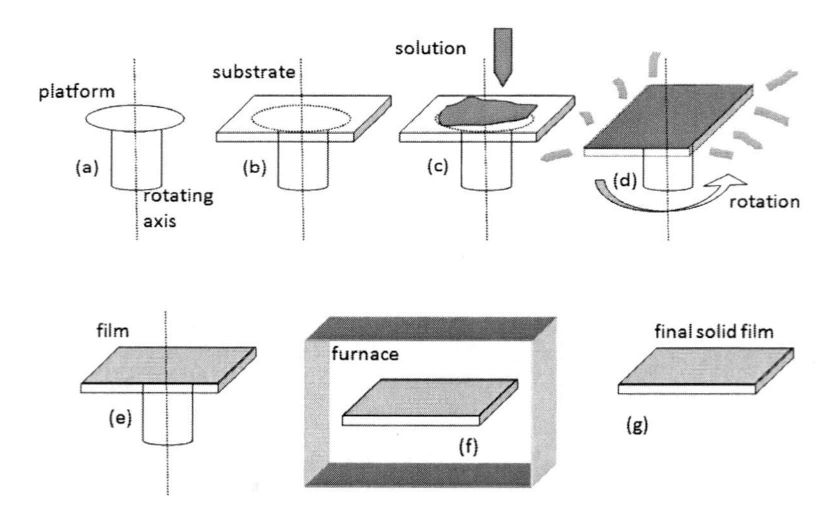

Figure 2.16 Sequential steps for producing a thin film by the spin-coating technique: (a) the support in the rotating axis; (b) a substrate is positioned and fixed by the low vacuum to the rotating axis; (c) a small amount of solution (0.5–0.8 ml) is dropped into the substrate; (d) the system starts rotating in pre-defined angular velocity and time; (e) a "wet" film is obtained; (f) the solvent is evaporated in the furnace; (g) the final film is obtained.

The main advantage of the spin-coating technique is its simplicity and low-cost; the drawbacks are the difficulty of defining parameters for reproducible and confident results, the low area deposited with good uniformity, and a limitation in the number of layers that use the same solvent. Moreover, the solution preparation for spin-coating deposition strongly depends on parameters such as concentration, viscosity, etc., implying a time-consuming previous work. However, this technique still usually produces average good results.

2.3.3 Ink-Jet Printing

A much more precise technique for liquid deposition of organic semiconductor thin films (viewed as a precise alternative to spin-coating technique) is ink-jet printing. As the name states, this technique is completely based on the similar and originally developed widely known, consumer-used ink-jet systems for paper printing.

The main advantage of the ink-jet printing process is the possibility of large area deposition with a micrometric control that allows, for instance, the fabrication of displays by making a RGB pixel in the micron scale. The technology, in a simple theory, is as follows: the printing head drops a very small amount of solution (usually less than 1 μl) in a substrate point (x, y) defined with precision by the system software. The "secret" resides in the printing head, specifically developed to handle solutions containing organic semiconductors dissolved in specific solvents. These liquid are a bit different from the conventional liquids for paper printing. Large deposited areas can be obtained, with the pattern defined by the user (by software) and specifically dependent on the final application.

Figure 2.17 shows an ink-jet printer. The system is quite complex due to the existence of several supporting parts that are needed for a correct work.

Much constrains mentioned before for the spin-coating technique are also applied to the ink-jet system. Moreover, an exceptional care must be taken in the solution preparation due to the fact that the smallest error can be fatal: the printing head degradation and even clogging, with the consequent definite end of service. In particular, viscosity and solvent are the most important factors.

Figure 2.17 A Xennia Inkjet printer with pico-liter drops with sub-micron resolution. [Courtesy: CeNTI—Centre for Nanotechnology and Smart Materials (www.centi.pt).]

As a simple scheme, Figure 2.18 shows the basic work of an ink-jet printer.

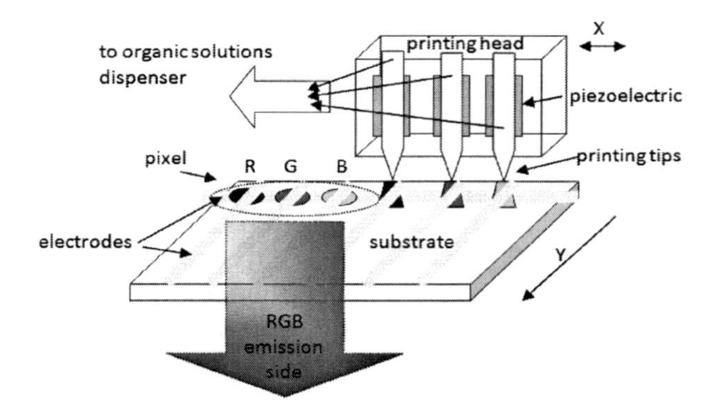

Figure 2.18 Simple scheme of the ink-jet printing technology.

Several problems are currently associated with this technique. One is the rheological question that appears when a polymer is deposited onto a surface in a very small amount each time (pico-liters). The surface tension tends to spread and drag the drop with serious consequence for the final film.

Another problem of using the ink-jet printing for OLEDs fabrication is naturally the cost. This is the consequence of the specific head development with relative low lifetime and all the chemical expenses needed for solution preparation and complete system cleaning. Also, the large amount of solution used (in order to allow a decrease in the cost and implies a large area/time of the deposition) with large wastefulness contributes to high cost. Clearly, this technique is only ideal in an "industry-like" manufacturer and must be avoided for laboratory research use (small areas, small amount of solution to be deposited) except naturally for the "tuning" of new materials deposition.

2.3.4 Self-Assembled Technique

Several other systems can be used for small and large area deposition of organic semiconductors. As they are not much used, only a brief description is given here. One of the most important for small areas is the self-assembled technique.

The self-assembled deposition technique is, perhaps, the most precise and efficient for an exact film molecular conformation. As the name indicates, the technique takes advantage of the property of some organic semiconductors (polymers) to, under specific chemical solutions environment, perform a "self" aggregation, on a molecular scale. The technique provides a very uniform film, even when the thickness is below the nanometer. On the other hand, it is strongly dependent on the accurate chemical samples and solutions preparation, and limited by the number of polymers that can be deposited. The ideal target are small sensors based on polymeric semiconductors, where a precise molecular architecture is necessary to improve the signal response but our some preliminary results in OLEDs have also shown its potential application specifically as hole injection and transport layer.

The principle is very simple: a substrate is submerged into a solution during a specific time, where, due to the electrostatic attraction force, a polymer is deposited. During immersion, the polymer is left to "self-organize" in the substrate surface. The substrate is then emerged from the polymer solution and immerged into another containing de-ionized water. The deposited material shows a structural organization at molecular scale, in a very thin film, usually denominated as molecular layer. It is possible to deposit multiple molecular layers intercalating a solution of an anionic polymer with a solution with another of a cationic polymer. This deposited bi-layer is held united by Van der Waals forces. Naturally the total of deposited bi-layers can be as many as desired depending, as expected, on the desired final application. Figure 2.19 shows a scheme of the process.

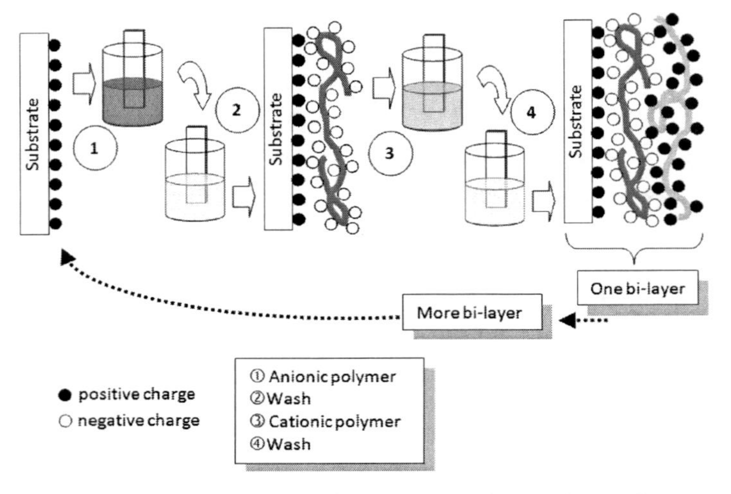

Figure 2.19 Simple schemes that show the steps for a polymer deposition by self-assembled technique. A complete cycle is shown to obtain a bi-layer. More "bi-layers" can be added by repeating all cycle as needed.

The advantages of this technique are the following:

(1) It is a very precise technique on a molecular scale;

(2) The final film shows excellent uniformity and homogeneity;

(3) Avoid the rheological problems of the ink-jet printing;

(4) Ideal for a nanostructured semiconducting polymeric film;

(5) When automatized, it is very simple to use.

On the other hand, there are some disadvantages that, in some cases, are crucial and can make the technique impracticable:

(1) The success of self-organization, strongly depends on a precise solution pH, in some cases is a bit hard to control;

(2) Multiple layers depend always on the "bi-layer" concept;

(3) Only a restrict number of polymers can be used, and it is not adequate for small molecules (that do not have a molecular chain);

(4) Only small areas can be deposited;

(5) The precise automatization is relatively expensive, taking into account the limited use.

2.3.5 Final Device Fabrication Considerations

The device deposition/fabrication techniques herein reported are, of course, not unique. There are many other techniques that can be used for specific situations, materials, and applications. Considering that these are not currently used for rare-earth or transition-metal organic complexes-based OLEDs, they will not be specifically described. These techniques are applied from small to large area device production and involve systems as different as screen-printing, doctor blade, roll-to-roll, etc.

As a final consideration, the choice of the best fabrication technique depends clearly on two distinct, but interconnected, issues:

(1) Physical and chemical properties of the organic semiconductor; and;

(2) The final device target (application, lifetime, area, patterning, and deposition precision, etc.)

Some others points, related with the both above-mentioned conditions, can be used as guidelines about material processing and device fabrication:

(1) Polymers and some coordination compounds cannot be processed by thermal evaporation (molecules degrade with temperature);

(2) Small molecules can only be processed by thermal evaporation, due to their incapacity to form a film from a solution;

(3) The last point can be overcome by embedding the small molecules into a conductive organic polymeric matrix, in order to process the material as a hole, from a solution.

All of these points will be further detailed in relevant examples throughout this book.

References

1. Singh, J. (1995) *Semiconductor Optoelectronics*, McGraw-Hill, USA.

2. Housecroft, C. E. and Sharpe, A. G. (2008). Inorganic Chemistry, 3rd edn., Prentice Hall, Italy.

3. Santos, G. (2010) Polymer Light Emitting Diode using PFT – poly (9,9'-n-dihexil-2,7-fluorenodiilvinylene-alt-2,5 thiophene), *J Nanosci Nanotechnol*, **10**, 2776–2778.

4. Silva, V. M. (2006) Effect of field cycling on the ac and dc properties of Alq3 device, *J. Non Cryst. Sol.*, **352**, 1652–1655.

5. Pereira, L. (2003) Green OLED's: electroluminescence and the electrical carrier transport, *MRS Symposium Proceedings Series: Organic and Polymeric Materials and Devices*, **771**, 99–103.

Chapter 3

Measuring the Electrical and Optical Properties of OLEDs

The measurement of the main electrical and optical properties on an OLED (usually known as figure of merit) is the most fundamental task in order to get a complete characterization and, sometimes assess the device viability. This chapter briefly describes the physical basics for the understanding of the measured data. Also, the most common electrical and optical parameters are explained, based on the former physical description. Finally, explanations are directed to the device viability and efficiency.

3.1 Introduction

Although the electroluminescence study of organic semiconductor-based devices uses knowledge from the inorganic ones, a special attention must be paid to the specific differences of the organic materials. In general, the physical approach starts using known models derived and applicable to inorganic devices, and then tries to adapt these models to the own properties of the organic semiconductors. The differences are not difficult to see; the problem is on how to adapt the current models.

Considering that most of the organic semiconductors (polymeric and small molecules) used in OLEDs form disordered amorphous thin

Organic Light-Emitting Diodes: The Use of Rare-Earth and Transition Metals
Luiz Pereira

www.panstanford.com

films, without a macroscopic crystalline lattice, it is not a good idea to adopt directly and a straightforwardly the physical mechanisms developed for molecular crystals. Due to the absence of localized states, the electrical carrier transport is usually not a coherent movement inside well-defined energetic bands, but a stochastic process of hopping between localized states, that is responsible, for instance, for the very low-carrier mobilities found in organic semiconductors. As a consequence, the excitations are located either in the individual molecules or in a small amount of monometric unities (if a polymer chain is considered). Usually, these have the largest ligation exciton energy (some dozens of eV). Furthermore, most OLED organic semiconductors have a bandgap of about 2–3 eV (and sometimes larger). This implies that the intrinsic concentration of the thermally generated free carriers is usually disregarded (less than 10^{10} cm^{3}) and, from this point of view, the organic semiconductors are more "insulators" than "semiconductors." On the other hand, and again contrary to the inorganic semiconductors, the organic materials exhibit a large number of intrinsic impurities that act like traps and not as extrinsic sources for the carriers, which are moveable. Some exceptions exist, but once again, are disregarded. The origin of the traps found in organic semiconductors can be the residual impurities remaining from the chemical synthesis but also structural traps due to molecular disorder and/or molecular conformation. In several situations, the environmental conditions (e.g., oxygen, humidity, etc.) are equally responsible for changes in the materials properties or (as cited in last chapter) in the device behavior.

In Section 2.1, a summary of the OLED electroluminescent process is given. For simplicity, Fig. 2.1 (and extensive to Figs. 2.8 and 2.12) shows only the spatial variation of the molecular levels inside a bandgap. But, we must remember that these kinds of semiconductors are disordered materials without a well-defined band structure.

In the absence of doping, the interfacial electrical dipoles, and other effects (assuming a vacuum leveling), the energy barrier for the carrier injection is, in a first approximation, given by the energy offset between the metals or TCO (cathode and anode, respectively) work-functions and the organic energy levels. Although with relatively good results, the experimental data can reveal a different situation [1].

The carriers injected into the organic semiconductor are transported across the material, from one electrode to another, by means of the external applied field (voltage). Due to the disordered transport, this process is usually described as a hopping among locations with different energies and distances (variable range hopping). Additionally, the carriers can be intermittently captured in energy levels inside the gap, with origin at impurities and/or structural traps. Adding to the low mobility due to the hopping process, the resulting real mobility for a carrier in organic semiconductor is very low (typically about 10^{-3} to 10^{-7} cm^2/Vs) at room temperature, and in many cases strongly dependent on temperature and applied electrical field. With these low mobilities and a neglected free carrier density, even without energy barriers to decrease the carrier injection, the metal–organic (or TCO) contact can inject a carrier density higher than the value in the thermal equilibrium inside the organic semiconductor (bulk). This effect leads to the formation of spatial charges that reduce the electrical field at the injecting contact and therefore "block" further carrier injection.

It must be noted that the necessity of using contacts with different work-functions to allow the double carrier injection, leads to the appearance of a built-in potential V_{bi}, across the organic layers, which in some cases is not neglectable. Disregarding the shift in the energy levels due to the interfacial dipoles, the V_{bi} is equal to the difference between the electrodes contact potential. Its physical importance is the reduction of the external applied voltage (V_a), and the consequence of a non-null balance of a drift electrical current in forward bias, if the V_a will be higher than V_{bi}. Thus, the knowledge of V_{bi} is crucial and must be considered in the equation that describes the OLED carrier injection and transport. In the development described in the next section, the effective potential V is defined as $V = V_a - V_{bi}$ and represents the potential effectively applied across the organic device bulk, when under forward bias.

3.2 Carrier Injection and Transport in Semiconductors

There are two regimes that can limit the electrical current in device operation: the injection limit and the bulk limit. The first is electrode–organic interface dependent; the second is bulk dependent. Let's now see the essentials of each one.

3.2.1 Injection Limit: The Schottky Barrier at Electrode–Semiconductor Interface

Consider for instance an n-type semiconductor [a]. What happens when a metal is deposited, with a physical contact with it? Consider for instance a metal (or a TCO instead) and a semiconductor (at this moment it's not important whether it is organic or inorganic), with an essential property: the metal work-function is higher than the semiconductor.

The basic physical parameters can be viewed in Fig. 3.1; and is the departure point for the discussion.

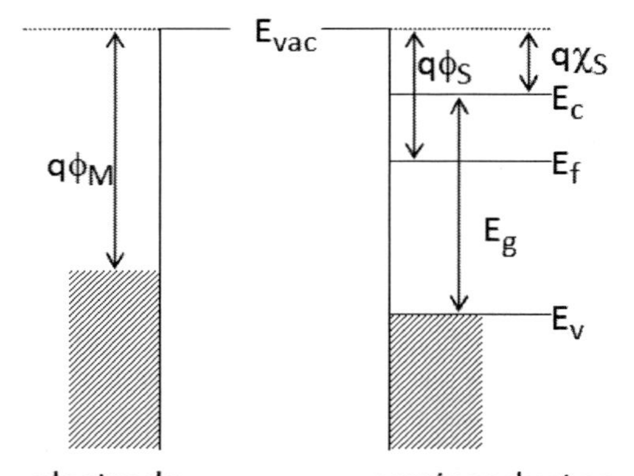

Figure 3.1 Basic physical parameters for a metal and a semiconductor when isolated.

In the figure above, E_F, E_{vac}, E_C, E_V denote the Fermi level, the vacuum level, the conduction band energy, and the valence band energy, respectively; χ_S is the semiconductor electronic affinity, q the elementary charge, and $\varnothing_M$ and $\varnothing_S$ the work-functions of the metal and the semiconductor, respectively.

---

a Similar description and conclusion can be made for a p-type semiconductor, regardless the differences in the band curvature (further along designed as band bending) but with the same discussion and applicable models.

When the metal and the semiconductor are joined through a common interface, several electrical, besides morphological issues will occur. The final result (considering the energetic physical modifications) is show in Fig. 3.2.

In this figure $q\o_D$ is the diffusion potential ($\o_D = \o_M - \o_S$).

Figure 3.1 shows the materials' physical parameters before the connection. The differences between them will determine the final property of the interface after the junction. When an n-type semiconductor is connected to a metal with a high work-function, some electrons migrate from the semiconductor to the metal, leaving positive sites near the interface (i.e., without electrons or *depleted*). This results from the need to align the Fermi level. The electron energy in the semiconductor bulk is now different and an electrical field arises in the semiconductor gap directed to the metal interface. This causes a negative charge to appear in the metal interface compensated, in turn by a positive charge in the semiconductor interface. Since the donor density in the semiconductor is much lower than the electrons in the metal, the non-compensated donors will occupy a region in the semiconductor interface, denominated *depletion width* (w_d). This is represented by an upward band curvature (upward *band-bending*).

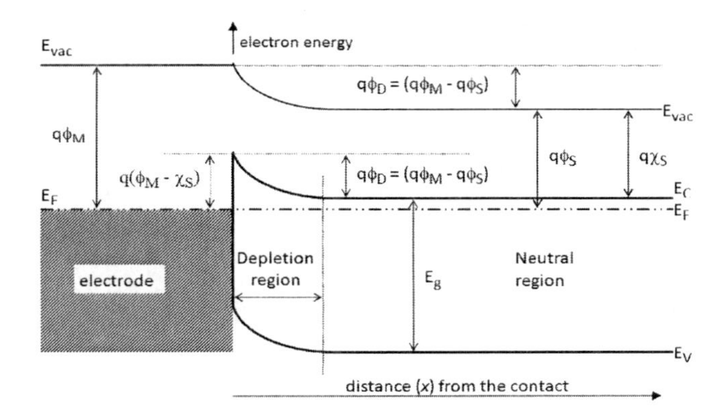

Figure 3.2 Formation of a Schottky barrier. In this example we have an n-type semiconductor and an electrode (for simplicity consider a metal) where ($q\phi_m > q\phi_s$).

The barrier height found by an electron when moving toward the metal (from the semiconductor side) – V_b, is high than $q_{\phi D}$ by $(E_C - E_F)/q$. This barrier can be expressed using the semiconductor electronic affinity and is given by:

$$V_b = \phi_M - \chi_S \tag{3.1}$$

In this simple model no interfacial states are considered. If a positive potential is applied to the metal, the energy levels in the semiconductor are increased and therefore more electrons flow to the metal increasing the current due. This case corresponds to a junction barrier decrease and is denominated *forward bias*. On the contrary, if a negative potential is applied to the metal, the junction barrier is increased and no more electrons can flow to the metal (at least). Only a small electrical current is measured, regardless of the applied voltage magnitude. This case is denominated *reverse bias* and the flowing current is due to the minority carriers. This electrical current is known as *saturation current*. When the electrical current is clearly dependent on the potential polarity, we are in the presence of a rectification.

The dependence of the electrical current I with the applied potential V can be done by [2]:

$$I = I_0 \left[\exp\left(\frac{qV}{kT}\right) - 1 \right] \tag{3.2}$$

where k is the Boltzmann constant and T the temperature. Considering that the current flows across the contact by thermionic emission [3] in both directions, the value for I_0 can be done by [4]:

$$I_0 = AA^* T^2 \exp\left(-\frac{qV_b}{kT}\right) \tag{3.3}$$

in which A is the contact area and A^* the Richardson constant, given by

$$A^* = \frac{4\pi m^* qk^2}{h^3} \tag{3.4}$$

m^* being the carrier effective mass.

Equation 3.2 is derived considering that V_b does not vary with the applied potential. This is not true. As a general case, V_b decreases as the external applied potential increases. Suppose that V_b is reduced by an amount of ΔV_b when an external potential V is applied. In this case the effective barrier is given by

$$V_e = V_b - \Delta V_b \tag{3.5}$$

and the Eq. 3.2 becomes

$$I = I_0 \left[\exp\left(\frac{qV}{nkT} \right) - 1 \right] \tag{3.6}$$

where n is called the ideality factor. This ideality factor measures the approximation between the experimental electrical junction behavior and the theoretical estimate. The farther n is from one, the less "ideal" the junction will be.

In a perfect metal–semiconductor interface, all processes are exclusively controlled by the work-function values as explained in last section. Nevertheless, the real situation is different. The conjugation of the so-called *image force lowering* with the electrical field tends to decrease (sometimes significantly) the Schottky potential barrier. Understanding this will allow us to have a much more realistic electrical behavior.

The explanation is somewhat simple, although the mathematical models are not so simple and will not be developed here. Let us consider, for instance, an electron inside the semiconductor but near the metal surface: its presence will induce a positive charge inside the metal (and near the surface) that tends to attract the electron towards the metal. The electrical potential near the surface decreases due to the appearance of its induced "charge image." This implies that an additional energy is necessary to repel the electron. The reduced electrical potential of an electron in the neighboring metal surface gives an extra semiconductor energy band bending in a region immediately adjacent to the metal surface.

The decrease in the potential barrier by the image force, under forward bias, is then given by [5]

$$\Delta V_b = \sqrt{\frac{q^3 E}{4\pi\varepsilon_r\ \varepsilon_0}} \tag{3.7}$$

where E is the electrical field.

A contact between a semiconductor and a metal is sometimes not so physically perfect as the predicted ideal model. The interface can present small structural defects that play an important role in the final macroscopic electrical behavior as these may provide levels of energy accessible to carriers.

The typical roughness of the surface of organic semiconductors and their molecular nature are easily affected when another material (particularly a metal) is deposited. In a few extreme situations, metal diffusion may occur, destroying the device. All these factors lead to a metal–semiconductor interface region with a non-neglectable defects density. This interface region is usually called *interfacial layer* [6] and the applied voltage V is now partially applied through the semiconductor and through the interfacial layer.

The voltage drop across the interfacial layer can be expressed by [6]

$$V_i = \frac{\delta}{\varepsilon_i}\left(\varepsilon E_{\max} - \frac{Q_s}{\varepsilon_0}\right) \tag{3.8}$$

where ε_i is the relative electrical permittivity of the interfacial layer. The electrical field in the semiconductor changes the value of V_i across that layer, and therefore, not all applied voltage is effectively applied across the depletion region. When the layer thickness is less than (or about) 50 Å, the more important carrier transport process across the interfacial layer is the tunneling.

3.2.2 Application of Schottky Barriers to Organic Semiconductors: The Richardson–Schottky Emission

The Schottky barrier considerations discussed in the last section are, of course, based on a model developed for inorganic semiconductors, in particular the crystalline ones. The real approach to a complex organic model is clearly quite difficult. In general, when an effective potential barrier exists, the model can be applied to organic semiconductors, as many authors have shown [7]. Although good results, two constrains may exists: the experimental fit may be obtained with a very high ideality factor (up to 4–5 in extreme) and with an unrealistic Richardson constant (A^*), completely different of the theoretical prediction that is 120 (m*/m) Acm^{-2}K^{-2}. But, in truth, excluding the very high ideality factors, the value for A^* for inorganic semiconductors is always a little unrealistic. Nevertheless, the existence of a complex interface between an electrode and an organic semiconductor, leads, in many cases, to a significant deviation from the theoretical model developed.

Based on the two distinct effects described before, new derived models are currently employed in metal–organic semiconductors. The first, that takes into account the image force effect, is the Richardson–Schottky thermionic emission; the second, considering the predominance of interfacial states, is the Fowler–Nordheim tunneling effect. The last will be discussed in next section. Lets for now look into the Richardson–Schottky emission. Basically, this emission follows the principle afore-mentioned for the Schottky barriers (a thermionic emission over a potential barrier) and the electrical current behavior can be given by

$$I = AA^*T^2 \exp\left(-\frac{qV_b}{kT}\right)\exp\left(-\frac{\sqrt{q^3/4\pi\varepsilon_r\varepsilon_0}\sqrt{E}}{kT}\right) \tag{3.9}$$

that simply results from subtracting the corresponding term of potential barrier reduction by image force effect to Eq. 3.7 under the thermionic model. It is clear that this equation is a more accurate representation of the metal–semiconductor interface.

Naturally other effects can be incorporated into the Schottky equation. Nevertheless, when the electrical behavior of an organic device is injection controlled, the Richardson–Schottky thermionic emission is the most commonly used model. Again, this model is specifically derived for inorganic semiconductors, where the valence and conduction bands (that results from the periodic crystalline lattice) are extended, yielding a large mean free path for the electrical carriers. Contrarily, on organic semiconductors, where the electrical conduction resembles the hopping between molecular orbital for which the carrier is more or less localized, we expect a very small mean free path (realistically speaking, it must be of molecular distance). Adding to that clear difference, the structural disorder typical of organic semiconductors (that don't have a structured crystalline lattice), certainly originates randomly distributed potential barriers, such structural defects that have their own local potential. This brings a higher difficulty for carrier injection into the organic semiconductor, comparatively to the inorganic crystalline model. As a result, we cannot expect such a simple behavior as described for the Schottky emission model, although we have a reasonable experimental fit. The main problem remains on the extracted Richardson constant because to fit the experimental current, an unrealistic value must be considered. Also, as the model depends on temperature and electrical field, additional problems can appear, as in some cases the expected dependence cannot be verified experimentally.

So, care must be taken when using the Schottky emission model to determine the carrier injection into an organic semiconductor. Naturally that model is mainly useful to describe the physical bases of the metal–organic semiconductor interface, giving, in a first approach an idea about the expected macroscopic behavior. Secondly, it's a good starting point for further refinements considering the specificities of the organic semiconductors. In general, the model fits reasonably when using a relatively low applied potential.

3.2.3 A Special Case: Ohmic Contact

Strictly speaking, the ohmic behavior in a metal–semiconductor interface is not exactly an "injection" because the electrical carriers don't need to overcome a potential (energy) barrier. Usually it's called ohmic contact. However, it is still an interface process and we need to study the physical process at the metal–semiconductor junction that is responsible for gives this behavior.

An ohmic contact is a low (or almost null) electrical resistance contact that allows a carrier to flow in both directions (metal → semiconductor and semiconductor → metal) depending only on the electrical field direction imposed by the applied voltage. The importance of this kind of contact is that, when having an ohmic contact behavior, the device can be bulk limited and, in specific situations, this is an advantage. Ideally, the ohmic contact resembles an electrical charge reservoir when a voltage is applied. In order to understand this phenomenon, let us consider an n-type semiconductor with a work-function higher that the metal used for the contact. For low-applied voltages when the electrical flow is controlled by the semiconductor bulk, we obtain a linear relationship between the electrical current and that applied voltage. For higher applied voltages, it is possible that the injected carriers from metallic electrode control the electrical current flow and therefore a space charge limited current (SCLC) can occur (a more detailed discussion will be given later).

Consider the metal–semiconductor interface when $\emptyset_m < \emptyset_s$. In order to achieve thermal equilibrium, electrons must flow from the metal to the semiconductor lower energy levels, implying an excess of negative charges in the semiconductor surface. If an external positive voltage is applied to the metal, there is no further barrier to the electron flux from the semiconductor to the metal; and under this positive voltage applied to the semiconductor, the barrier for the electrons flowing to the semiconductor is approximately equal to $\chi_S - \emptyset_S$, that is low, even for a non-heavily doped semiconductor. Therefore, electrons can equally flow to the semiconductor at ease, without a "real" barrier.

3.2.4 Injection Limit: The Tunneling Effect

The tunneling effect is another way to explain the carrier injection into semiconductors from a metal contact. The physical basic mechanism can be viewed (and treated) as a direct tunneling (when the carrier tunnel through the total barrier) or a Fowler–Nordheim model (when the carrier tunnel only a partial width of the barrier). For a brief explanation, we can use the main idea developed for the interfacial states in the junction. Although the tunneling effect is not dependent on Coulomb interactions, the presence of a significant density of interfacial states may promote the process.

The tunneling effect generally occurs when the metal–semiconductor interface band energy is "distorted" by the presence of the cited large density of interfacial states. Considering, again, an n-type semiconductor, the band bending shown in Fig. 3.2, is affected by the presence of much available levels of energy. From the diagram showed in Fig. 3.3, a new and simple representation of the band bending can be made, considering that the width of the interfacial layer (δ) is the physical location of the energy levels that can allow the effect. It must be noted that the tunneling effect is a pure quantum effect, in which the dual nature of the carriers (particle and wavelength) is taken into account.

The detailed analysis of the tunneling effect falls beyond the scope of the present work and is not given here. For more complete information refer to some excellent bibliographies [8]. Herein, we just concentrate on the phenomena description and the consequent model for electrical current derived for the inject carriers.

Consider, for instance, an electron in the metal surface, near the junction with the semiconductor. For this carrier (particle) a wavelength function is associated, describing, in the corresponding Schrodinger equation, the available energy levels. If the wave-function can be extended to the semiconductor surface near the junction, then a finite probability exists for the electron to cross the barrier and to further be located in the semiconductor region. It is not too difficult to understand that this probability increases with the interfacial density of states increase.

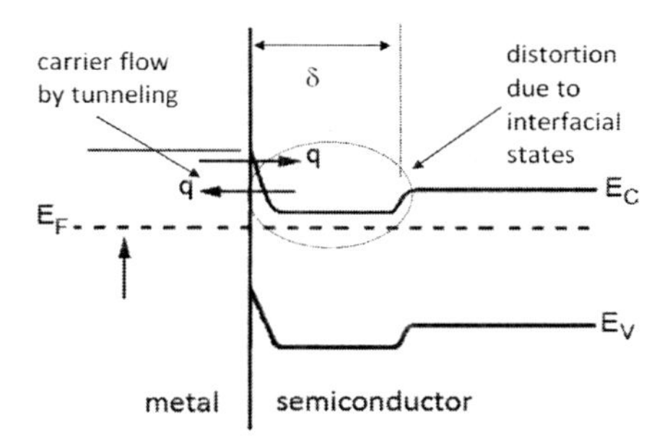

Figure 3.3 Energy band diagram for a metal–semiconductor (n-type) tunneling effect. The diagram is derived from the Schottky barrier in the presence of interfacial states of energy.

In Fig. 3.4, a simple diagram, derived from Fig. 3.3, is shown.

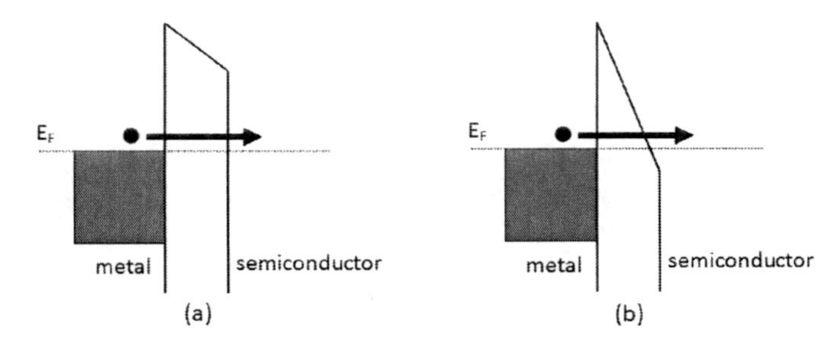

Figure 3.4 Simplified diagrams of energy bands for (a) direct and (b) partial barrier tunneling.

The tunneling effect has a strong dependence on the electrical field but is generally, not affected by the temperature.

Many equations have been derived for the tunneling effect, depending on the "geometry" of the barrier. The most common barriers types are square, triangular, trapezoidal and parabolic, giving, as expected, different tunneling models with different macroscopic electrical current behaviors [8]. Currently, the most used is the triangular barrier and we only focus this type.

It is also common to, distinguish between direct and partial barrier crossing as mentioned. This simple difference is shown in Fig. 3.4 for a triangular potential barrier.

The most common model is precisely the partial triangular barrier tunneling, also known as Fowler–Nordheim model. In this case, the electrical current behavior can be given by [9]:

$$I = A\frac{A^*q^2E^2}{V_b\alpha^2k^2}\exp\left(-\frac{2\alpha V_b^{3/2}}{3qE}\right) \tag{3.10}$$

with

$$\alpha = \frac{4\pi\sqrt{2m^*}}{h} \tag{3.11}$$

Although the process is expected to be temperature independent, very low temperature activation can be observed. As a final remark, it's commonly accepted that the tunneling effect is one of the most important processes involved in the high increase of the reverse bias current.

3.2.5 Bulk Limit: Space Charge

Another limit to the carrier regime in a device is the bulk limit. In this case, the electrical current is limited by the region inside the semiconductor instead of the electrode interface. The most well-known model for carrier transport inside a semiconductor is the SCLC, which can be affected by the existence of energy levels acting as carrier traps. Here we present a description of the basic physics of this model, either trap-free or trap-dependent scenarios. A brief description of the ohmic bulk conduction is also presented.

3.2.5.1 Ohmic conduction

Consider a metal–semiconductor junction under a forward bias. As the applied voltage increases from zero, the energy band bending decreases, until the voltage is enough to cancel the voltage drop over the junction. In this situation, the barrier to the carrier flow disappears and the process is "bulk-dependent."

In the simplest scenario, the semiconductor can be treated as a simple electrical resistance and the electrical current is plain given by:

$$I = Aqn\mu\frac{V}{d} \tag{3.12}$$

in which d the semiconductor thickness. In this regime, all the carriers injected into the semiconductor are immediately flushed out. Is the ohmic regime, where $I \propto V$. It must be noted that this regime is often only observed in relatively low applied voltages; and if the device is injection-limited, the most common scenario is an impossibility to observe the ohmic behavior in current–voltage data plot.

3.2.5.2 Trap-free SCLC

By further increasing the applied voltage, more carriers are injected into the semiconductor. If, at least, one of the electrodes is a very good carrier injector, it is possible that the injected carrier density flushed is higher than that of the thermally activated carriers inside the semiconductor. In this case the carriers are not immediately flushed out forming space charge regions.

As an immediate consequence, these space charge regions invert the normal band bending (of the non-compensated carriers) due to the carrier excess. As a consequence of that, the carriers injected into the active region are attracted to the injection electrode by the "new" electrical field. Therefore, the total electrical current is given by the carrier diffusion, i.e., flows to the high free carrier density gradient. The relationship between electrical current and applied voltage has been extensively study by use of an insulator model. This model, [10] known as Mott-Gurney law, is valid for a unipolar trap-free semiconductor considering that the carrier mobility is independent of the electrical field. With these constrains, this law can be expressed as follows:

$$I = \frac{9}{8} A\varepsilon\mu \frac{V^2}{d^3} \tag{3.13}$$

Following the former above discussed ohmic conduction, a transition from this behavior to the Mott-Gurney law must occur when the applied potential increases. This specific applied voltage (V_Ω) can be easily obtained, by combining Eqs. 3.12 and 3.13. The solution is the following:

$$V_\Omega = \frac{8qnd^2}{9\varepsilon} \tag{3.14}$$

The immediate application is for the determination of the density of states n, when knowing the others parameters.

3.2.5.3 Trap-dependent SCLC: shallow levels

Semiconductors in general are, physically speaking, much more complex that the simple model of defect-free perfect bands. In many cases, and perhaps for all organic semiconductors, the morphological differences at the molecular level, caused by growth and device fabrication lead to structural defects that may be electrically active, i.e., influence the carrier transport. This is globally accepted for organic materials, in which the local conformation of relatively independent molecules in a thin film originates a very complex and non-crystalline semiconductor. Some of these intrinsic defects may have energy levels near the Fermi level and are called shallow levels. The SCLC model requires a new approach when these shallow levels acts as carrier traps.

Trapping carriers in semiconductor bulk shallow levels will change the total electrical charge density (ρ_{total}) because we must add the amount of charge trapped (ρ_{trap}) to the free charge (ρ_{free}) i.e., $\rho_{\text{total}} = \rho_{\text{trap}} + \rho_{\text{free}}$. The relationship between the free charge (available for electrical current) and the total charge is simply given by:

$$\theta_0 = \frac{\rho_{\text{free}}}{\rho_{\text{free}} + \rho_{\text{trap}}} \tag{3.15}$$

Using this equation the electrical current model for this new situation is given by [11]:

$$I = \frac{9}{8} A\varepsilon\mu\theta_0 \frac{V^2}{d^3} \tag{3.16}$$

The factor $\theta = \mu\theta_0$ is known as the *effective mobility*. Clearly, this effective mobility is lower than that obtained for the trap-free SCLC. The trap levels, in a first consequence cause a decrease in the carrier mobility.

When the regime changes from ohmic to SCLC shallow-trap dependent the corresponding voltage (V_Ω) can be obtained:

$$V_\Omega = \frac{8\theta q n d^3}{9\varepsilon} \tag{3.17}$$

A further analysis of the current–voltage behavior can be made. At low applied potentials, an ohmic behavior is expected, i.e., $I \propto V$. Increasing the applied voltage, the electrical current will depend on a V^2, i.e., $I \propto V^2$, independent on trap-free or shallow-trap SCLC. In this last case, only the magnitude of the electrical current is decreased by θ_0. As the applied voltage increases, the shallow levels become occupied by trapped carriers, and do not contribute to the total flowing current. When all shallow levels become occupied, a saturation regime is expected, with the consequent electrical current abrupt increase. The applied voltage for this scenario is called *trap-fill limit voltage* (V_{TFL}) and is given by [12]:

$$V_{TFL} = \frac{q n_t d^2}{2\varepsilon} \tag{3.18}$$

Finally, when the equilibrium is achieved, the electrical current will again depend on a quadratic way of potential because all shallow trap levels are occupied and does not influences the electrical carrier flow. Figure 3.5 shows the schematics of the expected current–voltage. The logarithmic scale is useful to show the expected slopes in the different regimes.

3.2.5.4 Trap-dependent SCLC: deep levels

A scenario in which the trap energy level is relatively far from the Fermi level, is consider as being a *deep level*. These deep levels interact with the electrical carrier, by a different behavior which can be observed for the measured current–voltage data.

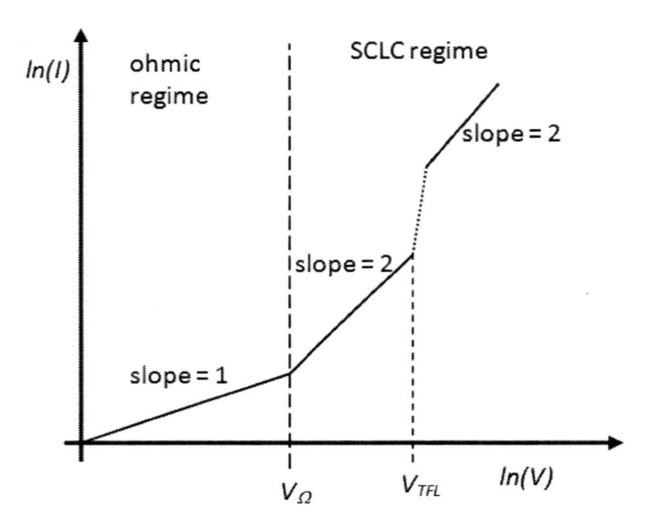

Figure 3.5 Scheme of expected electrical current dependence on applied voltage in a SCLC trap-free or shallow trap-dependent. The ohmic regime is show at low applied voltages.

Assuming that the difference between the energy level of the deep trap and the Fermi level is of some amounts of kT the V_{TFL} now, is given by [13]:

$$V_{TFL} = \frac{qn_{tr}d^2}{2\varepsilon}\exp\left[-\frac{q(E_F - E_t)}{kT}\right] \tag{3.19}$$

with n_{tr} representing the trap density at room temperature.

The existence of discrete trap levels is not the most common situation. Considering the intrinsic disorder in organic semiconductors, a much more realistic situation would be to consider that the trap levels are distributed around a deep energy value. The most useful distributions are the exponential and Gaussian.

In the exponential distribution of trap deep levels and assuming that the carrier mobility is electrical field independent, the electrical current is given by [13, 14]:

$$I = An\mu q^{(1-m)}\left[\frac{2m+1}{m+1}\right]^{m+1}\left[\frac{m}{m+1}\frac{\varepsilon}{n_{t0}}\right]^{m}\frac{V^{m+1}}{d^{2m+1}} \tag{3.20}$$

where $m = T_t/T$ and $m > 1$. This implies a supra-quadratic dependence of the electrical current on the applied voltage. The expression for V_Ω is now given by:

$$V_\Omega = \left(\frac{n}{N}\right)^{1/m}\left[\frac{qn_t d^2}{\varepsilon}\right]^{m+1} \tag{3.21}$$

with N being the band density of states. The term $E_t = kT/q$ is the characteristic energy of the exponentially states distribution and can be obtained from the temperature dependence of m. The plot of m vs. $1/T$ gives a linear function with zero intercept.

Instead an exponential deep trap distribution, a Gaussian distribution is also commonly considered. In that case, the electrical current is then [14]:

$$I = An\mu q^{(1-m)}\left[\frac{2m+1}{m+1}\right]\left[\frac{m}{m+1}\frac{\varepsilon}{n'_{t0}}\right]^{m}\frac{V^{m+1}}{d^{2m+1}} \tag{3.22}$$

with

$$m = \left(1+\frac{2\pi\sigma_t^2}{16k^2T^2}\right)^{1/2} \quad \text{and} \quad n'_{t0} = \frac{n_t}{2}g\exp\left(\frac{E_{tm}}{mkT}\right) \tag{3.23}$$

And again we have $m > 1$.

It can be easily observed that both equations (3.20 and 3.23) express the same current dependence on applied voltage, i.e., $I \propto V^n$ where $n > 2$. We this expect, regardless of the deep trap density profile, that, after the ohmic regime observed at low applied voltages, the current will increases with supra-quadratic dependence for $V > V_\Omega$. In a similar way to that described for shallow traps when the applied voltage achieves the V_{TFL} values, all the deep traps become occupied and a quadratic dependence of the current over the voltage must occur. Figure 3.6 shows a schematic of the expected current–voltage behavior in this situation.

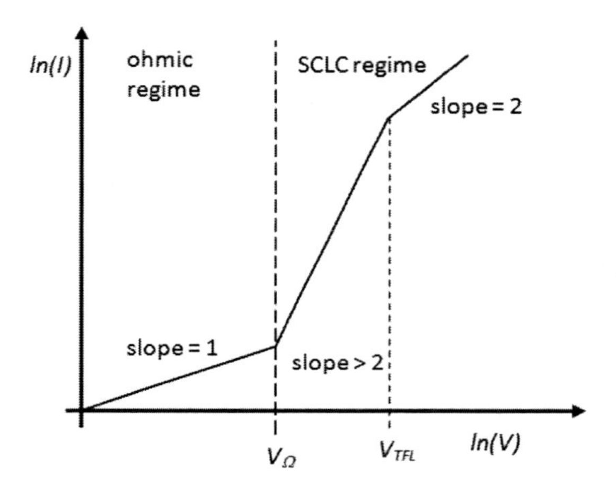

Figure 3.6 Scheme of expected electrical current dependence on applied voltage in a SCLC deep trap-dependent. The ohmic regime is shown at low-applied voltages.

3.2.6 Bulk Limit: Poole–Frenkel Effect

The Poole–Frenkel effect is a thermionic process that may promote a carrier to be located in a deep energy level due to an effective decrease of this deep level. This may occur by work of an external electrical field (applied voltage). In fact, a high electrical field near a localized carrier can change the "effective depth" of the energy state. This model has been tentative attributed to organic semiconductor bulk electrical conduction with, relative success in fitting the experimental data [15, 16].

To better explain this phenomenon, let us consider a carrier in a deep state at the depletion region formed by the metal–semiconductor interface. Applying an external field changes the local potential, as viewed from the metal surface towards the semiconductor region. The electrical potential maximum shifts from the equilibrium point to a different distance of the metal, as schematized in Fig. 3.7.

The effective depth of the energy level can be quantified as follows:

$$\Delta V = -\sqrt{\frac{q^3 E}{\pi \varepsilon}} \tag{3.24}$$

and an increase of free carriers is obtained.

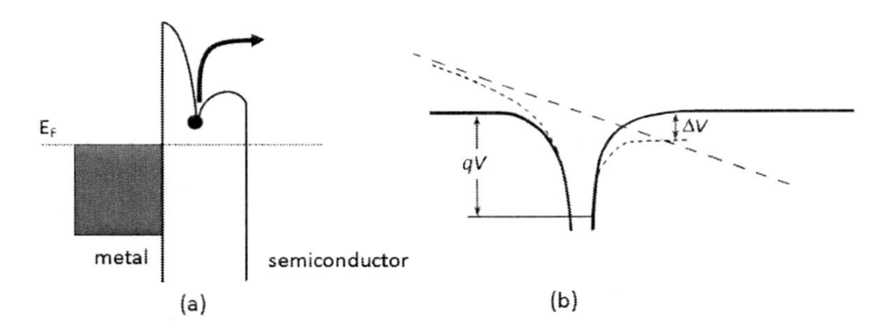

Figure 3.7 (a) General scheme of the Poole-Frenkel effect and (b) a more detailed energy diagram, showing the change (ΔV) in the effective "depth" of a deep state, by application of an external field of energy qV.

An analytical relation between the electrical current and the applied potential can be obtained [11]:

$$I \cong A\varepsilon\mu\theta_0 \frac{V^2}{d^3} \exp\left[\frac{q}{kT}\sqrt{\frac{qV}{\pi\varepsilon d}}\right] \tag{3.25}$$

The main difference is that the mobility now depends on the electrical field, according the following relationship [11]:

$$\mu(E) = \mu_0 \exp\left(\beta\sqrt{E}\right) \tag{3.26}$$

in which μ_0 is the SCLC trap-free mobility (also called zero-field mobility) and β a constant. It must be noted that, although the Poole-Frenkel effect is a bulk effect, some of the experimental data is sometimes interpreted as an injection effect.

3.2.7 Bulk Carrier Transport: The Hopping

The herein presented electrical models fit reasonably to the experimental data; nevertheless, a real approach to the organic semiconductors behavior is quite impossible. These electrical models can be used to help understand the organic device macroscopic electrical performance (and therefore are of a surplus importance) but from a physical point of view, some doubts exist about their "perfect" applicability.

All in all, there is much controversy on which path to follow, in part due to the general knowledge that the organic semiconductors do not have a real energy band; but also due to the fact that many results are contradictory *per se*. Usually, organic semiconductors are considered as disordered systems. On the other hand part of the macroscopic behavior of organic semiconductors electrical transport resembles that of inorganic ones [17, 18]. Finally, some authors points that the electrical carrier injection and transport that even conjugates polymers cannot be perfectly modeled to the inorganic known process [19, 20]. It is natural that the above described models cannot be completely adjusted to organic semiconductors, even because these are usually bi-polar.

Regardless of the injection mechanism, the electrical carrier transport in the organic semiconductor bulk, from the molecular point of view, is considered as a hopping transport process. This hopping transport is usually determined by a thermal-activated carrier jump form a localized site to another. Taking into account the general form of the Poole–Frenkel mobility (Eq. 3.26) the hopping process is characterized by an increase of the β factor accompanied by a decrease of μ_0, when the temperature drops. This hopping mechanism can be described under two distinct models: a modified Poole–Frenkel model [21] or a Gaussian disorder model [22].

The modified Poole–Frenkel model is based on the assumption that the electrical carrier jumps between two localized sites by a de-trapping process assisted by the electrical field and the temperature. It is considered that the carrier is trapped by a Coulomb potential in a charged trap. In this case, the carrier mobility is given by [21]:

$$\mu = \mu_{PF} \exp\left(-\frac{\Delta E_a - \beta_{PF}\sqrt{E}}{kT_a} \right) \tag{3.27}$$

where $(1/T_a) = (1/T) - (1/T_0)$, with T_0 an empirical factor as for $T = T_0$ the mobility is equal to μ_{PF}. ΔE_a is the activation energy and $\beta_{PF} = (q^3/\pi\varepsilon)^{1/2}$.

In turn, the Gaussian disordered model assumes that the hopping process is the result of a carrier jump between sites subject to a Gaussian distribution of energies (energy disorder – σ) or a Gaussian distribution of positions (positional disorder – Σ). In this case, the carrier mobility is given by [22]:

$$\mu = \mu_G \exp\left[-\left(\frac{2\sigma}{3kT} \right)^2 \right] \exp\left\{ C\left[\left(\frac{\sigma}{kT} \right)^2 - \Sigma^2 \right]\sqrt{E} \right\} \tag{3.28}$$

where μ_G is the mobility high temperature limit and C a parameter.

It is obvious that, regardless of the hopping model, the macroscopic electrical current will be different from that obtained by the classical models given before, as the carrier mobility is different. The best way to test these hopping models is to calculate the carrier mobility under temperature. In the modified Poole–Frenkel hopping model, and from Eq. 3.26, the plot of $ln(\mu_0)$ and β and versus T^{-1} must present a linear fit, from which the useful parameters of Eq. 3.27 are extracted (μ_{PF}, T_0 and β_{PF}); in the Gaussian disordered model, the plot is now of $ln(\mu_0)$ and β

and versus T^{-2} and a linear fit must also be obtained. Further parameters extraction from Eq. 3.28 can be performed (μ_G, C, σ, and Σ) taking into account that the parameter C is usually previously determined from numerical simulation.

A more detailed analysis can be viewed, for example, in [23]. Some corrections have been added to the Gaussian disordered model and the numerical solutions are, in some cases, in agreement with the experimental date. Unfortunately, there are no perfect models. The mobility temperature dependence on hopping models is in most cases, not verified in specific ranges, suggesting that the electrical carrier transport in organic semiconductors will be a "mixture" of several effects and physical mechanisms, each one is explained by one of the models given.

3.2.8 Overview About Electrical Properties of Organic Semiconductors: An Attempt to Use the Common Models

We can now discuss the electrical injection/transport of carriers in organic semiconductors, considering the models/discussions given before.

The physical specifications of the organic semiconductors pose a lot of questions (and problems) when we try to use these models (derived under specific conditions) in organic materials. First of all, the organic semiconductors do not have a "perfect" crystalline lattice, like the inorganic ones, for which the models was derived; secondly, the presence of disordered molecular structures on organic semiconductors creates additional problems, not addressed by the cited models; and finally, there are a lot of different results from several authors, even for the same organic semiconductor, leading to a more or less incapacity to establish the exact model for it. If some explanations for this last point appear to be realistic (based on different semiconductor thin film deposition parameters) they only stress the difficulty in understand the real nature of the electrical behavior of organic semiconductors.

A common problem when analyzing an experimental current–voltage data, is to have a relatively good fit to a formerly described physical model, that upon a more consistent analysis (in special with temperature and electrical field dependence) reveals inapplicability: in some cases only for a short region, thus being considered consistent; but for others cases the experimental data deviation from the theoretical model is extended in a large region and the doubts increases. In general one is aware that no perfect models exist (even in inorganic crystalline semiconductors) and the main idea is to choose the "best adjusted" model. This necessity to try to describe the device electrical model is crucial for the device development as some important figures of merit can be improved when a good feedback it's obtained by fitting the data with the selected model. So, even some difficulties, an attempt to model the device electrical behavior is fundamental.

Several methods are used to get some insight about the electrical device nature. As a simple procedure, we can start by observing the electrical field dependence on the electrical current. The aforementioned equations (either for injection or transport) can be re-written in simplified ways. Two analyses can be made: the direct dependence and the normalized differential conductivity. This (normalized) differential conductivity can be expressed as

$$\alpha = \frac{\partial(\log I)}{\partial(\log E)} \tag{3.29}$$

Table 3.1 summarizes the analysis for all the common models.

The analysis of the electrical behavior given by the specific plot and/or the normalized differential conductivity change with applied field as indicated in Table 3.1 is an exceptional tool to establish a possible electrical model. Once again, it must be noted that these plots can help understand the electrical behavior; but it's sometimes very difficult to assert with precision the electrical model because most probably a superimposition of different carrier injection/transport phenomena.

Table 3.1 Physical dependencies of electrical conduction models with the electrical field and normalized differential conductivity

Model	I(E)	log-log plot	α(E)
SCLC/ohmic	$I = I_{SCLC/TCLC}E^n$	$\ln(I) \propto \ln(E)$	N
Poole–Frenkel	$I = I_{PF}E\exp(\beta_{PF}\sqrt{E})$	$\ln(I) \propto \ln(E) + E^{1/2}$	$1 + \frac{1}{2}\beta_{PF}E^{1/2}$
Richardson–Schottky	$I = I_{RS}\exp\left(\beta_{RS}\sqrt{E}\right)$	$\ln(I) \propto \ln(E^{1/2})$	$\frac{1}{2}\beta_{RS}E^{1/2}$
Fowler–Nordheim	$I = I_{FN}E^2\exp\left(-\beta_{FN}E^{-1}\right)$	$\ln(I) \propto 2\ln(E) - E^{-1}$	$2 + \beta_{FN}E^{-1}$

The simplified equations are derived from equations of the respective models: Richardson–Schottky (3.9), Fowler–Nordheim (3.10), SCLC/ohmic (3.12, 3.13, 3.16, 3.20, and 3.22) and Poole–Frenkel (3.25).

The main difficulty arises from the fact that, for most cases, the experimental data resembles one of the most conventional models, which in a first approach, can be viewed as the correct application of the model to organic semiconductors. On the contrary, there are some cases in which a known model from inorganic materials cannot be applied at all. In a specific point of view, the more conventional models are generally good approaches to extract the main device macroscopic properties as they can be correlated with light emission in a relatively simple way. If the mobility is known to clearly change with electrical field and/or temperature, the Poole–Frenkel model can be a first correct choice. Despite the good approximation, the best known approximation for a strong temperature and electrical field mobility dependence, the hopping models are the best choice. But curious, they do not give a simple and straightforward macroscopic understanding of the real magnitude of the device figures of merit. The classical models for carrier injection/transport have the handicap of failing to fit the device behavior under temperature to a greater or lesser extent depending on the organic semiconductor, the morphology, the device geometry and design and the methods (and parameters) for the thin film fabrication. For instance, conjugated polymers tend to exhibit a typical Schottky interface while small molecules have a SCLC or Poole–Frenkel behavior. The best way to surpass these problems is to start by analyzing the device experimental data under the common models (using, for example, the information summarized in Table 3.1) in order to check if the fit results are physically acceptable or not. If no believable results are obtained, more complicated ways must be taken into account. Next section focuses on the relation between electrical conduction and the expected electroluminescent behavior. In a simple description, we will see that, for many known and practical devices, the simple models are quite useful. For further practical interesting examples, see [24–27].

3.3 Carrier Recombination and Electroluminescence

3.3.1 Carrier Recombination

Although conceptually simple, the electron–hole recombination is not so simple as expected being a particularly complex problem in organic semiconductors. Naturally, many simplifications can be made to help understand the device behavior, but, in a similar way to that found for the electrical carrier models, there is not a unique solution.

The main theory that supports the electron–hole recombination is the Langevin theory, where it's assumed that the mean free carrier path is smaller than the electrical capture due to the Coulomb interaction, i.e.,

$$r = \frac{q^2}{4\pi\varepsilon kT} \tag{3.30}$$

In a simplified view, we can assume there is carriers drift (electrons and holes) in an electrical field towards a recombination zone. From the original model and considering that the charge concentration is proportional to the electrical field ($n \propto E$), i.e., for electrical currents controlled by the organic semiconductor bulk, a recombination zone can be expressed by [28]:

$$W_{eh} \cong \left[\frac{2\mu}{3(\mu_e + \mu_h)} \right] d \tag{3.31}$$

In this equation d is the film thickness and μ, μ_e and μ_h are the majority carrier mobility, the electron mobility and the hole mobility, respectively. Naturally, the recombination point is determined by the mobility ratio. Also the electroluminescence region can be viewed as the recombination zone of the faster carriers.

Considering an injected limited device, the recombination zone can be given by [28]:

$$\begin{aligned} W_{eh} &= \mu_{eh} \varepsilon E^2 / d \\ W_{eh} &= \mu_{eh}^2 \varepsilon E^2 / (\mu_e + \mu_h) d \end{aligned} \tag{3.32}$$

The first equation is derived when the electrical charge increases with the electrical field E and the second when a decrease is observed.

As shown by some authors, the recombination efficiency under SCLC is electrical-field independent (due to the electrical field independence of W_{eh}) and can be given by [28]:

$$P_{rec}^{eh} = \frac{1}{\left(1 + W_{eh}/d\right)} \tag{3.33}$$

implying that the efficiency increases as W_{eh} decreases (low recombination zones). In the limit (when W_{eh} equals the layer thickness d), the efficiency is equal to 0.5.

3.3.2 Electrical Current and Electroluminescence

The first presuppose for an efficient OLED behavior is the correct conversion of electrical current into photons. The simplest model considers a kinetic process in which singlet excited states are homogeneously produced throughout the organic film [28].

A simple equation can be derived assuming that both electrons and holes have the same injection mechanism. The result is given by [29]:

$$\Phi_{EL} \propto \frac{\left(I_i^{e,h}\right)^2}{\mu_h(E)\mu_e(E)} \tag{3.34}$$

in which Φ_{El} is the photon flux per unit area and I the electrical injected current (for both electrons—e and holes—h). Note that for this equation, the carrier mobility is considered as electrical field dependent. Equation 3.34 is derived assuming the thermionic injection mechanism described before. In this model, where the carrier mobility is independent on electrical field, the Φ_{EL} depends on I^2. However, as already noted, the mobility in organic semiconductors can be dependent on the applied electrical field and sometimes in a strong way. In these cases, Φ_{EL} the former equation is no longer valid. The new electrical current dependence of Φ_{EL} is now specific of the electrical injection/transport model considered as expected. For instance, when considering the Poole–Frenkel model we have [28]:

$$\Phi_{EL} \propto \left(I^{e,h}\right)^n \tag{3.35}$$

with

$$n = 2 + \frac{\beta_\mu^h + \beta_\mu^e}{a_t} \quad \text{and} \quad a_t = \left(\frac{1}{kT}\right)\sqrt{\frac{\varepsilon_r^3}{4\pi\varepsilon}} \tag{3.36}$$

The carrier transport behavior is correlated with the value of n. For example, in the Fowler–Nordheim tunneling mechanism the follow relationship is expected [28]:

$$n = 1 + (b_h + b_e) \tag{3.37}$$

with b_h and b_e being constants that depends on the hole and electron mobilities and are obviously dependent on the carrier injection properties. This simple dependence on Φ_{EL} with power laws, in the real experimental data, is sometimes not completely observed. It must be noted that here is given a simple insight about the electro–optical relationship of an OLED. The purpose is to illustrate the differences of physical models and to show how the different electrical carrier injection/transport can affect the electron–hole recombination. More detailed information can be viewed in [30].

3.3.3 Light Generation

In general, the light emission properties of organic semiconductors can be studied with base on the individual molecule properties. This is especially truth for small molecules, due to the weak intermolecular interactions. In this section a brief and schematic review of the principal molecular de-excitation mechanisms is presented.

3.3.3.1 Electronic spectroscopy: the basics of singlet–triplet organic molecular systems

The intrinsic electronic process that occurs in a molecule when subject to and external excitation source (optical or electrical) is a fundamental tool to understand the molecular levels interaction and the light emission. The most common process is the photo-physical study consisting of exciting a material by an optical source and studying its absorption/emission (photoluminescence). In organic semiconductors, based on π–π^* systems, the optical spectroscopy plays an important (and probably unique) role in predicting the expect behavior given by the molecule use into an electroluminescent device.

The total energy molecule is given by its electronic, vibrational and rotational levels. Figure 3.8 shows a simple scheme of the energy levels division.

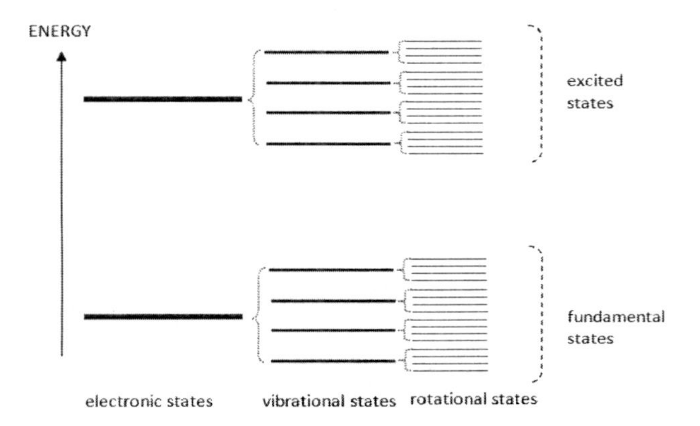

Figure 3.8 Simple scheme of the energetic levels of a molecule (electronic, vibrational, and rotational).

As the scheme depicts shows, an electronic level is divided into vibrational levels, each of them further divided into rotational levels. The total molecule energy is, therefore, the sum of the all energies. The optical study of the molecule give us an analysis of the photoluminescence (PL) that occurs after the absorption (ABS) of incident photons that, if having the required quantum energy (resonant with the electronic levels) can be absorbed, promoting and electron from the fundamental level to an excited level. After the successful photon absorption several de-excitation mechanisms can occur involving the energy levels of the molecule. These mechanisms are dependent on the spin multiplicity of the involved energy states and of the local molecule symmetry. The electron transition probability between two energy levels can be obtained by the result of $\langle \Psi_a / r / \Psi_b \rangle$, where Ψ_a, Ψ_b and r are the wave-function of the actual energy level, the wave-function of the final level and r the electrical dipole quantum operator, respectively.

The basic electronic level of an organic conjugated system is formed by two kinds of spin multiplicity, namely the singlet state (represented by S with spin multiplicity equal to 0) and the triplet state (represented by T with spin multiplicity equal to 1). For a normal energetic level representation in a molecule we can write S_n and T_n where $n = 0$ for the respective fundamental level and $n > 0$ for an excited level. In the singlet case, the electrons have non-paired spins in the lowest energy level which is the common situation of a fundamental energy level of an organic molecule; in the triplet case, the electrons spins are paired in the highest energy levels and are only observed in excited levels.

The relation with the π levels formed by the molecular orbitals as described in Chapter 2, can now be expanded for a better understanding of the forthcoming explanations. Figure 3.9 shows, in a simplified way, the path from the LUMO and HOMO levels to the final singlet and triplet levels (in this case only S_0, S_1 and T_1). Considering the electronic charge in the HOMO level (having a spin configuration $\pi\pi^*$), four levels can be achieved depending on the total spin number (S = 0 or 1) and the spin moment ($M_S = 0, \pm S$). In the first case, we have $S = 0$ and $M_S = 0$, giving the first singlet excited level S_1; in the second case we have $S = 1$ and $M_S = 0, 1, -1$, giving the first three excited triplet levels, T_1. In both cases, those excited levels have a spin configuration $\pi\pi^*$ (both HOMO and LUMO levels as represented in Fig. 3.9). Although simple, the final energy levels are, in the excited state obtained from the LUMO level of the diagram band. While the first singlet excited level becomes above the LUMO level, all the three first triplet excited levels becomes below the LUMO level. This result from electrostatic interactions at molecular orbitals, giving two amounts that are responsible for those differences: the Coulomb Integral (J) and the Exchange Integral (K). The excited triplet level with $M_S = 0$ is located at LUMO $- J$ while the excited singlet level is located at LUMO + $(2K - J)$. The detailed explanation of these quantities will not be given here because we are only interested in the final distribution of the excited levels but a much complete description can be obtained in [31].

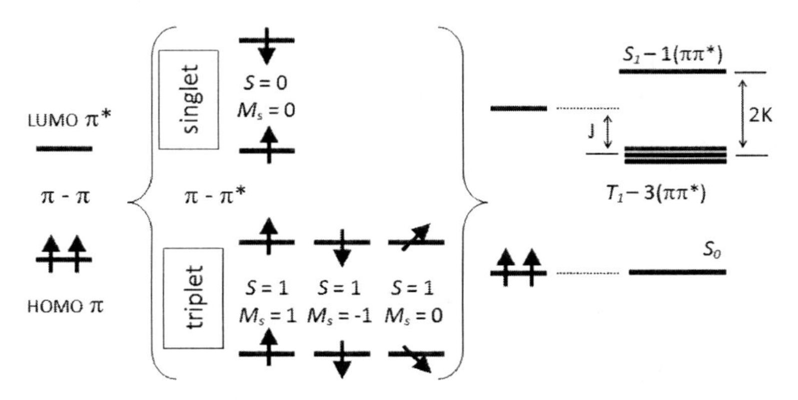

Figure 3.9 Distribution of the first excited levels singlet and triplet from the HOMO–LUMO diagram. The different spin orientations are given for the all four possible levels.

When an external energy source is used to excite an electron from its singlet S_0 fundamental state to an excited state (singlet S_n or triplet T_n, with $n > 0$) the molecule can undergo several internal and/or external physical mechanisms in order to allow the electron relaxation. If this relaxation is achieved by the electronic recombination with the hole that remain in the fundamental state we can have an exciton recombination (electron–hole) called as *radiative* with a corresponding photon emission; on the contrary, if the recombination is made by successive vibrational relaxations with semiconductor bulk phonons, the energy is dispersed by the molecular semiconductor structure and *non-radiative* emission occurs (without photon emission). The first mechanism is the basic principle of luminescence; the second is non-luminescent. These simple notions must be taken into account when synthesizing an organic semiconductor or when a device is fabricated. All non-radiative recombination must be avoided (at all costs) as it only contributes to the loss of device efficiency as it does not account to light emission.

Concerning the transitions lifetime, it's given as the reciprocal of the transition probabilities. The complete study is not within the scope of this book. For further reading some good bibliography can be consulted [32]. The fundamental idea about the transition lifetimes is, in spite of the complexity of processes associated with it, important to understanding the light emission and molecule mechanisms. Usually electron promotion to the excited level (by means of an optical excitation—and further photoluminescence—or by means of an applied electrical field—given the electroluminescence) is very fast ($\approx 10^{-15}$ s). The subsequent de-excitation processes have distinct lifetimes depending on the relaxation mechanism. For instance, the rotation and vibrational transitions are very fast (about 10^{-12} to 10^{-14} s) because these are high probability transitions and are considered as the first process experimented by the electron after the energy absorption. Due to their very low energy separation and very fast transition lifetimes, the rotational energy levels are often depreciable. When reach to the lowest vibrational exited energy level, the electron can relax by means of an electronic transition, between levels with the same spin multiplicity (i.e., $S \rightarrow S$ and $T \rightarrow T$). This transition can also be very fast (generally about 10^{-7} to 10^{-10} s) whereas when it's between levels of different multiplicity (i.e., $T \rightarrow S$) are slow (generally about 1 to 10^{-6} s, typically in the 10^{-3} s range) due to the spin forbidden transition with low probability. In the first case (same spin multiplicity) we are in the presence of a *fluorescence* process; in the second case (different spin multiplicity) we have a *phosphorescence* process.

As can be easily understood, several others mechanisms are responsible for the final molecule internal efficiency, and these may significantly alter the final device performance. These mechanisms involve the energy changes between different excited states (of same or different spin multiplicity) and/or further absorption to other excited states. Also the possible interaction between neighbor molecules may affect the energetic relaxation of an exited carrier. This last case is very common in polymers in which the polymer chain can have different conformations in a solid state film, allowing orbital overlapping among different polymer chains and modifying the orbitals interaction of each polymer. This effect is rare in small molecules although is not impossible. Sometimes can appear when a semiconducting polymeric matrix is used to embed coordination complexes of rare earths or transition metals small molecules.

In Fig. 3.10 the possible de-excitation mechanisms are represented over the general energy level diagram, known as Jablonski diagram [33].

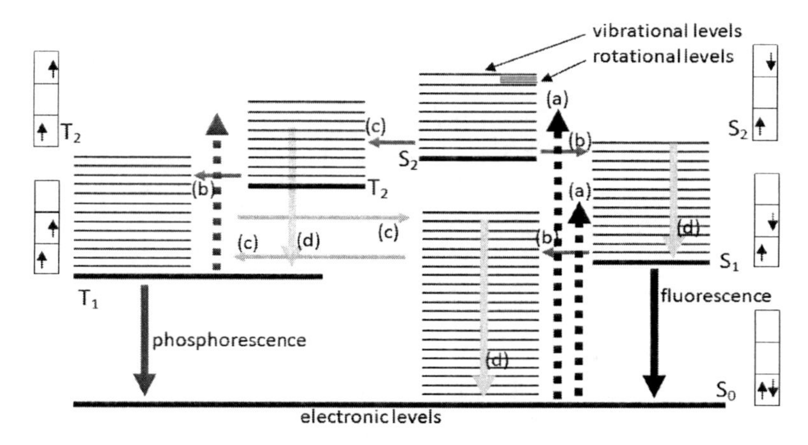

Figure 3.10 Jablonski diagram showing the available intermolecular processes in organic molecules with singlet and triplet energy states under electronic excitation and subsequent relaxation. The lateral boxes show the electronic spin orientations associated with the energy levels. The principal excitation/relaxation mechanisms are indicated: (a) excitation, (b) internal conversion, (c) intercrossing system, and (d) vibrational relaxation. The radiative emissions are also represented (phosphorescence and fluorescence).

In a first observation, the Jablonski diagram stipulates the fundamental energy level as S_0 (the lowest singlet level) and the electron promotion by excitation from this level to the S_1 and S_2 levels. Considering that the S_0 level is, in the molecular system, the lowest energy configuration (and not T_0) and that the excitation is very fast corresponding to an allowed spin transition, is not difficult to understand that the principal excitations are $S_0 \rightarrow S_{1,2}$.

In these complexes the intermolecular processes are the main focus of study, despising the influence of neighbor molecules. After excitation to the $S_{1,2}$ energy states corresponding to a transition $\pi \rightarrow \pi^*$, the electron relaxes to the singlet state of lowest energy (typically S_1) by phonon activation. This process is designed as internal conversion (Fig. 3.10). Also, depending on the transition probability, another process can occur when the carrier is in the excited energy level: the relaxation to a excited triplet state (with lower energy). This mechanism called of intersystem crossing (Fig. 3.10). As mentioned before, levels S_n and T_n, $n > 1$, having different spin multiplicity and consequent energy degradation, may occur when the electron inverts its spin by the spin–orbit interaction [34]. Nevertheless, the triplet state is more energetically stable due to the lowest electronic repulsion (lowest energy electronic configuration as given by Hund rule [34]). This means that the intersystem crossing favors energy stabilization of the system, and of course, occurs quite frequently considering the molecular demand for the lowest energy/stable configuration.

After this first relaxation to the lowest excited energy singlet or triplet sate, the recombination with the hole in the fundamental state will occur. There are three possibilities for this (Fig. 3.10): fluorescence ($S_1 \rightarrow S_0$), phosphorescence ($T_1 \rightarrow S_0$) and non-radiative relaxation ($S_1 \rightarrow S_0$). The first two processes give (at different lifetime as referred) light emission; the last one is phonon-assisted and no light, i.e., radiative emission occurs. De-excitation by one of these two ways, radiative and non-radiative, clearly depends on the molecular structure and must be taken into account when the molecular structure is defined. Molecules prone to non-radiative relaxation between the excited and fundamental singlet levels are clearly inadequate for light emitting devices as obviously.

Extending the concepts about carrier excitation and further radiative emission by recombination, it is not difficult to predict the expected excitation/emission band taking into account the previous discussion. To understand this idea a very simple model can be used, when only two electronic levels are considered (fundamental and exited) each one with its intrinsic vibrational levels.

In fact, when considering the excitation from the lowest energy singlet level to the excited one, each of them having its vibrational levels, there are many allowed transitions that may determine the excitation and emission dependence on energy. Naturally transition probabilities are not equal and depend on the specific molecular nature (in an electronic point of view) but for now, we only consider the energy difference between a fundamental state and an excited one. A simplest case of two levels, each of them with four vibrational levels is shown in Fig. 3.11.

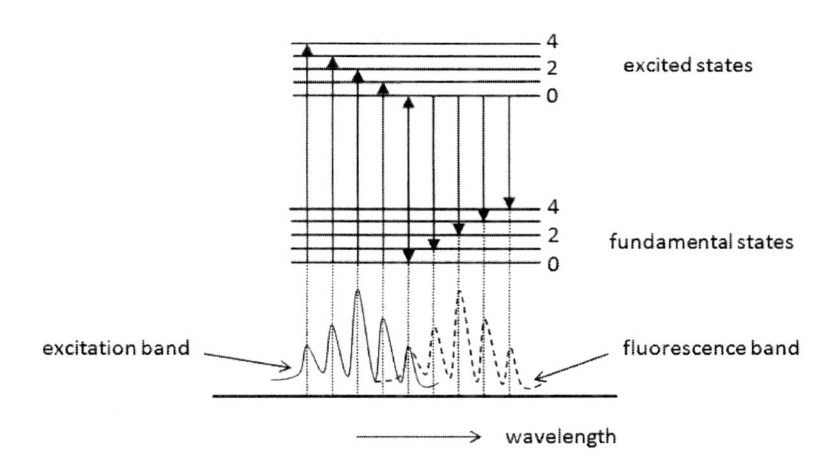

Figure 3.11 Simplified model of expected exited and emission bands profile depending on the electronic–vibrational transitions between a theoretical molecular system with one fundamental electronic energy level and one exited energy level.

In a limit case where exists a lot of vibrational energy states (as typically in organic semiconductors) a Gaussian shape bands is expected.

A slightly different approach is used when for example a rare-earth organic emitting molecule is employed as the emissive layer to fabricate an OLED. Although the organic molecular counterpart follows the behavior expected for a singlet/triplet energy level system, the "molecular engineering" idea is to populate the excited levels of the coordinated rare-earth ion so that emission takes place not at the organic molecular fragments but only at the lanthanide ion. Further application of the principles explained here are described in the next chapter; the main part will comprise the design of the organic ligands for the preparation of an efficient and useful emitting molecule. Considering all de-excitation mechanisms, the emitting efficiency will depend completely on the organic ligands ability to efficiently deliver energy to the ion. The same principle will be used to increase emission of the small molecule emission by a direct energy pumping from a polymeric matrix. All these cases are discussed later.

Regarding the emitting metal–organic coordination complexes, two kinds of electronic charge can occur and are briefly described here as an introduction for forthcoming chapters. For a better understanding of the process we must consider two entities, the organic ligands and the coordinated ion. The process by which the electronic charge is transferred (clearly dependent on the ion type and organic ligand) determines the molecule intrinsic behavior. Also, this behavior depends on the relative location of the energy levels of the organic ligands and the ion's orbitals.

The most common acceptable mechanism for rare earth coordinate ions in an organic matrix (in a similar way described before) is the energy transfer where the electronic charge is moved from the organic part (usually the lowest excited triplet state T_1) to the excited level of the coordinated ion, with further rare earth ion emission. The emission is clearly dependent on the organic molecular part that plays the fundamental role on the excitation and energy transfer to the coordinated ion.

Two other more specific cases are the so-called *metal to ligand charge transfer* (MLCT) and the *ligand to metal charge transfer* (LMCT) that are, in the case of transition metals ions, electronic charge process involving *d* orbitals and ions coordinated to organic ligands. As for the rare earth complexes, here is only presented a simplified description of these phenomena. In the MLCT system, the system has a partial transition of electrons from the metal to an orbital of the ligand. In general this process occurs for metals with highly filled *d* orbitals that can acts as electron donors into the anti-bounding (π orbitals) of the organic ligand. The non-bonding *d* organic orbitals must match the anti-bounding organic orbitals in size, shape and symmetry. This is particularly truth if the organic ligand belongs to the aromatic molecular structures that are much used in light emitting organic complexes. In the LMCT charge transfer (also involving the *d* orbitals of transition metals) one excited electron in the organic ligand orbital is transferred to an empty metal-based orbital (also called LMCT orbital). This implies a partial electronic transference of electrons from the organic ligand to the coordinated metal ion. Both systems (MLCT and LMCT) involve *d* orbital transitions (we still consider the transition metals ions) and can be studied by optical spectroscopy experiments. As is easily understood, these processes are not simple and present a very high complexity. Usually depend on molecular symmetry and obey the spin ($\Delta S = 0$) and orbital (or Laporte – $\Delta l = \pm 1$) rules. Under specific conditions these rules may be less strict. Chapter 5 briefly describes these situations along with the MLCT mechanism and formation of the molecular energy levels.

The important point to retain for now is that these light emitting transition metal coordination complexes exhibit a high colored emission and therefore are effectively very good candidates for OLEDs.

3.3.3.2 Luminescence advantages of metallo–organic complexes: the internal efficiency maximization

Metal–organic complexes addressed in this book, both rare earth (RE) or transition metal (TM) complexes are considered small molecules with small conjugation. The difference from the major part of organic light emitters is the existence of a coordinated metallic ion in to an organic matrix around it (giving the called coordination spheres). The way an excited carrier relaxes to the energy ground state, although similar in scheme, is quite different in practice. This difference is the surplus of organic–metal complexes.

In a pure organic emitter molecule, the traditional spin forbidden emission from the lowest triplet excited state has generally a low or even near zero transition probability. This means that, in general, there is no phosphorescence and the carriers de-activate by non-radiative transitions, involving only the phonons in the molecular structure. As a first consequence, and observing the Jablonski diagram in Fig. 3.10, the unique radiative emission expected is the fluorescence from the first singlet excited level to the ground state ($S_1 \rightarrow S_0$). Thus, and according to the conventional energy levels π–π^* obtained from the *sp* orbitals, we have four excited levels (one singlet and three triplets) but only one, singlet, is involved in the radiative emission. This means that the maximum internal efficiency is only 25% due to the non-contribution of the triplet states. All the luminescent organic complexes with energy states based on carbon molecular orbitals have this constrain, in special some dyes and conjugated polymers. Moreover, the intercrossing system (i.e., singlet S_1 to triplet T_1) has low probability on those materials.

A different pathway can be found in metal–organic complexes. Regardless of the charge transfer mechanism, which usually comprise a direct de-excitation by intercrossing system, the internal efficiency of that metal ions complexes is near 100% because the probability of the transition is very high (as are discussed in next chapters). This is the main advantages of using coordination complexes as emissive layers in an emitting light device. It is obvious that the structure of the coordination complex must also be optimized for a maxima energy transfer to the radiative emitting levels, taking advantage of the high internal efficiency intrinsic nature of these materials. The molecular "design" rules must thus comprise a relaxation process (after charge excitation) that minimizes or ideally eliminates any non-radiative relaxation in the organic ligands. This strategy will allow us to obtain an excellent OLED based on that coordination complexes.

The *pros & contras* of using coordinated metal–organic as the emitting materials for OLED active layers can be summarized in the following topics:

- Different approach to device color modulation and allowing the use of a polymeric conductive matrix where the organic–metal complex is embedded;
- They are relatively easy to process, even under thermal evaporation systems;
- For rare earth based complexes have quasi "monochromatic" emission;
- Transition metal complexes have strong brightness;
- Very high internal efficiencies (a tremendous surplus face to the traditional pure organic complexes).

3.4 Characterization of the OLED Light Emission

3.4.1 Introduction and Light Emission Drawbacks

An OLED is a typical 2D light surface emitting, with an area that can be sometimes large. The main difference from the other sources of light is that it exhibits a large viewing angle, it's very thin (the organic layers have a total thickness of 100 to 200 nm typically), it's "self-emitting," i.e., does not need a backlight like the conventional LCD, and it can be fabricated on different kind of substrates. Such wide range of advantages (concerning on the light emission) makes the choice of the methods for correct light emission characterization a difficult task. Also it will be not easy to properly define and implement a technique to measure the efficiency [35]. Currently, there are several methods/techniques (and definitions) to characterize the OLED light emission. None of them is perfect and distinct research groups used also distinct processes which raises an additional problem when a comparison is needed. In the next pages, a simple (and summarized) explanation is given, for better understanding on how to characterize the OLED light emission and to measure its efficiency.

As a simple idea on the difficulties measuring the OLED light emission profile, Fig. 3.12 shows an OLED lighted under different viewing angles.

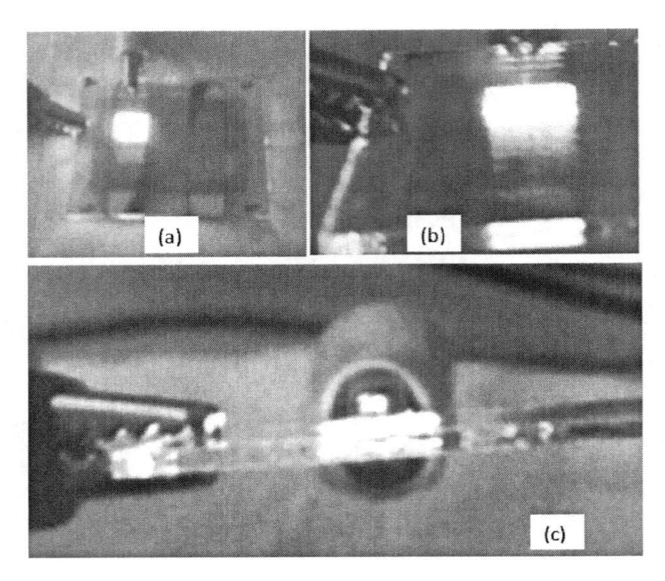

Figure 3.12 Pictures of an OLED lighted. Viewing angles of 20, 50, and 180 degrees. Work done at Department of Physics of University of Aveiro.

As may clearly be seen, even under a viewing angle of 180 degrees, there is still OLED emission coming from the "side" of the glass substrate.

3.4.2 OLED Light Measurements

When measuring a light, there are essentially two kinds of data sets we can collect, depending on the specific application programmed for such light. Each of them has on proper scientific nature and interpretation and the main problem in the difficulty on establish the correspondence among them.

The first type comprises the radiometric measurements, based on the fact that light is a form of radiation; the second type are the photometric measurements, based on the ability of light to illuminate for visual purposes, so therefore it can be measured considering its perception by the human eye. This photometric measurement is the most commonly accepted by the industry. As easily understood, the efficiency measurements will depend on the light measured type.

An important and currently indispensable tool to characterize light emission is the determination of so-called coordinates of chromaticity, where visible light is characterized by two coordinates (x, y) in a 2D diagram that represents the human eye perception. These coordinates are related to the colorimetric theory and are based on the CIE (Commission Internationale de L'Eclaire) models [36]. The most widely used was created in 1931 and is called two degree model [36][b]. This model considers three illuminants called of A, B, and C (A is related with the black body emission at 2856 K, B represents the sunlight with a color temperature of 4900 K and the C represents the daylight with a color temperature of 6800 K). The model defines also a color matching function, which gives the relative contribution of a light of wavelength λ to the CIE tristimulus values X, Y and Z. This color matching function gives the average color perception of a human eye observer in the all visible light wavelength, from near 380 nm to near 780 nm, (i.e., from the near ultra-violet to the far red). Naturally this color matching function represents a color as the result of the sum of contributions of each X, Y, and Z function. To represent a specific color, a simple calculation can be made following the equation:

$$xX + yY + zZ \tag{3.38}$$

where the x, y, and z are the chromaticity coordinates and are defined as

$$\begin{aligned} x &= \frac{X}{(X+Y+Z))} \\ y &= \frac{Y}{(X+Y+Z))} \\ z &= \frac{Z}{(X+Y+Z)} \end{aligned} \tag{3.39}$$

[b] This 1931 model was object of further transformations in 1960 and 1976. Moreover, adding to the 2 degree standard observer, a more complex standard observer with ten degrees was introduced in 1964. In this book only a simple description of the two degree 1931 model is given, as explained before.

and of course $x + y + z = 1$. As z can be easily obtained from x and y, $z = 1 - (x + y)$, the chromaticity coordinates are usually given simply by (x, y). Figure 3.13 shows the CIE 1931 (two degree observer) diagram.

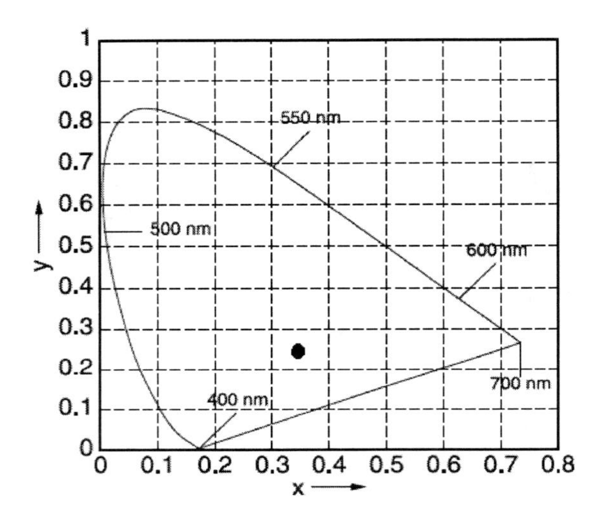

Figure 3.13 Chromaticity diagram CIE 1931 (two degree observer) with the typical wavelengths (from 400 nm—blue to 700 nm—red). The border in curvature with the wavelengths represents the so-called spectral colors, whereas the straight line in the diagram bottom represents the non-spectral colors (see the definition in the text). The point at $(x, y) = (0.33, 0.33)$ marked in the diagram, is the pure white color.

In order to calculate the (x, y) color coordinates, we must make a convolution of the OLED electroluminescence spectrum to the X, Y, and Z tristimulus functions. The corresponding tristimulus intensity is available from the CIE in the 380–780 nm range with 5 nm step. Several software routines are available for a simple and fast calculation of (x, y).

More than the mere color representation, the chromaticity diagram offers the possibility of obtaining two important definitions for OLED (in our cases) light characterization. From the diagram, we can extract the dominant wavelength (λ_d) that can be further used in the external quantum efficiency (next section) and the color purity, a very important parameter. Figure 3.14 shows how to determine those parameters.

In Fig. 3.14 we arbitrated that an OLED has color coordinates values of about (0.45, 0.45) represented by the open circle. In order to determine the dominant wavelength we simply must connect the pure white coordinate point of (0.33, 0.33) in black, to the one indicated above and extrapolate the straight line to the diagram border. This border point (which of course represents the spectral colors) has an associated wavelength that corresponds exactly to the dominant wavelength. In our simple exercise, the value of λ_d is approximately 578 nm. The color purity is obtained by dividing the corresponding length in the diagram scale, that is $A/A + B$. The closer the ratio is to one, the more pure is the color. Reporting again to our exercise, the ratio is about 0.25/0.35, i.e., approximately 0.71. In this system, we have a yellow–orange color with a poor purity. As a simple indication, rare earth and transition metals based OLEDs usually exhibit good color purity in the green, yellow, orange and red spectral regions.

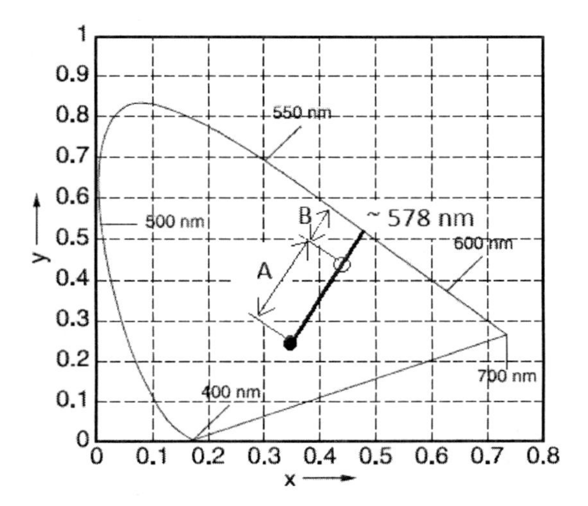

Figure 3.14 Exemplificative method for the extraction the dominant wavelength (λ_d) and the color purity from the electroluminescence spectra. Refer to the text for a complete explanation.

3.4.3 OLED Efficiency Measurements

At least five different parameters are defined for OLED efficiency measurements [37]. The first problem is the experimental conditions for OLED light output measurements. For instance, and only by comparison, one of our polymer-LED structure gives a brightness of 1200 cd/m^2 or 12000 cd/m^2 depending on a direct (normal to the surface emitter) measurement or using an integrating sphere.

First of all, we must take into account the light emitter application. If it is conceived for a display, only the photometric (human eye) response is important; for a general illumination system, and even against the industry's desire, a radiometric measurement is acceptable. Also, this measurement is simple, direct and allows immediate comparison of the power efficiency, an important parameters when the energetic efficacy is the ultimate goal of the OLED. Secondly, some approximations must be considered if a more simple measurement is to be considered, in special assuming that the OLED surface emitter is of a Lambertian type (that in many cases are clearly wrong), i.e., the emitter area is isotropic emitting with equal radiance into any solid angle from the surface (is discussed later). Finally, some previous definitions related to the light measurement (both radiometric and photometric dependent) must be taken into account, as sometimes is not easy to understand what physical amount is considered. A detailed description concerning the radiation pattern of a light-emitting device can be easy found in published literature. For those are interesting in more accurate models, see, for instance [38].

Starting from this last point, the radiometric units usually employed are the radiance ($W/sr\ m^2$), radiant efficiency ($W/sr\ A$) and the power efficiency—W/W—also called "wall plug efficiency"; the photometric "correspondent" units of measure are the luminance (or brightness) (cd/m^2), the luminous efficiency (cd/A) and the luminosity lm/W, also called luminous power efficiency. *W*, *sr cd* and *lm* denotes watt, steradian, candela, and lumen, respectively.

The following pages define, in a simple discussion, the most important quantities, related to the radiometric and photometric measurements [39, 40].

The first definition of light emitter efficiency takes into account the relationship between the number of generated photons and the number of electrical carriers. Cleary is a radiometric measurement. If we consider the OLED optical power integrated spectrum (P_{OLED} (λ)) over all emitting wavelengths, we can obtain the total number of photons generated by time unity, which is given by:

$$N_{photons} = \int_0^\infty \frac{P_{OLED}(\lambda)}{hc} \lambda d\lambda \qquad (3.40)$$

where h is the Plank constant and c the light velocity. The total number of electrical carrier that flows throughput the semiconductor, can be easy obtained from the electrical current I, because $N_{carriers} = I/q$. So the relationship between both quantities (η_{ext}) is given by:

$$\eta_{ext} = \frac{N_{photons}}{N_{carriers}} = \frac{q}{hcI} \int_0^\infty P_{OLED}(\lambda)\lambda d\lambda \qquad (3.41)$$

This relation, called *external quantum efficiency* can be corrected in order to accommodate the specific conditions of the experimental set-up used. This efficiency has no units.

Considering a much simpler radiometric measurement, direct power efficiency can be obtained when considering the ratio between the total OLED optical emitting power and the electrical input power. The optical power (W) can be written as:

$$L = \int_0^\infty P_{OLED}(\lambda) d\lambda \qquad (3.42)$$

and obviously the electrical power (W) is given by $P_{ele} = IV$. This means that the *power efficiency*, or *Wall Plug*, represented by η_{WW}, is given by:

$$\eta_{WW} = \frac{L}{IV} = \frac{\int_0^\infty P_{OLED}(\lambda)d\lambda}{IV} \tag{3.43}$$

and also has no units. Although with some constrains in the experimental set-up needed for its measurements, the Wall Plug efficiency is relatively less exposed to calculation simplifications.

Leaving the radiometric efficiencies, in spite of their wide spread used in laboratories, we now direct our attention to the photometric data for which two main definitions must be referred.

The base is the human eye perception. In order to obtain numerical information about this perception, the CIE created a photopic response in 1924, for which the maximum intensity (equal to 1) occurs at a wavelength of 555.17 nm, corresponding to the maximum response of the human eye[c].In Fig. 3.15, the 1924 CIE model of the photopic response function is showed. This function also has no units.

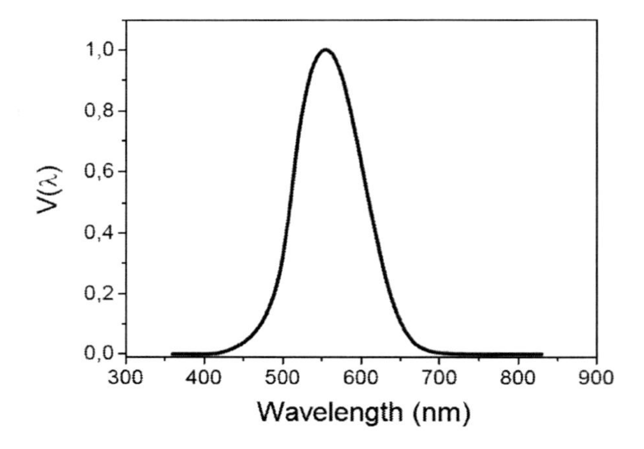

Figure 3.15 Photopic response V(λ) as defined in the CIE 1924 model.

[c] There are changes in the human eye spectral response due to the effects of the two kinds of receptors, the rods and the cones. This implies that, in addition to the photopic response, there is a scotopic response. For the aim of this book, these questions will not be explained as the photopic response in function during daytime or in well-lighted environments is the most usually employed.

In order to obtain (under photometric responses) the total luminous flux Φ_{total} we need to obtain the photopic response. The expression of Φ_{total} is given by:

$$\Phi_{\text{total}} = \int_0^{\infty} K(\lambda)P_{\text{OLED}}(\lambda)d \tag{3.44}$$

$$K(\lambda) = V(\lambda) \times 638$$

The of Φ_{total} and $K(\lambda)$ are given in *lm* and *lm/W*, respectively. It is now easy to define the *luminous power efficiency* (η_{lp}), calculated from the ratio between the total luminous flux and the total electrical power:

$$\eta_{lp} = \frac{\Phi_{\text{total}}}{IV} = \frac{\int_0^{\infty} K(\lambda)P_{\text{OLED}}(\lambda)d\lambda}{IV} \tag{3.45}$$

This luminous power efficiency (η_{lp}) is expressed in *lm/W* and is one of the most important photometric efficiency parameters.

Another photometric efficiency parameter, is the simple *luminous efficiency* (η_l), expressed in *cd/A* and defined as the ratio between the luminance and the OLED electrical current. The major problem in obtaining this parameter is to determine the luminance, because it depends on the emissive area S and on the luminous intensity (dependent itself on the emissive angle). The most used approximation is to consider the OLED emissive shape as a Lambertian one. The use of this method requires the acquaintance of some previous notions. If we consider a light flux in a circular shape with a radius r, arising from an infinitesimal area ΔS (for now let us consider a ΔS in a perfectly flat OLED surface from an emissive area S with a completely homogeneous photon emission) it is possible to suppose that, for a quite large distance d so that $d >> r$, we shall have a theoretical single-point source of light. The correspondent light detection (basically the photon flux by stereo-radian) is defined precisely as the *luminous intensity* I, that depends on both the polar (θ) and azimuthal (φ) angles, being $d\Omega$ the infinitesimal change in the emissive angle. The total emitted luminous flux is then defined as:

$$\Phi_{\text{total}} = \int_0^{2\pi} \int_0^{\theta/2} I(\theta,\phi)\sin(\theta)\,d\theta\,d\phi \tag{3.46}$$

The first approximation, is to consider only the photon flux as function of θ for which the referential makes a 90 degrees angle with the emitting surface. This is the definition of an emitting Lambertian surface where the luminous intensity can be expressed as:

$$I(\theta) = I_0 \cos(\theta) \tag{3.47}$$

and the luminous flux is given by:

$$\Phi_{\text{total}} = 2\pi \int_0^{\pi/2} I_0 \cos(\theta)\sin(\theta)\,d\theta \tag{3.48}$$

In both equations, I_0 is the maximum luminous intensity perpendicular to the emitting surface. The solution of Eq. 3.48 is very simple: πI_0. This means that for a Lambertian surface, the luminous intensity of 1 *cd* in the normal direction (of the emitting surface) corresponds to a total luminous flux of π *lm*. Also, in that emitting profile, the luminous intensity (from the observer's perspective) is directly proportional to the cosine of the angle between the observer and the normal direction of the emitting surface. Figure 3.16 shows a general and a Lambertian case.

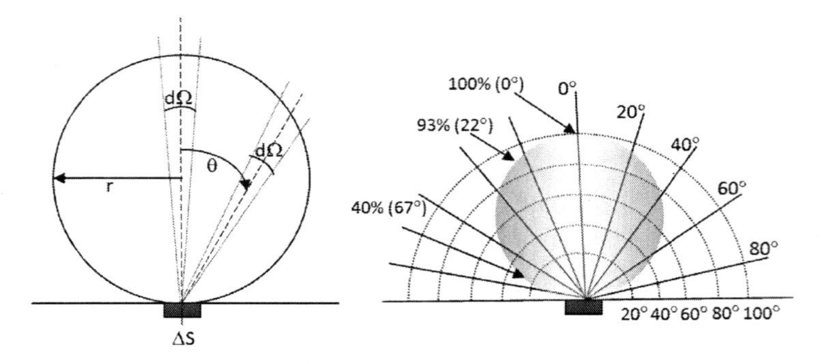

Figure 3.16 (a) Circular shape of a point light emitter and (b) a light emitter with a Lambertian radiation pattern.

Some considerations must be made about these approximations, in particular for the Lambertian one, which unfortunately is relatively far from the OLED reality. As stressed before the typical device structure used in an OLED includes both organic and inorganic layers, each one of them with different optical properties, introducing significant interference effects with strong influences on the emitting spectrum and pattern, that in some cases departs clearly from the conventional Lambertian "shape." Another important aspect comprises, for instance, the OLEDs build by the spin coating technique where the alignment of the polymeric matrix (the emissive or for energy transfer as well used when coordination complexes are embedded in them) implies that the azimuthal angle φ cannot be left aside in the Eq. 3.46.

In spite of these considerations, many authors are still using the luminous efficiency to characterize OLEDs and a obtaining final relation. Considering that the luminous intensity for θ is equal to zero degrees (and cumulatively to the Lambertian approximation) the luminance L (cd/m^2) is easily obtained as follows:

$$L = \frac{I(\theta = 0)}{S} \tag{3.49}$$

Considering also that the electrical current density is $J = I/S$, the luminous efficiency η_l (cd/A) is then given by:

$$\eta_l = \frac{L}{J} \tag{3.50}$$

From the definitions given before, a simple relation can be established to extrapolate the luminous power efficiency from this luminous efficiency. The result is simply given by:

$$\eta_{lp} = \frac{\pi}{V}\eta_l \tag{3.51}$$

A special case is the indirect determination of the *external quantum efficiency* (η_{EQE}) considering the luminous efficiency and the photometric magnitudes like the photopic response V(λ). A simplified expression for η_{EQE} is given by:

$$\eta_{EQE} = \frac{\pi q}{K(\lambda)hc}\eta_l \approx \frac{3.7\times 10^5}{V(\lambda)}\lambda_d \eta_l \tag{3.52}$$

where λ_d is the dominant wavelength (as described in last section). All these indirect data must be carefully used as their inherent error measurements can be lead to significant mismatch with the reality.

Finally, from the physical point of view, some authors try to estimate the *internal quantum efficiency*, defined as the ratio between the total numbers of phonons generated *within* the OLED structure to the total number of carriers injected. In general, this measurement can only be done with an integrating sphere and is not of absolute importance for assessment of the effective efficiency of an OLED. There are many sources of errors and only in a chemical molecular design point of view can be clearly useful. Sometimes the OLED design structure can (in many cases of systems with not so distinct internal quantum efficiencies) overcome those differences. So, the specific explanation is not given here.

In a main overview of the OLED efficiency measurements, two methods are widely used: the power efficiency (wall plug), a radiometric measurement, and the luminous power efficiency and/or the luminous efficiency, which are photometric measurements. Depending on the experimental set-up, the error associated with each case can be minimized in order to give a realistic idea about the OLED efficiency figure of merit.

As a final note it's important to highlight that besides the overall differences in efficiency measurement methods, the many different experimental set-up may also pose serious problem. For instance, and recalling my personnel experience on the development of Terbium based OLEDs, I measured, in a very green device, an efficiency rounding 20 *cd/m*^{2} where another group measured almost a four times higher value with a similar device structure and the exact same instrumentation and procedure. The difference was in the environment conditions: I performed the measurements in normal atmosphere and my colleagues measured them under high vacuum conditions.

3.4.4 OLED Dynamic Emitting Region

Independently of methods for OLEDs efficiency measurements, special considerations must be considered for clearly obtain a figure of merit that corresponds to the real emitting device behavior. And naturally, it's understandable that the wide range of available measurements methods make it complicated to attempt a scientific comparison or published results.

A traditional tool for device characterization which also constitutes a figure of merit is the dynamic operating region. Why? It is very simple to understand. Consider, for instance, and OLED with a luminous efficiency of 10^4 *cd/A*, and another one with an efficiency one order of magnitude lower. But what would happen if the first one could only operate regularly, e.g., in an applied voltage within 10–11 V, and the second one could operate in a voltage range between 10 and 30 V? It is obviously that the second OLED offers a much more useful applicability (unless for specific applications, which is very rare).

In all the efficiency measurements described before, the specific plots are determinant in the assessment of the so-called OLED dynamic operating region, i.e., the applied voltage region, where the output (optical power, brightness, etc.) is a linear function of the applied voltage. In this dynamic operating region the OLED response is directly proportional to the electrical voltage input which determines the region of practical applicability. Currently that dynamic operating region is one of the main drawbacks to extensive application of OLEDs, because of the chemical electrical degradations that limit the OLED light output range. This region is limited at low applied voltages (when the OLED does not start to operate) and at high applied voltages (when the OLED saturates, i.e., further increase in the applied voltage does not correspond to an equivalent increase is the light output—and may correspond to an irreversible degradation). Figure 3.17 shows an example for a Terbium-based OLED [41].

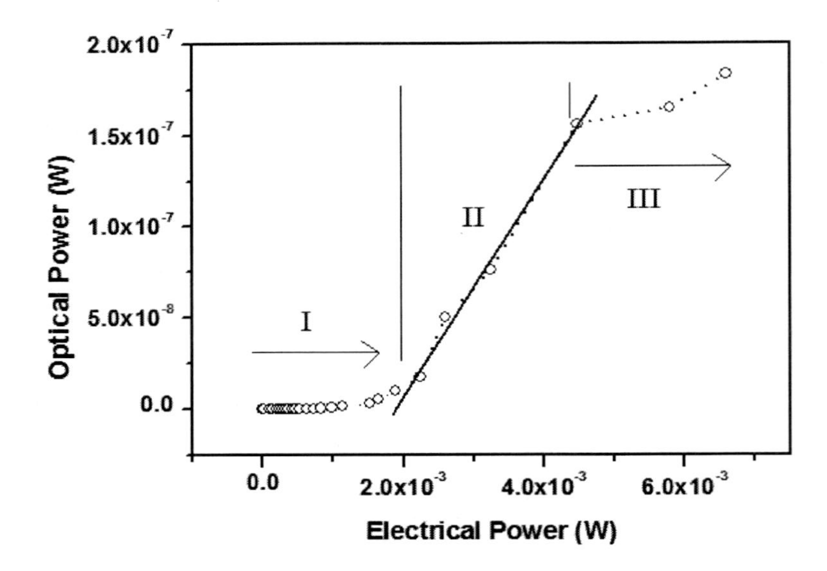

Figure 3.17 Optical power vs. electrical power of a Terbium based OLED. The optical power was measured only in the strongest line of the electroluminescence spectrum (546 nm). The graph shows three distinct regions: (I) OLED not working, (II) OLED dynamic operating region (for which the power efficiency can be obtained from the showed linear fit), and (III) OLED in saturation regime.

References

1. Brütting, W. (2001) Device physics of organic light-emitting diodes based on molecular materials, *Org. Electron.*, **2**, 1–36.
2. Tuck, B. and Christopoulos, C. (1986) *Physical Electronics*, Eduard Arnolds, UK.
3. Sze, S. M. (1985) *Semiconductor Devices, Physics and Technology*, John Wiley & Sons, USA.
4. Allison, J., (1989) *Electronic Engineering Semiconductors and Devices*, McGraw Hill, UK.
5. Heinisch, H. K. (1984) *Semiconductor Contacts*, Clarendon Press, UK.
6. Hwang, K. (1981) *Electrical Transport in Solids with Particular Reference to Organic Semiconductors*, Oxford Pergamon, UK.

7. Tagmouti, S. (1999) Effect of water vapor on poly (3-methylthiophene)/Mo Schottky diodes, *Synth. Metals*, **2**, 109–115.

8. Braun, D. (2003) Electronic injection and conduction process for polymer devices, *J. Polym. Sci.: Part B: Polym. Phys.*, **41**, 2622–2629.

9. Sze, S. M. (1981) *Physics of Semiconductor Devices*, Wiley, USA.

10. Mott, N. F. and Gurney, R. W. (1940) *Electronic Processes in Crystals*, Oxford University Press, USA.

11. Murgatroyd, P. N. (1970) Theory of space-charge limited current enhanced by Frenkel effect, *J. Phys. D: Appl. Phys.*, **3**, 151–156.

12. Juan, T. P. (2004) Temperature dependence of the current conduction mechanisms in ferroelectric $Pb(Zr_{0.53},Ti_{0.47})O_3$ thin films, *J. Appl. Phys.*, **95**, 3120–3125.

13. Kao. K. C. and Huang W. (1981) *Electrical Transport in Solids*, Pergamon Press, UK.

14. Mark, P. (1962) Space-charge-limited currents in organic crystals, *J. Appl. Phys.*, **33**, 205–215.

15. Shafa, T. S. (1992) Observations of Schottky and Poole-Frenkel emission in lead phthalocyanine thin films using aluminium injecting electrodes, *Int. J. Electron.*, **73**, 307–313.

16. Li, L. (2009) Electric field-dependent charge transport in organic semiconductors, *Appl. Phys. Lett.*, **95**, 153301/1–3.

17. Ma, D. (1999) Electron and hole transport in a green-emitting alternating block copolymer: space-charge-limited conduction with traps, *J. Phys. D: Appl. Phys.*, **32**, 2568–2572.

18. Mu, H. (2008) Temperature dependence of electron mobility, electroluminescence and photoluminescence of Alq3 in OLED, *J. Phys. D: Appl. Phys.*, **41**, 235109/1–5pp.

19. Steiger, J. (2002) Energetic trap distributions in organic semiconductors, *Synth. Metals*, **129**, 1–7.

20. Ogawa, T. (2003) Numerical analysis of the carrier behavior of organic light-emitting diode: comparing a hopping conduction model with a SCLC model, *Thin Solid Films*, **438–439**, 171–176.

21. Gill, W. D. (1972) Drift mobilities in amorphous charge-transfer complexes of trinitrofluorenone and poly-n-vinylcarbazole, *J. Appl. Phys.*, **43**, 5033–5041.

22. Bässler, H. (1993) Charge transport in disordered organic photoconductors—a Monte Carlo Simulation Study, *Phys. Stat. Sol. (b)*, **175**, 15–56.

23. Hertel, D. (1999) Charge carrier transport in conjugated polymers, *J. Chem. Phys.*, **110**, 9214–9223.

24. Burrows, P. E. (1996) Relationship between electroluminescence and current transport in organic heterojunction light-emitting diodes, *J. Appl. Phys.*, **79**, 7991–8006.

25. Nguyen, P. H. (2001) The influence of deep traps on transient current-voltage characteristics of organic light-emitting diodes, *Org. Electron.*, **2**, 105–120.

26. Blom, P. W. (2005) Tickness scaling of the space-charge-limited current in poly(p-phenylene vinylene), *Appl. Phys. Lett.*, **86**, 092105/1-3.

27. Walker, A. B. (2002) Electrical transport modeling in organic electroluminescent devices, *J. Phys.: Condens. Matter.*, **14**, 9825–9876.

28. Kalinowski, J. (1996) Electrical and related properties of organic solids, (eds. Munn, R. W., Miniewicz, A. and Kuchta, B.), *Electron Process in Organic Electroluminescence*, Kluwer Academic Publishers, The Netherlands, pp. 167–206.

29. Kalinowski, J. (1996) *Organic Electroluminescence Materials and Devices*, eds. Miyata, S. and Nalwa, H.S., Chapter 1 "Electronic Process in Organic Electroluminescence" (Gordon & Breach, UK) pp. 167–206.

30. Kalinowski, J. (1996) Injection-controlled and volume-controlled electroluminescence in organic light-emitting diodes, *Synth. Metals*, **76**, 77–83.

31. Ramachandran, K. I., Deepa, G. and Namboori, K. (2008) *Computational Chemistry and Molecular Modeling: Principles and Applications*, Springer-Verlag, Germany.

32. Kemp, W. (1991) *Organic Spectroscopy*, 3rd edn., Macmillan Education, USA.

33. Rabek, J. F. (1987) *Mechanisms of Photophysical Processes and Photochemical Reactions in Polymers*, John Wiley & Sons, Sweden.

34. Guillet, J. (1987) *Polymer Photophysics and Photochemistry*, Cambridge University Press, UK.

35. He, Y. (2000) Light output measurements on the organic light-emitting devices, *Rev. Sci. Inst.*, **71**, 2104–2107.

36. Fortner, B. and Meyer, T. E. (1996) *Number by Colors: A Guide to Using Color and Understand Technical Data*, Springer-Verlag, USA.

37. Forrest, S. R. (2003) Measuring the Efficiency of Organic Light-Emitting Devices, *Adv. Mater.*, **15**, 1043–1048.

38. Moreno, I. (2008) Modeling the radiation pattern of LEDs, *Optics Express*, **16**, 1808–1819.

39. Wolf, W. L. (1998) *Introduction to Radiometry*, SPIE — The International Society for Optical Engineering, USA.

40. Santos, G. (2008) *Estudo de Dispositivos Orgânicos Emissores de Luz Empregando Complexos de Terras Raras e de Metais de Transição*, PhD, University of São Paulo, Brasil, in Portuguese.

41. Santos, G. (2007) Electro-optical measurements, stability and physical carrier behavior of rare-earth based organic light emitting diodes, *Proc. SPIE—The International Society for Optical Engineering*, vol. 6655, 6655U/1–6.

Chapter 4

Rare Earth Complexes: The Search for Quasi-Monochromatic OLEDs

The advanced highly internal efficient rare earth emitting complexes have been launched with their application in active layers of OLEDs. Besides the internal efficiency that theoretically can be about 100 percent, the relative simple (and low cost) chemical synthesis allied with the novel simple device structure make these complexes very attractive. In addition, the relative "quasi-monochromatic" emission, typical of rare earth coordinated ions, definitely triggered the research on these materials. In this chapter, a wide overview to rare earth-based OLEDs is described from the chemical nature to the internal energy process mechanisms related with the specific nature of these emitting materials. The specific applications to active OLED layers are focused, addressing the benefits and assets, and also the common scientific barriers and difficulties to overcome for a further efficient technological application of these OLEDs.

Organic Light-Emitting Diodes: The Use of Rare-Earth and Transition Metals
Luiz Pereira

www.panstanford.com

4.1 Introduction—The Choice of Rare Earth Complexes

In the earlier chapters, importance of the correct development of RE- and TM-based OLEDs was briefly explained. Over passing the justifications made and detailing how on the rare earth organic complexes specific world, a new framing is developed.

In contradiction with their "name," rare earth materials belong to the most abundant in nature and are also known as lanthanides because they belong to this group in the periodic table. The main problem is the amount of then that are "ready to use." The extraction, processing, and preparation for specific applications (e.g., as precursors for instance for further development of new luminescent complexes) remain yet a problem although new chemical routes, more efficient and simple, have been developed. In spite of these points, the work on new RE organic complexes is on an incredible progress in last two decades.

The RE complexes are currently viewed as the simplest to improve the OLED production technology due to, in part, their high internal efficiency when compared to the conjugated polymers or "pure" organic small molecules. Moreover, the conjugated polymers and small molecules strictly based on π-π^* orbitals (as described in last chapter) are relatively instable from a chemical point of view: easy oxidization and high susceptibility to environmental conditions—temperature and atmosphere composition, leading to a well-known fast degradation and short lifetime. This way, new emitting complexes for use in OLED active layers are welcome, especially if they can help overcome the drawbacks and simultaneously simplify the device final structure avoiding sophisticated and expensive protection layers development. Naturally that these new materials are only interesting if, beyond any improvement on the pure organic light emitters, they can also be scientifically viewed as useful substituents. Regarding the electroluminescent spectrum, the pure organic materials are known to produce a featureless emission band (typical from the several π-π^* orbitals emissions) and therefore are not suitable for sharp (and color specific) emission. RE complexes, due to the molecular internal mechanisms (explained later), have a spectral emission exhibiting a Full Width at Half Maximum (FWHM) of about some nanometers, which may obviously be related with "color purity." Secondly, a very large color spectrum emission may also be obtained, although not efficient enough for technological application, in special organic complexes based on europium (red) [1–4], terbium (green) [5–7] and thulium (blue) [8, 9], allowing the fabrication of the basic RGB pattern with the enormous advantage of high "pure colors." There is also some evidence of yellow emission with dysprosium organic complexes [10, 11], which further broadens the range of potential applications.

Luminescence of metallo-organic emitting complexes (and not only RE based) is completely governed by the molecular structure of the coordination sphere that the metal is incorporated into. This means that we have a high degree of control over the luminescence properties, working on the organic counterpart; the same does not occur in the "pure" organic emissive complexes. Moreover, replacing a pure organic emitter with an optically active metal ion coordination complex usually increases the chemical, and more importantly, the electrical stability. Taking into account all these considerations, it is natural that the interests in the metallo-organic emitting coordination complexes in general, and in RE complexes in particular, has strongly increased.

4.2 The Basics of Rare Earth Complexes: Application to Visible Spectrum

In general, the RE complexes that emit on visible spectral region are usually found in a valence of 3^+, i.e., RE^{3+} or RE(III), in coordination with an organic part. In fact RE ions have a higher stability under a tri-valence oxidization and thus most of them lie in that situation. This stability requirements is also an asset because it assures all of them are of interest due to simplicity/low cost/efficiency/sharp lines emission. Moreover, the RE ions have a higher probability to make chemical bonds with some chemical elements including in a decreasing order of "preference" due to the electro-negativity, fluorine, oxygen, nitrogen, and sulfur, that are simply the basic compounds (excluding the ubiquitous carbon) of the most common organic ligands. Still the exceptional properties of the RE organic complexes is the so-called "blindage effect."

4.2.1 The Structure of a Coordinated Rare Earth (III) Ion

A rare earth (lanthanide) element has an electronic configuration based on the xenon element, i.e., $1s^2\ 2s^2\ 2p^6\ 3s^2\ 3p^6\ 3d^{10}\ 4s^2\ 4p^6\ 5s^2\ 4d^{10}\ 5p^6$ hereinafter called [Xe]. That configuration is the core of all the RE coordinated ions in coordination complexes that we have described herein.

A quick look at the periodic table shows that the *4f* orbitals are only partially filled (exception to the lutetium and ytterbium that are a part of the elements not considered in this book). Table 4.1 shows the external orbitals' configuration of the RE elements mostly used in emitting complexes.

Table 4.1 Electronic configuration of the most-used emitting rare earth elements

Element	Chemical symbol	Electronic configuration
Samarium	Sm	[Xe] $4f^6\ 6s^2$
Europium	Eu	[Xe] $4f^7\ 6s^2$
Terbium	Tb	[Xe] $4f^9\ 6s^2$
Dysprosium	Dy	[Xe] $4f^{10}\ 6s^2$
Thulium	Tm	[Xe] $4f^{13}\ 6s^2$

[Xe] denotes the electronic configuration described in the text.

An important question is, why are some RE(III) elements are suitable (and so desired) for use on OLED active layers? The answer lies in a combination of two distinct properties: the "blindage effect" and the nature of the organic ligands of the RE(III) ion first coordination sphere.

In fact not all organic ligands are suitable to form a RE complex that is of use as emitting molecule. In order to achieve a high luminescence yield, the RE(III)–organic complex must fulfill two simple rules:

- The energy location of the RE(III) resonant excited level must obviously be located at lower energy than the organic excited triplet level (T_1 as described in last chapter); the goal is to allow a high transition probability from the organic ligand to the RE(III) excited resonant level;
- On the other hand, the non-radiative deactivation probability of that resonant level should be lower than the radiative one. The goal is that all the energy transferred from the organic ligand can be emitted as photons, i.e., radiative deactivation.

When both of these conditions are addressed and taken into account, not only the several chemical considerations (and constrains) related to the organic ligand synthesis but also the each RE(III) resonant excited energy levels and the luminescence properties of that elements can be fitted into three general groups, that will determine their usefulness for OLED active layer application as follows [12]:

1. Non-visible and near IR RE(III) spectral emission: La, Gd, and Lu;
2. Without radiative emission due to the fast non-radiative transitions between the very proximal RE(III) energy levels (although the ion receives energy from the organic ligand excited triplet state): Pr, Nd, Ho, Er, Tm and Yb. A special appointment must be made for Thulium (with some near blue emission) and for Erbium (with some emission in IR region that is under investigation for optical communications applications);
3. Finally, RE(III) ions with strong light emission in the visible region and without (or very weak) molecular emission (considering the organic whole molecule): Sm, Eu, Tb, and Dy.

Recalling that the electronic configuration for all RE of the interest for OLEDs is [Xe] $4f^m\ 6s^2$. One can clearly see that all neutral RE element have the same *6s* configuration and that the *4f* is variable. Nevertheless, when in the oxidized 3^+ state, the RE(III) ions exhibit a change in the *4f* orbitals ($4f^m$ with *m* ranging from 6 to 13, although in the overall elements it ranges from 1 to 14) leading to a gradual increase of occupancy of the orbital's levels. Though these processes are not surprising, the most interesting point is that, the probability of finding an electron belonging to *5s*, *5p*, and even *5d* and *6s* orbitals at a position *r* from the atomic nuclei is higher than the observed to an electron belonging from the *4f* orbitals, as observed from Fig. 4.1 [13].

This effect gives origin to the well-known "blindage effect" where the *4f* electrons are "shielded" by the *5* (*s*, *p*, *d*) and *6s* electrons. The advantages of such system are enormous: as the electronic transition in the RE(III) ions involves the *4f* electrons, there is no influence of the organic ligand on the radiative RE(III) emission (the electronic transitions also become "shielded") and therefore, this kind of rare earth based materials exhibits sharp and strong spectral lines, corresponding only to the internal ion emission. This system provides them with a "quasi-monochromatic" emission of a naturally high interest.

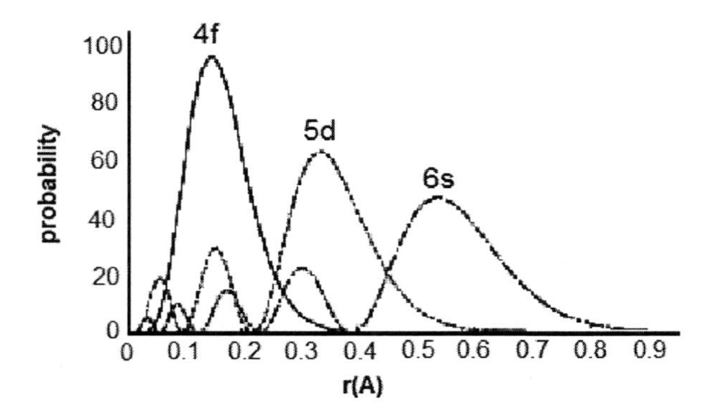

Figure 4.1 Probability of finding an electron at a distance r from nuclei, showing specifically the orbitals *4f*, *5d*, and *6s*.

The determination of the possible resonant levels of the RE(III) ion is sometimes complex. Here, we present a simple description in order to help understand the forthcoming analysis of the luminescence properties that are showed.

As expected, the many intrinsic RE(III) interactions will give different energy levels, which belong to the same original configuration. Regardless of the number of electrons in each ion (and configuration) they can have several distributions between the *4f* orbitals (responsible for the electronic transitions). The total number of *4f* orbitals is 7 so a very complex framing is predicted to appear due to the many possible different interactions. These interactions can be divided into Columbian, spin–orbit coupling and small electrical and/or magnetic interactions like the Stark and Zeeman effects, respectively. The Columbian interaction is usually the strongest, as it represents the electronic repulsions in the orbitals that give origin to the first energy levels separation (in order of 10^4 cm^{-1}). Due to high weight of the rare earth nucleus, the spin–orbit coupling is also relatively strong (in order of 10^3 cm^{-1}). In these situations, the spin and orbital moments, S and L, cannot adequately describe the RE(III) ion sate; in order to give a more realistic description the angular momentum j for each electron must be included. The combination of the individual j momentum yields the total angular momentum J, known as the j–j coupling. S, L and J can be interrelated by recurring to the Russell–Saunders schemes where $J = L + S$. Thus, for a free ion such as happens in our case because the RE(III) shielding induces a ion-like behavior, the energy levels can be described by the states $^{(2S+1)}\boldsymbol{L}_J$, where $2S+1$ represents the total spin multiplicity. Representation of $\boldsymbol{L}$ depends on its numerical value: L = 0, 1, 2, 3, 4, 5, gives $\boldsymbol{L}$ = S, P, D, F, G, H, respectively. So, for instance, an energy level given by $^5D_2$ represents $S = 2$, $L = 2$ and $J = 2$. It must be noted that as the J is equally quantized, the result from the vectorial addition between S and L allows J to ranges from $|L+S|$, $|L+S|-1 \ldots |L-S|$. For instance, considering a Russell–Saunders level ^7F, we will obtain the following levels from J: $^7F_6$, $^7F_5$, $^7F_4$, $^7F_3$, $^7F_2$, $^7F_1$ and $^7F_0$, i.e., seven different levels.

Finally, the small electrical (Stark) and/or magnetic (Zeeman) effects can be significant as the ion is coordinated (surrounded by a crystal or organic ligands). The several J levels can also be separated in $2J+1$ levels (with small separations in order of 10^2 cm^{-1}) and can be observed as very sharp lines in each (already sharp) spectrum "bands" (main lines!). These effects (called as crystalline field interactions) are only of mater in a pure physical study such as photoluminescence because the original *J* level cannot be more used to a "good" quantum number and the ion energy levels are now described by the irreducible representations of the point group to which the ion belongs. In our case, the electroluminescence spectrum is much more weak than the photoluminescence, so, even using an exceptional equipment for the measurement, these new energy sates are usually undetectable and, of course irrelevant for an RE(III) based OLED emission, where the color purity, carrier confinement, efficiency and CIE coordinates depend only of the main spectrum lines, i.e., on the *J* levels.

Figure 4.2 shows the predicted most important levels for a Eu^{3+} coordinated ion, giving the effect of each interaction inside the ion. To get a more complete idea about all the energy levels that arises from different metal ions (even under different oxidizing state) refer to [14].

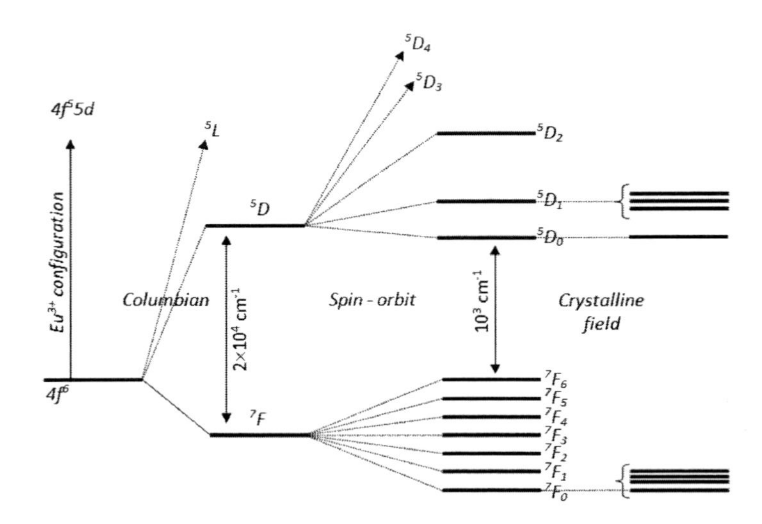

Figure 4.2 Energy level splitting for a Eu^{3+} coordinated ion due to Columbic and spin–orbit interactions. The crystalline field interaction is represented to provide a more widened notion of the phenomena and will not be used in RE(III) based OLEDs studies.

It is important to refer that the energy states arising from the Russell–Saunders description, have, obviously, different energy values. In order to establish an energetic order, we can apply to the Hund's rules that easily help on determine some fundamental points on the derived energy states. Here are the three Hund rules [15]:

1. The lowest energy state, i.e., the ground or fundamental state, has the largest spin multiplicity (given, as referred by $2S+1$);
2. If two energy states have the same spin multiplicity, the lowest energy level is the state for which the L is larger;
3. If an electronic sub-shell is less that half-filled, the state with the lowest J value has the lowest energy; and, if an electronic sub-shell is more than half-filled, the state with the highest J value has the lowest energy.

Simple application of these rules correctly predict the ground state; the others states are sometimes more difficult to predict but for all the matter covered in this book about RE(III) emission in OLEDs, the rules apply and corresponds to the main visible emitting scheme.

Although focused on the Eu^{3+} coordinated ion, the physical process showed in Fig. 4.2 is the "main route" for every other RE(III) ion energy level splitting. Starting always with basic ion configuration (in this case, configuration $4f^6$) we then apply the rules for Columbian and spin–orbit coupling, taking into account the Hund rules to scale the energy levels. Obviously, that the precise numeric value of the energy levels cannot be empirically obtained and must be estimated by theoretical calculations or determined from experimental spectroscopy data. For the aim of this book and for those working with OLEDs, only the energy levels involved in the visible spectrum region matter. This is the reason why the levels with energy higher than that at the visible region are omitted in Fig. 4.2 and only their reference is given. An important notion to highlight is the wide range of splitting of the RE ions energy levels leading in extreme scenarios, to complex and difficult to analyze diagrams. Using Eu^{3+} ion as a typical example, over 110 energy levels are obtained after taking into account the spin–orbit coupling effect. Once again, we focused only the visible spectral region.

After establishing the RE(III) energy levels, the radiative and non-radiative transitions can be explored. The luminescence of RE(III) ions is due to the *4f–4f* transitions that, following the Laporte rule (described in last chapter) are forbidden by electrical dipole. In spite of this constrain, the Laporte rule relaxes in molecules without an inversion center as are RE(III) complexes [16], because the parity is not completely defined [16]. For a free ion, only the magnetic dipole transitions are allowed and the rule is $\Delta J = 0, \pm 1$, although the transition $(J = 0) \leftrightarrow (J' = 0)$ is forbidden (J' denotes an excited state). This explain the absence of the ${}^5D_0 \rightarrow {}^7F_1$ in the emission of Eu^{3+}, for instance. If the crystalline field effect is taken into account, the picture gets even more complex due to the fact that this effect can mix configurations of odd parity, and therefore the main spectral lines will be the result of a forced electric dipole or can even be the result of a double characteristic: both electrical and magnetic dipoles. That situation can modify the line intensity (compared to the theoretical predicted free ion) and, in some cases, add or suppress other spectrum lines (for instance ${}^5D_0 \rightarrow {}^7F_1$ emission is sometimes observed, although weak, implying a low transition probability). For RE(III) complexes used in OLEDs, the difference in practice is not so extreme because the organic ligands does not imposes a strong crystalline field effect, thus making its contribution neglectable in the performance of OLEDs with active layers based on these RE(III) materials. As a matter of fact, previous results obtained in our group with Eu^{3+} [17], on a study of the influence of changing the organic ligand in these complexes, showed that the only noticeable modification was the change in some nanometers of the main emission spectral line position, corresponding to the ${}^5D_0 \rightarrow {}^7F_2$ usually located near 612 nm in most Eu^{3+} complexes with organic ligands.

4.2.2 Excitation and Luminescence Mechanisms of Rare Earth Coordinated Ions

4.2.2.1 Basic principles of the luminescence of coordinated rare earth ions

As reported before, the electronic transitions involving coordinated RE(III) ions, of interest for OLEDs development are the *4f–4f* transitions. In theory it's possible to excite an electron directly into those orbitals and then obtain the respective luminescence when the carrier deactivates in a radiative way to the ground state. Although simple, this idea is impracticable due to the very low absorption of the RE(III) free ions. This is where the ligands enter the picture.

In order to obtain the strong visible luminescence from the *4f–4f* RE(III) transitions, it's absolutely necessary to populate the resonant energy levels of the ligands coordinated to the ion. In complexes with organic ligands, the molecular structure of the ligand is chosen to do just that.

The basic process involves, in a similar way to that described in Chapter 3, a previous charge promotion to the organic ligand's excited energy levels (excitation) with a subsequent transfer to the resonant excited levels of the RE(III) ion. The process aims to "pump" to the RE(III) ion, overcoming its weak direct excitation, and therefore yielding a final strong light emission. Naturally that the careful choice of the organic ligand is fundamental (and will be discussed later) because the energy levels involved and the electrical conductivity requirements. This energy transfer mechanism is commonly known as *antenna effect* [18], as it works by a similar way as an usual antenna that collects the energy and transmits it to the system for further application. One important issue that can currently be posed is the difficulty in assessing if the energy transfer in this antenna effect process is generated on the S_1 or the T_1 organic exited level, although some previous work [19] showed that the transfer $S_1 \rightarrow$ RE(III) is not the most important (and in many situations can be despised). So we keep in the assumption that the energy transfer $T_1 \rightarrow$ RE(III) is the main source of the antenna effect. An empirical rule, called the energy-gap rule, deems that the total luminescence quantum yield (briefly described in the next section) decreases due to an energy back transfer from RE(III) to T_1 when the ΔE between the T_1 and RE(III) resonant exited emitting level is low [20]. For instance, applying this empirical rule to Tb^{3+} and Eu^{3+} coordinated ions, leads to a high luminescence quantum yield achieved if the energy gap between the T_1 organic ligand state and the $^5D_4$ Tb^{3+} excited level is less than about 0.23 eV; a value that ranges from 0.31 eV to 0.59 eV in the case of Eu^{3+} [21]. Regardless of different arguments on this issue, it is widely accepted that the molecular design of an organic RE(III) emitting complex must guarantee that the T_1 organic ligand level is located slightly above the exciting resonant emitting levels of the RE(III). Naturally, this assumption *per se* does not suffice to obtain a very efficient emitting complex but it is at least warrants the antenna effect and, therefore, the emission efficiency.

In organic complexes based on RE(III) ions the "antenna" is of course the organic part of the molecule (we stressed its importance from the chemical point of view) and must efficiently transfer the energy (electrical charge in our OLED discussion) to the final resonant RE(III) ions. The organic part must also comprise two molecular requirements: (1) balance the charge (+3) of the RE(III) ion in order to guarantee the molecular charge neutrality and (2) allows the molecular (as a whole system) chemical stability in order to avoid fast chemical degradation and/or easily molecular ligations broken that will produce with the subsequent OLED fabrication procedure, a molecular inhomogeneous active layer composed by a mix of the desired emitting complex and their molecular fragments.

Currently, there are three mechanics that lead to the antenna effect on the RE(III) complexes with organic ligands, all based on the intermolecular energy transfer as described in Chapter 3. The final process, that is, RE(III) emission is the result of contribution from all of them and manifests as the intrinsic *4f* electronic transitions. The mechanisms can be summarized as follows:

1. An excitation ($S_0 \rightarrow S_1$) leads to intersystem crossing ($S_1 \rightarrow T_1$) followed by an energy transfer from the first excited triplet state (T_1) of the organic ligand to the lowest resonant energy level of the RE(III) ion;
2. Upon excitation, the organic ligand goes into a singlet excited level (S_1) and this level transfers the energy to the resonant levels of the RE(III) ion;
3. Upon excitation, the organic ligand goes again into a singlet excited level (S_1) but the energy transfer is then done to a high excited resonant level of the RE(III) ion; after this process, the RE(III) high excited level transfers the energy back to the first excited triplet level of the organic ligand (T_1) that in turn, transfer the energy again to the lowest resonant level of the RE(III) ion.

It is not difficult to understand that these mechanisms depend entirely on the structure of the organic ligands and their energy levels alignment with the resonant levels of the RE(III) ions. Sometimes, one of the cited mechanisms is chosen not particular for the cases of coordinating a particular ligand to the RE(III) ion, but for the final efficiency purpose, aiming at the maximization of the internal quantum efficiency. On the other hand we must pay much attention to the subtle differences that may be found between optical and electrical excitation, which, in some specific cases, bring down the feasibility of an exceptional theoretical idea as experimental results come to show no success. As an implicit advice, and about application of RE(III) coordinated ions in active OLED layers, experimental tests that guarantee the correct charge confinement, are always critical, if no previous computational predictions of the electrical charge transfer in the organic–RE(III) complexes are available. And, of course, after the correct molecular structure description is confirmed, for instance, by X-ray crystallography. In the Fig. 4.3 a simple diagram of the three mechanics for antenna effect is shown. Of those depicted in the diagram, the most usually mechanisms are (1) and (2), although the last one (3) cannot be excluded. For an OLED it's not of vital importance to know which mechanism is the dominant one but the molecular internal efficiency may be affect by this issue and implicitly so well the OLED efficiency.

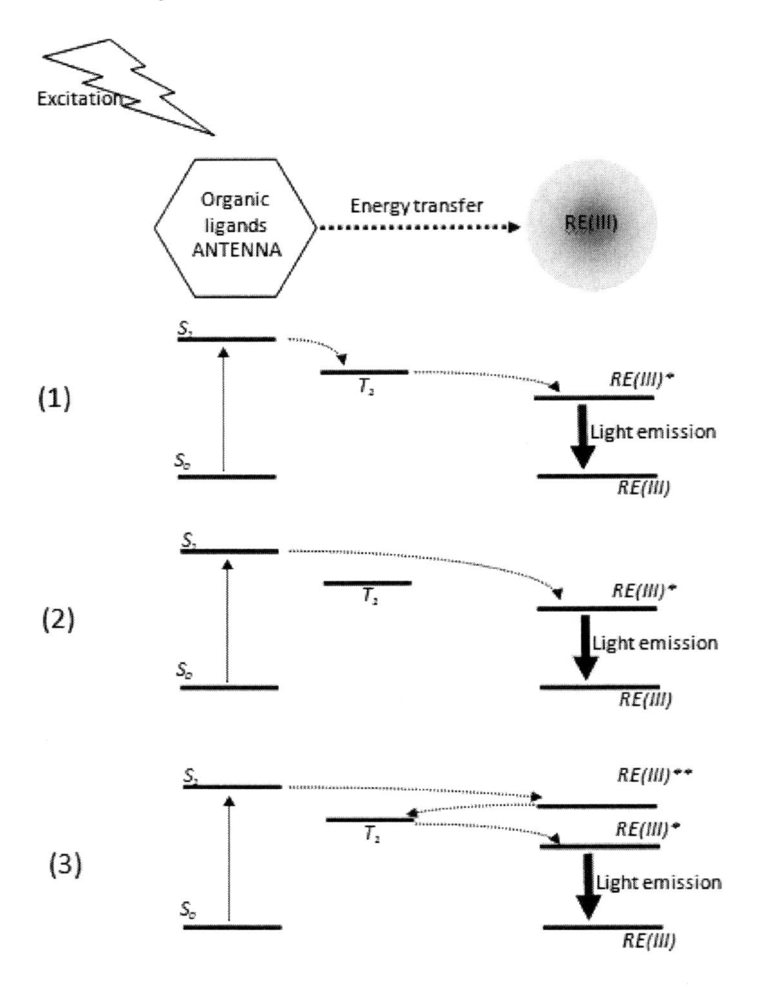

Figure 4.3 Different mechanics for antenna effect. RE(III)* and RE(III)** denote RE(III) excited states.

By looking to the molecular structure we will now see how the antenna effect works on the different molecular fragments. A RE(III) organic complex typically comprises, besides the coordinated ion itself, a group of three organic molecular ligands each with one negative electronic charge (a total of −3 to compensate the +3 from the RE(III) ion) and an electronic neutral organic ligand that completes the coordination sphere around the ion and may also contribute to the electrical carrier conduction. In a more general designation, we have, in a RE(III) organic complex, the *central ligand* (typically a set of three organic anionic ligands that balance electronic charge of the RE(III) cation) and a *neutral ligand*. It is not difficult to understand that this designation—central and neutral—has a strong connotation with the geometry of the first sphere of coordination (we will return to this idea later) and with the electric charge, respectively.

The luminescence mechanisms of the RE(III) complexes are similar in the light emission but differ a little under optical or electrical excitation. In photoluminescence, the organic ligands responsible for the electronic charge conduction, absorb a phonon (from the optical excitation source) and then, if the absorbed energy from the excitation phonon is sufficient, promotes an electron from the ground singlet state to an excited energy state, singlet or triplet, as already mentioned. This part of the molecule part is organic "antenna." Then, the de-excitation takes place by one (or more) of the referred process. Under electrical excitation (to induce electroluminescence) the role of the molecular "antenna" is to harvest electrical charge from the transport layers of the OLED structure, by a recombination process involving an electron and a hole, which yields the excitation of the T_1 state. The external applied voltage will inject further carriers as many as those needed for the process in a similar way to that found for the excitation by a light source. Following the excitation of the T_1 state, the process proceeds in the same way to that described for the photoluminescence. The differences in the initial step of the excitation is usually responsible for difficulties in obtaining an efficient OLED even when employing in its active layer an extremely efficient photoluminescent RE(III) complex. So, it is not excessive to stress, once again, the importance of the designing OLED structure, by a pondered choice of the best organic electrical carrier transport and of the interface barriers, which must allow the carrier flow, but also have to act as blocking barriers to increase the electrical carrier confinement inside the active layer.

4.2.2.2 Considerations about luminescence efficiency and quenching of coordinated rare earth ions

The RE(III) organic complexes luminescence process is intrinsically related with two important molecular features: the quantum yield and the luminescence quenching, indicate the efficiency of the energy transfer from the organic triplet excited state to the RE(III) excited resonant ions states.

The quantum yield, Q, is usually defined as the ratio between the emitted and the absorbed photons. The value of Q can be related to the RE(III) intrinsic quantum yield Q_{RE}, which in turn, depends on the depopulated ratio of the excited level, denoted by K_d and on the radiative rate K_r [22]:

$$Q = \eta_O \eta_I Q_{RE} = \eta_T Q_{RE} = \eta_T \left(\frac{K_r}{K_d} \right) = \eta_T \left(\frac{\tau_d}{\tau_r} \right) \tag{4.1}$$

where τ_d and τ_r denote the lifetimes of the excited state and of the radiative decay, respectively, and are related to the respective rates (K_d and K_r) as the inverse, $\tau = 1/K$; η_O and η_I, represent the overall and intrinsic quantum yield (depends on the organic ligand) and can be simply represented by η_T, the total efficiency of energy transfer from the ligand to the RE(III) ion. It is thus not surprising to observe that the maximum efficiency energy transfer to the RE(III) ion can be obtained from Eq. 4.1 by the maximizing η_O and η_I. This information is of vital importance for the choice of the ligands as a rational molecular design.

The experimental determination of Q is not easy and requires specific equipment. The most common route is to determine, rather than the absolute quantum yield, a relative value obtained by direct comparison with the Q value of a known reference material.

Concerning the intrinsic quantum yield a simple definition is that it corresponds to the quantum yield of the RE(III) ion luminescence when a direct excitation of its *4f* levels occurs. That value is intrinsically related to the non-radiative de-excitation process that occurs in both the inner and the outer coordination spheres of the RE(III) ion. The K_d rate expression can be thus summarized as follows:

$$K_d = K_r + \sum_n K_{nr}^n \qquad (4.2)$$

where K_{nr} denotes all non-radiative de-excitation processes. These non-radiative de-excitation processes involve vibrational induced mechanism, and other non-radiative energy losses such as photon induced charge transfer.

Others mechanisms can contribute to this non-radiative decay. In a simplified way, these can be divided into vibrational and electronic processes. The first are, traditionally more important that the second ones although both contribute to the K_{nr}. In a vibrational process, some molecular bonds can, under specific conditions, promote energy dissipation in the RE(III) ion coordination sphere and therefore lead to the quenching of the RE(III) luminescence. A particular case of this situation we must note that some neutral organic ligands are known to, under specific circumstances, promote non-radiative de-excitation by vibrational mechanisms. In addition, some central organic ligands may follow the same mechanism. Naturally, excluding these few problematic organic ligands (both neutral and central), the non-radiative quenching by vibrational de-excitation can be strongly related to the molecular structure (and RE(III) ion).

Electronic de-excitation processes comprise (i) the energy back transfer, in which the energy transferred from the organic ligand excited state to the RE(III) resonant excited state, is sent back to the excited organic energy level; and (ii) the photon induced transfer of the organic ligand to the metal ion with a further reduction of the RE(III) to RE(II) and the consequent quenching of the RE(III) luminescence.

Detailed explanations on quantum yield calculation from optical spectroscopy measurements can be found, for instance, on the following references [23, 24]. Though this is a fundamental topic in RE(III) luminescence behavior its in-depth description is not within in this book's objectives. For those is interested in this question we recommend reading details on the suggested literature.

4.2.3 The Molecular Structure of a Rare Earth Organic Complex and the "Antenna" Effect

Minding the before mentioned molecular structure definition for RE(III) organic complexes, composed by (i) the emitting RE(III) ion, (ii) the central ligand and (iii) the neutral ligand, we can now establish a more depth notion on the importance (and influence) on each part in the excitation/emission process involved in the antenna effect. For that, and to keep it simple, let us consider for instance, a europium organic complex.

Figure 4.4 shows the *tris (dibezoylmethane)-mono (4,7-dimethylphenanthroline) Europium(III)* complex, commonly abbreviated as Eu(DBM)$_3$phen. The RE(III) emitting ion is the Eu^{3+}, the central ligand DBM (*dibezoylmethane*) and the neutral ligand phen (*phenantroline*). This emitting molecule is one of the most commonly used europium based organic complexes.

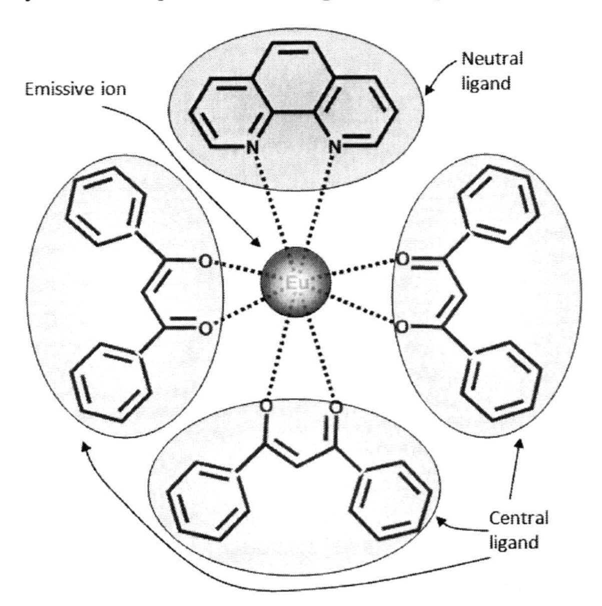

Figure 4.4 The molecular two-dimensional structure of the RE(III) emittingcomplex,*[tris(dibezoylmethane)-mono(4,7-dimethylphenanthroline) Europium(III)]* commonly known as Eu(DBM)$_3$phen. The shadowed areas represent the different parts of the molecule as indicated.

In this molecular structure, the ligands must be responsible for energy delivery into the excited level of the Eu^{3+} ion. This can be accomplished by the neutral and/or by the central ligand, regarding the respective excited levels (singlet or triplet). In photoluminescence, the photon absorption can occur on both ligands, depending on the transition probabilities, but when we look into the electroluminescence process, some differences exist.

In Fig. 4.5 this complex's general photoluminescence process is shown.

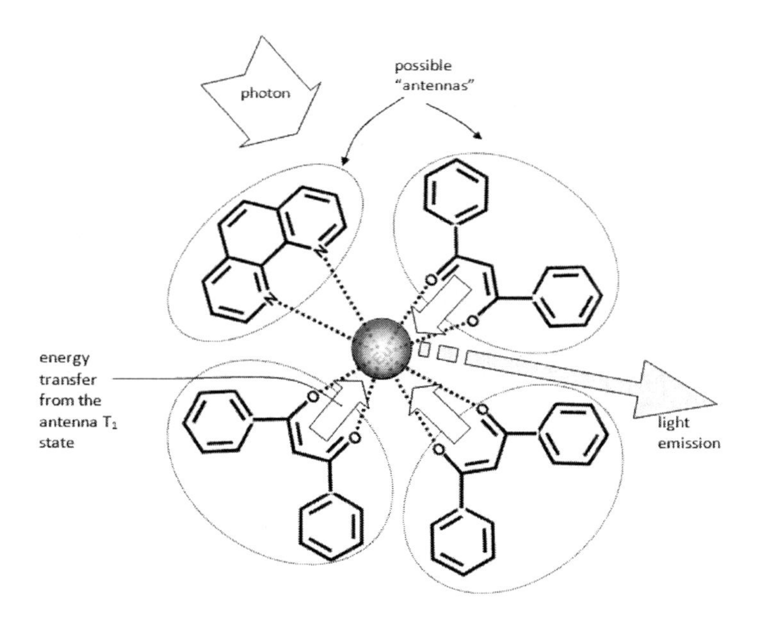

Figure 4.5 A typical photoluminescence process for the $Eu(DBM)_3$ phen molecule. Upon photon excitation, the organic ligands (either the central or the neutral, depending on the optical absorption properties) transfer the energy for the lowest excited triplet state T_1 of the ligands to the resonant excited levels of the Eu^{3+}, with the consequent light emission.

It must be noted that sometimes it is very difficult to establish which is the actual part of the organic ligands that is working as antenna. Usually the spectroscopic absorption band of the RE(III) organic complexes shows a large and featureless band (traditionally in the UV spectral region) that cannot be easily correlated with any of the employed ligands (central or neutral). The desired model for the molecular structure, does not apply exhaustively for that questions as the main know organic ligands used in the RE(III) organic complexes efficiently absorbs energy. As mentioned before, the main features to be considered are precisely the non-radiative mechanisms.

Some authors, however, claim a mixed-ligand process [25] with the introduction of different neutral donor ligands, which can be responsible for important changes in the RE(III) ions photoluminescence; others [26] point to a possible energy transfer from the central ligand triplet state to the neutral ligand triplet state followed by transfer to the resonant levels on the ion. But at the moment this theory has little experimental evidence to support it.

Regardless of these questions, (some of them to be further detailed in special cases of RE(III) organic complexes in the forthcoming sections) the main approach to the design of an efficient emitting complex lies in managing the energy difference between the main ligand excited triplet state (T_1) and the resonant level of the RE(III) ion. In many cases, the later will be the ^5D_J level but some ^4F_J, ^4G_J or others levels may also be found (we will consider the generic $^{(2S+1)}L_J$). We have already see the energy levels of Eu^{3+}, and for the remaining RE(III) ions, the energy levels will be presented whenever they are relevant for the discussion. This difference, ΔE, will influence the rate of the direct energy transfer from the organic ligands to the RE(III) ion (and the back transfer) and also determine the probability of the non-radiative ligand relaxation [27].

The energies associated to a RE(III) excited level is not difficult to learn, as already mentioned, as it can be found in several databases (we stress again that the crystalline field may slightly modify these values albeit without but special relevance for the RE(III) complexes here studied). Determining of the energy of the T_1 level however, is sometimes more complicated. A typical solution for the problem is to replace the emitting RE(III) ion with the gadolinium ion in the same organic framework, i.e., to synthesize the parent Gd^{3+} complex with same organic ligands. The spectroscopic studies with this model complex, allows to estimating of the T_1 energy levels of the organic ligands. This process is used for two main reasons: first, in Gd^{3+} organic complexes the ratio between phosphorescence and fluorescence quantum yield is higher than 100 allowing a better singlet–triplet study; and second, the Gd^{3+} does not have any resonant excited level and its resonant level (in high *P* levels) is higher than the triplet levels of organic ligands [28] thus allowing us to study only the organic ligands. Once again the data obtained is still an estimate but it can provide important information about the T_1 energy location and the usefulness of an organic ligand for a particular RE(III) ion.

Considering the electroluminescence mechanism, and still using the $Eu(DBM)_3phen$ as example, the electrical excitation process will be considerably different. If an active layer is composed solely by this complex and the OLED can emit reasonably to feature the typical molecule spectrum, it is fair to assume the organic part must have a bipolar characteristic, i.e., it has the ability to transport either holes either electrons. In our example, *phenantroline* is considered both electrically bipolar and more an electron transporter. Before extending the discussion about these issues, it is important to mention that the general concept for electroluminescence is that electrons are injected into the active layer—$Eu(DBM)_3phen$—from a cathode or an electron injection/transport layer, while holes are transported into this layer trough the HTL after anode injection (we run on the assumption that the hole mobility is much higher than the electron mobility, as mentioned in last chapter). Then, excitation of the organic ligands is carried by the before mentioned exciton formation (electron–hole pair), with it further transference (after being converted to triplet excitons) to the excited energy to Eu^{3+}. Figure 4.6 represents an electrical excitation process for the $Eu(DBM)_3phen$ molecule.

The described model can be considered physically valid if, naturally, the RE(III) complex exhibits a bipolar electrical nature. If not, we must include into the active layer an organic semiconductor with an electrical transport nature (electrons or holes) that can compensate the absence of any transport carriers type in the organic ligands of the RE(III) complex. As a simple example, some complexes when employed in the OLED active layers, are embedded into an organic semiconducting polymeric matrix. Although this strategy also comprises an energy transfer from the polymer to the RE(III) complex, it can also be used for the electrical charge transport as we will be discussed later.

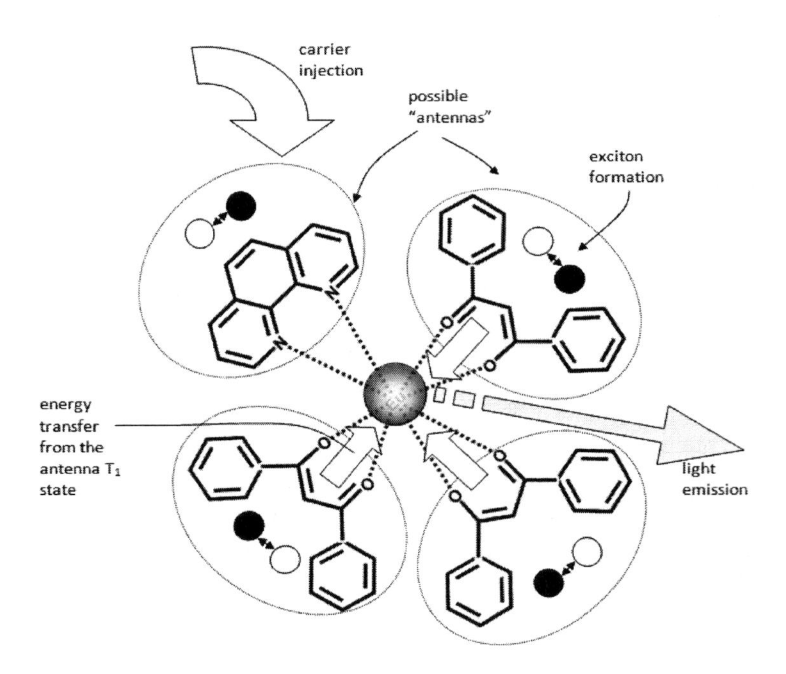

Figure 4.6 Typical electroluminescence process for the Eu(DBM)$_3$phen molecule. Under an applied voltage, the carriers (electron and/or holes) are injected into the organic ligands (either the central either the neutral, depending on the electrical transport properties) followed by exciton formation. The resulting energy is then transferred for the lowest excited triplet state T_1 of the ligands to the resonant excited levels of the Eu^{3+}, with subsequent light emission.

The carrier transport nature of the RE(III) organic complexes is unfortunately not easy to understand. Hall Effect measurements can evidently, be made, simultaneously determining the carrier mobility. But the different OLED fabrication processes, sometimes using different material chemical synthesis (with strong differences in the electrical behavior much more than the observed in the photoluminescence studies) and device structures, leads to an impossibility to asset an unique and universal model for a specific complex. This means that, though the reproducibility of the device and of the synthesis of the EL is assured, each OLED architecture is unique and follows a specific pathway, with its own results. Obviously, the only common factor within all these variables is that the intrinsic electrical nature of the RE(III) complexes remains similar.

Looking in further detail to the molecular structure of the ligands, it's obvious that these must present π–π^*, which is a synonym of delocalized orbitals and the consequent ability to allow an electrical carrier hopping to yield an electrical current. Although this assumption appears to be easy, in practice this strategy is not so simple mostly due to the local conformation of the coordinated organic ligands, strongly dependent on the electrostatic repulsion. In our example, these changes would be observed for the coordination angles and bond length of the three DBM parts and the *phen*: for a detailed case study refer to Chapter 6. Others factors include the intermolecular interactions among organic ligands (i.e., the influence of the proximity/overlap of two or more ligands belonging to distinct molecules). This last point, that is related, among other factors, to the OLED assembly process and thus depends on the methodologies and parameters it uses, is currently the object of several research projects that try to correlate the assembly methodology with all the variables involved in the local molecular conformation in a very thin film.

It is evident that for a few known ligands it will be relatively simple to estimate what will happen under an electrical carrier injection. For instance, a DBM derivate known as c-DBM exhibits a hole conduction, whereas a derivate of *phenantroline*, with the abbreviation *bath*, allows an electron conduction. However, these examples cannot be used to create a general rule and we must again stress that predicting the electrical behavior of a molecule is not as straightforward as we might wish.

An interesting observation, has resulted from our own research on the electrical AC and DC behavior of some RE(III) organic complexes, based either in europium or in terbium, and used to fabricate single layer devices by thermal evaporation. That all of them exhibited a relatively high electrical conductivity but non electroluminescence was also observed (and we were not expecting that). For the current discussion on the electrical molecular behavior of such RE(III) organic complexes, two conclusion can be easily drawn: (1) the complexes clearly have a bipolar nature (otherwise a very low electrical current would have been be measured with rectification) but (2) the electron–hole recombination that would lead to excitation of the triplet states of the organic ligands is very low or does not exist. This means that, although these materials are bipolar, they must possess significant unbalanced carriers (electrons and holes) at the molecular level or that the organic fragments of several adjacent molecules are overlapping, creating a channel; a third possible explanation is that electrons and holes flow on the delocalized states that behave similarly to polymer chains. All those hypothesis clearly warrant a more detailed investigation. The improvements on the knowledge of electrical conduction phenomena in RE(III) organic complexes processed into a thin films, will guarantee new developments on such based OLEDs.

In the sequence of this schematics molecular architecture and its implication to the excitation and luminescence behavior, a simplified diagram can be presented showing both the photon and the electrical carrier excitation/de-excitation process. We will consider the singlet and triplet states of the ligand without specifying whether this is a central or a neutral ligand, in agreement with the previous comments. It must be noted that both the energy transfer and the de-excitation (either radiative either non-radiative) have their own particular proper probability rate, which further stipulates the efficiency of the complex. In a general overview we can regard the complex as a whole and single molecular entity yielding, upon excitation, a final luminescence (resulting from the sum of all the internal competitive mechanisms). This simple scheme is present in Fig. 4.7 .

The figure omits some pathways, (in particular the energy back transfer), because these are implicit, following the Jablonski diagram showed in Fig. 3.10. The direct excitation of the RE(III) resonant levels from the organic ligand S_1 is also not shown because it usually is considered as not efficient when compared to the $T_1 \rightarrow$ RE(III) energy transfer [29]. The figure thus aims to represent the most common energy mechanisms observed in such complexes.

Concerning the energy transfer from the organic ligand to the resonant excited levels of the RE(III) ion, it can be explained by specific mechanisms. In particular, two well known mechanisms are traditionally involved in this transference: the Dexter and Förster [30, 31].

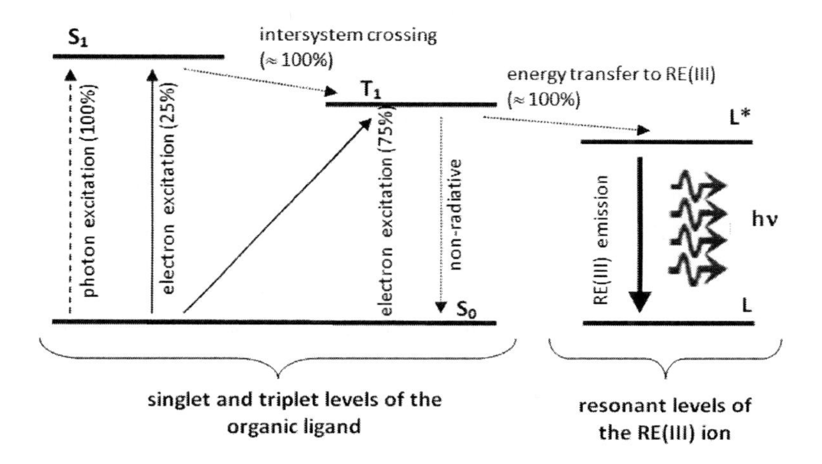

Figure 4.7 Diagram the for energy transfer from the organic ligand to RE(III) ion under both photon and electrical carrier excitation. The resulting light emission will be, respectively, the photoluminescence and electroluminescence. In RE(III) ions, the fundamental and first excited levels are the *L* and *L**.

The first or Dexter mechanism, (also known as the exchange mechanism) involves a double electron transfer and requires a strong overlapping of the organic ligand and RE(III) orbitals; thus a strong dependence on the donor–acceptor distance r is expected. The second mechanism, Förster's (also known as dipole–dipole mechanism) is based on the coupling between the dipole moment of the organic triplet state and the dipole moment of the RE(III) *4f* orbitals. This mechanism depends on the overlap integral (denoted by *J*) between the donor emission spectrum and the acceptor absorption spectrum (respectively, the organic ligand and the RE(III) ion). Both mechanisms are dependent on the donor–acceptor distance: e^{-r} for Dexter and

r^{-6} for Förster. The last one has "long" range dependence (typically about 50–80 Å) whereas the Dexter mechanism depends on short distances (typically about 10–15 Å). A very simple scheme for both mechanisms is shown on Fig. 4.8.

As an observation from the Fig. 4.7, while the photon excitation is obtained only between singlet states (spin multiplicity rule) the electrical excitation is obtained between singlet–singlet and singlet–triplet in a proportion of 1:3 respectively. This occurs not only due to the number of singlet and triplet states (as referred in Chapter 3) and but also because the electrical excitation depends on the previous electron–hole recombination, which, in delocalized orbitals, occurs in both kinds of spin multiplicity levels. In a first approach, one may think that the electrical excitation will be more efficient than the photon one; but unfortunately it is the opposite due to the difficulty in having efficient and high-density electron–hole recombination in the organic ligands. Some specific OLED structures are then developed specifically to help increase this efficiency.

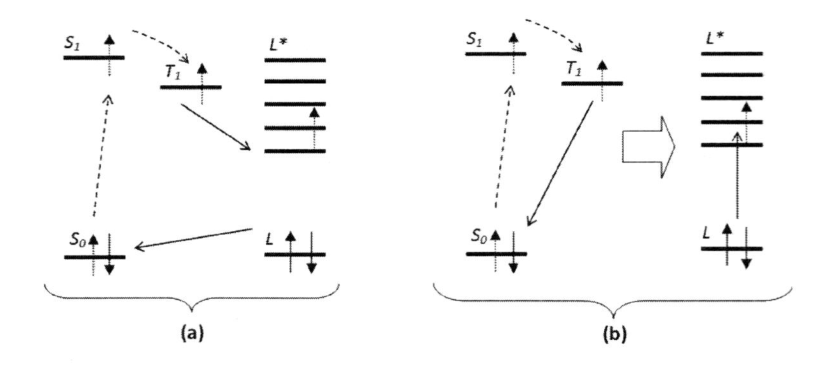

Figure 4.8 Simple schematics for the (a) Dexter and (b) Förster mechanisms for energy transfer between an organic ligand excited state (T_1) and the resonant excited state of the RE(III), a L^* level.

4.3 Rare Earth Based OLED: The Material Chemical–Physical Properties

When developing a RE(III) organic complex based OLED (in a process that is much different from those of other organic emitting complexes) much work must be done, starting with the molecular design and ending with an efficient OLED structure. The fundamental points we will explore are the following: (1) the molecular structure chemical-physical properties, (2) absorption and photoluminescence studies, (3) first device structure choice and fabrication process and (4) improvements on device structure and/or fabrication process.

4.3.1 Considerations of the Chemical–Physical Properties of the Rare Earth Organic Complexes

When we are looking for a particular RE(III) complex to start our work in the preparation of an OLED, there are some important chemically and structural data that must be known, as these are of extreme relevance from the physical point of view. In particular, the choice of the organic ligands that are coordinated to the RE(III) ion is the fundamental and perhaps the first that a physics and/or material researcher in this field will look into. The choice is clearly related with the ligands energy levels location (excited singlet and triplet).

Considering the previous discussion, there will not be much surprise in starting the need to have the RE(III) ion coordinated with a ligands that follows this one last rule: the energy difference between the lowest organic excited state (typically the T_1) and the resonant RE(III) excited level must be positive in order to optimize the energy transfer from the first to the second.

The typical structure of a RE(III)-organic complex for application into an OLED comprises a β-diketonate chelate (like the DBM used in last section) in conjugation with a second organic N-coordinating chromophore (like the previously shown *phenantroline*). This chromophore most often features two available sites of coordination with the RE(III) ion, being called a bidentate ligand as it forms two coordination bonds. In general, the β-diketonate chelate is anionic and the chromophore is electrically neutral (as referred before). Although this is the most widely used molecular structure, the use of an anionic chromophore has been reported to improve the chemical molecular stability, giving a more efficient device [32]. Therefore some of the formerly called neutral ligands are now becoming not electrically neutral and the name is not much realistic. For those ligands, the usual chemical designation (chromophore or bidentate ligand) will thus be used.

4.3.1.1 Rare earth common β-diketonates and organic neutral ligands

β-diketonates are the most studied and widely used organic ligands to coordinate to RE(III) ions. Although was first synthesized more than one century ago, only in last twenty years becomes important for organic electroluminescent complexes.

The three main groups of β-diketonates for RE(III) are the tris-complexes, ternary β-diketonates (also known as Lewis base adducts of the tris-complexes) and tetrakis-complexes. The basic molecular structure basic of each group can be easily described as follows:

1. In the first group, tris-complexes are formed by three β-diketonate ligands for one RE(III), giving a final electrically neutral molecule with a general formula of RE(III)(β-diketonate)$_3$;
2. The second group is a consequence of the fact that, the coordination sphere of the RE(III) is usually unsaturated in the former six-coordinate complexes and can expands its coordination sphere by means of an oligomer formation, either by bridging with the β-diketonates ligands either by adduct formation with the so-called Lewis basis (representing the neutral ligand L); its general formula is RE(III)(β-diketonate)$_3L$;

3. The final group is the less explored in RE(III)–organic complexes and is formed by a RE(III) with four β-diketonates ligands it a general formula of [RE(III)(β-diketonate)$_3$]$^-$, where the minus signal denotes an anionic complex with an overall molecular charge of −1. To achieve the electrical neutrality, a cationic counterion is needed, usually a protonated organic base, a quaternary ammonium ion or an alkali–metal ion.

Being the most widely used in active layer RE(III) OLED fabrication, this section will focus on the second β-diketonates group. In the specific RE(III) OLEDs, other groups may be cited as research examples.

The basic chemical structure of a β-diketonate comprises two acetyl groups linked by a single CH_2. This is the simplest of these structures, called *acetylacetonate*, represented by the acronym ACAC. Further chemical derivatives can be obtained from this basic ACAC structure by the substitution of the methyl groups by others moieties [33]. Also, the ACAC basic structure the methyl can be a substituted by an alkyl, a fluorinated alkyl, and aromatic or heteroaromatic groups. In Fig. 4.9 the most common β-diketonate ligands are shown.

Some important aspects must be noted about the molecular structure of β-diketonates and the relation to the physical luminescent behavior and the physical processing of the RE(III) complex. For instance, long alkyl chains are known to increase the solubility in organic solvents, so these are the best candidates for complexes for wet processing, allowing a good film formation [34]; on the other hand, fluorinated alkyl β-diketonates increase the volatility and therefore are the best candidates for film processing by thermal evaporation [35], giving higher thermal stability and lowering the rate of oxidation. Finally, β-diketonates with aromatic substituents have a higher absorption when compared to the β-diketonates bearing only aliphatic substituents. It is also understandable that, sometimes, the method of processing a RE(III) organic complex film in the OLED fabrication, depends on the β-diketonate ligands it bears.

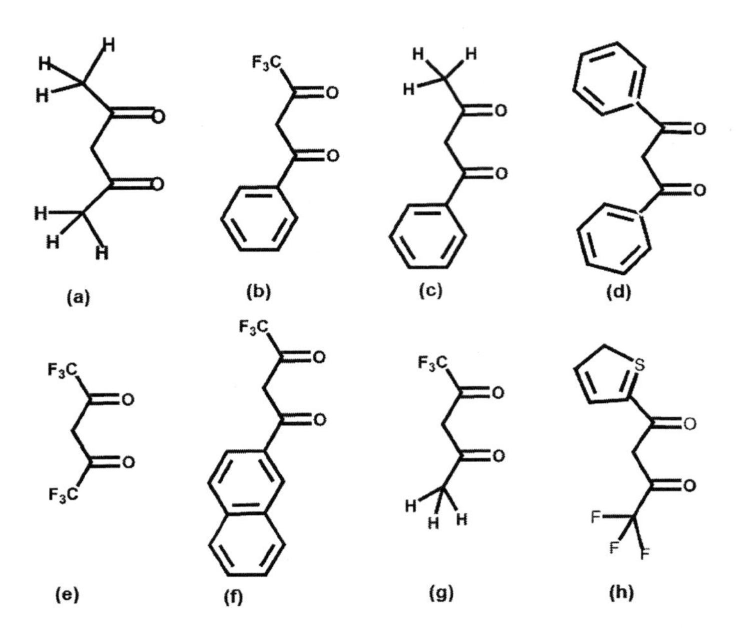

Figure 4.9 Usual β-diketonates structures employed in RE(III)–organic emitting complexes. The common designations are: (a) ACAC, (b) BTA, (c) BZA, (d) DBM, (e) FAC, (f) NTA, (g) TFA, and (h) TTA.

In addition, the energy levels and in particular the location of the first triplet excited state T_1 (the most important on the ligand to RE(III) ion energy transfer) are clearly dependent on β-diketonate ligand. This implies (knowing the resonant excited L^* levels of the RE(III) complex) that not any β-diketonate can be used with a specific RE(III), or at least, it is not advisable, because they may break the basic important rule: the T_1 level energy must be higher than that of the RE(III) resonant excited level. We will further detail on this further ahead.

As mentioned, besides the β-diketonates, RE(III) organic complexes used in the basic active layer of an OLED also posses, in the vast majority of cases, a forth ligand in the coordination sphere, usually electrically neutral. Incorporation of the neutral ligand brings major advantages, as the higher stability of an eight coordinated chemical structure improves the physical processing of the material.

The RE(III) organic complex formed by a tris-β-diketonate and a forth neutral organic ligand can increase the molecule emission efficiency as the neutral ligand replaces the coordinated solvent molecules traditionally present in tris-β-diketonate complexes. Based on the assumption [36] that RE(III) ions act as hard Lewis acids, the tris-β-diketonate complexes will bind preferentially with nitrogen or oxygen donors of an organic counterpart that are considered as Lewis basis.

There are several neutral ligands obeying the Lewis-bases character. When specifically aimed at incorporation into a RE(III) emitting complex, some are clearly preferential due better results in the energy transfer efficiency to RE(III). Some of them are showed in Fig. 4.10 .

A special case of neutral ligand is water. Some RE(III) organic complexes bear two water molecules on the site of the neutral organic ligand. These are usually formed when the RE(III) β-diketonate complex is synthesized in water, although the internal efficiency is lower than those obtained with others ligands. The problem is that the excited energy level of water is highly resonant with the excited energy levels of the most employed β-diketonate ligands and de-excitation by non-radiative decay trough intercrossing system of β-diketonate → water has a high probability. Still, for some specific photoluminescence applications this issue isn't a problem and RE(III) complexes with water as neutral ligands are thus not so rare. For application in OLEDs active layers, these complexes present some more trouble.

The incorporation of a neutral ligands on the β-diketonate RE(III) complex helps build a molecule that is simultaneously luminescence efficient and, chemically more stable for an active OLED layer. Unfortunately these molecules are still far from ideal efficacy in a real-case scenario, due to HTL and ETL mismatch at the desired energy levels, or to chemical and electrical instability. In fact, there is current consensus among the several research teams, naturally including the industry, that much scientific work must be still done to promote the RE(III) based OLEDs up to their expected applications. The present knowledge will be an essential platform for the future.

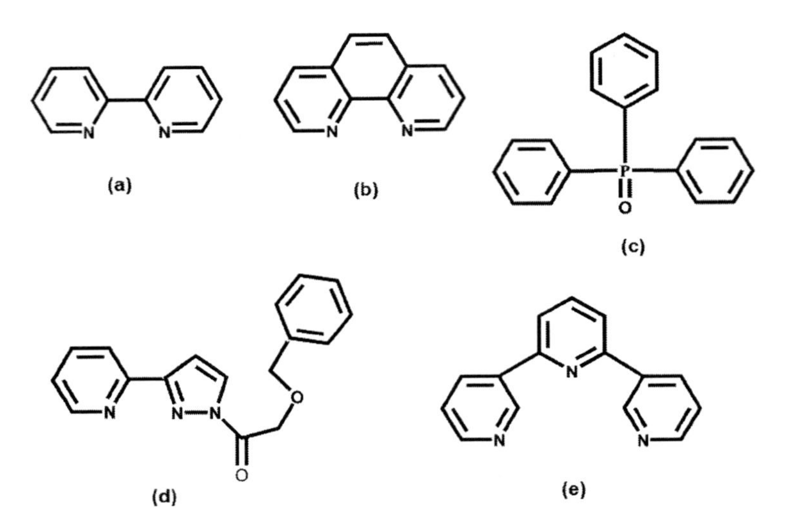

Figure 4.10 Molecular schemes of the most usual neutral ligand of RE(III)–organic complexes; (a) bipy, (b) phen, (c) tppo, (d) pypzB, and (e) terpy. Note that ligand (c) only acts as a monodentate because it features a single donor group (the oxygen atom) while ligands (d) and (e) may acts as tridentate as they possess several electron donor (i.e., Lewis base character) sites, two N and one O (for (d)) and three N (for (e)).

4.3.1.2 Excited energy levels of organic ligands and rare earth ions: looking for the match

Knowing that the energy difference ΔE between the organic ligand T_1 state and the resonant excited states of the RE(III) must be positive, i.e., $E(T_1) > E(RE(III)_{resonant})$ some important considerations must be taken into account to design efficient a RE(III) organic complex.

Before starting a "design" and synthetic work of a RE(III) complex, some simple questions must implicitly be taken into account. For instance, we must asset where are resonant excited levels of the RE(III) ion. For this, let us focus on Fig. 4.2 where the ^5D_J exited levels of the Eu^{3+} are shown. They start from low to high energy level, at $^5D_0$ and proceed to $^5D_1$, etc. Since the decay from each of these energy levels may occur by radiative or non-radiative processes (depending on the probability transition based on the spin and/or symmetry rules) we must first know which of these de-excitation processes is taking place. Fortunately, in most cases (including those relevant for OLEDs), the $^{(2S+1)}\boldsymbol{L}_J$ levels decay preferentially in a non-radiative way to the corresponding $^{(2S+1)}\boldsymbol{L}_{(J-1)}$ level immediately bellow and/or in a radiative way to an level in the RE(III) ground state. In general, the non-radiative decay from an excited state to a ground state is not competitive with the two described processes, unless promoted by the molecular structure, by influence of an organic ligand; however, this molecule would be useless for light emission. Considering the two mechanisms cited above, we can now see that the organic triplet excited state T_1 must be located above any of the $^{(2S+1)}\boldsymbol{L}_J$ RE(III) levels that can decay and promotes non-radiative energy transference to the corresponding lowest excited energy level. This level further decay radiatively to the ground state level. Another possibility is to have a $^{(2S+1)}\boldsymbol{L}_J$ level radiative decay to the ground state. So, the most important is to guarantee that the T_1 level becomes already above the lowest $^{(2S+1)}\boldsymbol{L}_J$ level. An exception is a molecular structure dependence on the ligands excited levels or on the RE(III) excited levels. The resonance idea will be further adapted.

The first consideration to be taken into account is the distribute ion of the energy levels of the RE(III) coordinated ion (assuming that the blindage effect occurs). This means that we consider the RE(III) coordinated ion as a free ion and therefore the corresponding energy levels distribution is applicable.

Figure 4.11 shows the energy level distribution for the RE(III) ions most commonly used in complexes for OLEDs active layers. The physical procedure to obtain such energy levels is the same used and described before when the Eu^{3+} energy levels are showed (Fig. 4.2). This includes the splitting energy levels from the fundamental configuration considering the Columbian and spin–orbit. As reported for the Eu^{+3} ion splitting due to the crystalline-field effect is negligible. So we finished the "energy resolution" at the spin–orbit splitting.

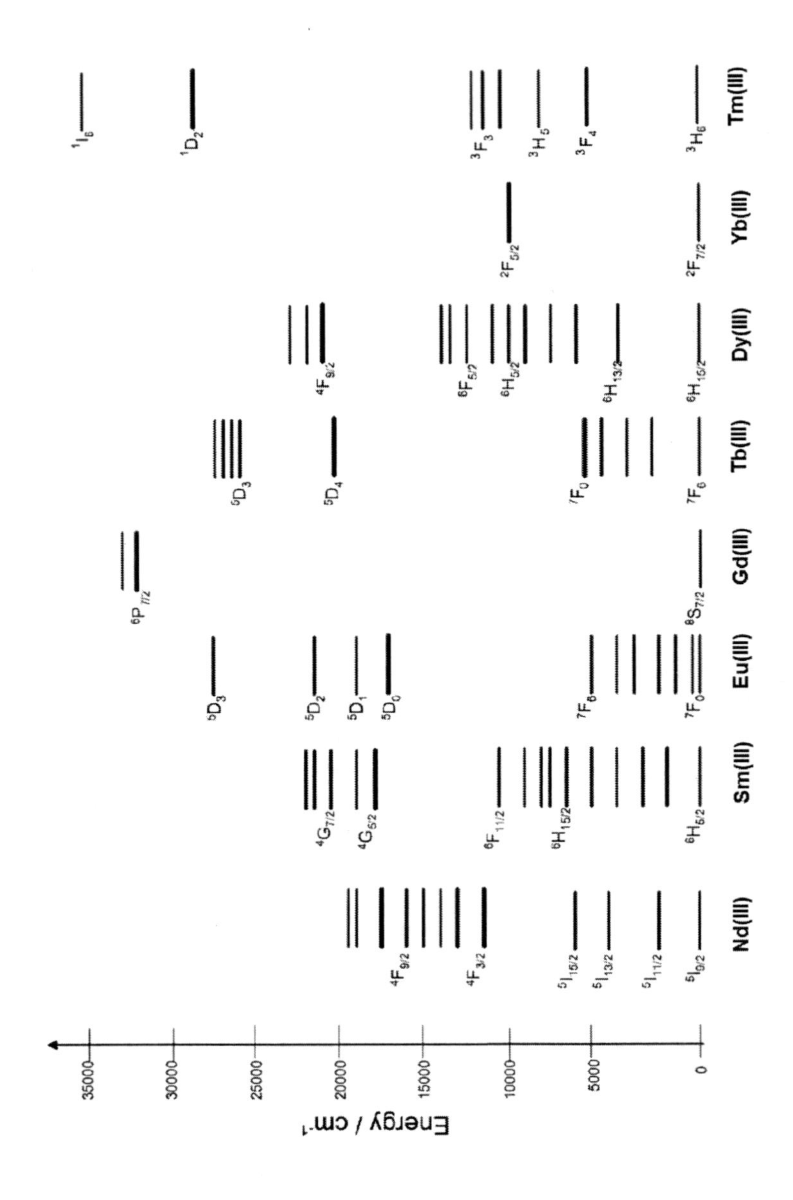

Figure 4.11 Ground energy levels and first excited levels of some common RE(III) ions.

The RE ions energy levels are usually considered in cm^{-1} units (independent on valence) and we follow this "rule." Further conversion in eV is very easy, so both systems can be applied.

Concerning Fig. 4.11, we show only the first excited levels because none of the known organic ligands has a T_1 state with an energy location high enough to populate more RE(III) excited states. At least the lowest $^{(2S+1)}D_{J'}$, $^{(2S+1)}F_J$ or $^{(2S+1)}G_J$ (depending on RE(III) ion) must be considered when selecting an organic ligand. In the shown cases (Fig. 4.11) and considering the most used RE(III) for light emitting molecules, we must highlight the Sm^{3+} $^5G_{5/2}$ level, the Eu^{3+} $^5D_0$ level, the Tb^{3+} $^5D_4$ level, the Tm^{3+} $^1G_4$ level and the Dy^{3+} $^4F_{9/2}$ level. Table 4.2 summarizes the information about these RE(III) ions.

Table 4.2 Lowest excited level characteristics of the most used emitting RE(III) ions.

Element	Lowest $^{(2S+1)}L_J$ level	Energy (eV)
Europium	$^5D_0$	2.17
Samarium	$^5G_{5/2}$	2.23
Terbium	$^5D_4$	2.54
Thulium	$^1G_4$	2.55
Dysprosium	$^4F_{9/2}$	2.63

Note: The energies given are approximated values.

More information can be obtained about the energy of the RE(III) excited levels. Also we must recall that although the lowest excited energy level is not the unique criterion for combining a RE(III) with an organic ligand, it cannot be ruled out. Others excited levels can also be involved in the radiative emission (later will present some practical examples), in agreement with the selection rules. For instance both $^5D_0$ and $^5D_1$ excited levels of Eu^{3+} can originate electronic transitions with any of the ^7F ground levels; in Sm^{3+}, only the $^5G_{5/2}$ excited level can originate electronic transitions to the any ^6H and ^6F ground levels except for the $^6F_{1/2}$;

for Tb^{3+}, both $^5D_3$ and $^5D_4$ excited levels can originate electronic transitions to any of the ^7F ground levels; for Tm^{3+} both $^1D_2$ and $^1G_4$ excited levels can originate electronic transitions with any ^3H and ^3F ground levels, except the $^3F_4$ and $^3H_4$; and finally (considering only the RE(III) referred in Table 4.2) for Dy^{3+} only the excited level $^4F_{9/2}$ can originate electronic transitions with any of the ^6H and ^6F ground levels, except the $^6F_{11/2}$ and $^6H_{9/2}$. A selection rules relaxation is possible, but not only they are very rare but also (and usually) they cannot be observed in electroluminescence due to their very weak intensity. In regard of the others RE(III) ions less used for OLED emitting complexes, similar considerations can be made.

It is also necessary to take into account the energy of the organic ligands excited levels. Besides the former consideration that the most important transference process is from the triplet T_1 level to the resonant RE(III) exited level, we will now consider also the first excited singlet state (S_1) because there is a finite probability of energy transfer from that state. In general, and despite the fact that the correct values for the T_1 and S_1 correspondent energy level are not easy to obtain, optical spectroscopy measurements in the UV–visible spectral region, may help assess the expected location of these levels. Simultaneously, some research works are the resource to valuable theoretical calculations that afford the results that, apart from few divergences, are in general, relatively close to the experimental values. Both determination methods have collateral problems not addressed by the mathematical approximations used in the calculations. The major problem is the influence of the neutral ligand (as previously mentioned) and the organic part of the molecule behaving as a whole, which may lead to a very different situation. For instance, the same β-diketonate organic ligand in conjugation with different neutral ligands (e.g., *phenantroline* or *bipyridine*) may show a noticeable (± 10%) difference on the S_1 and T_1 energy location levels. So, the absolute concept of the β-diketonate organic ligand excited energy levels cannot be used. The molecular configuration of the ligands is obviously responsible for those deviations and it is widely accepted that some of them cannot be completely evaluated. In practice, we accept that a common values for S_1 and T_1 can be "statistically" obtained from combining several experimental and theoretical results, in order to have a useful information for designing a RE(III) organic complex. In Table 4.3, the S_1 and T_1 energy level values of the most usual β-diketonate organic ligands are shown. Further information about others organic ligands can be obtained from several sparse published papers requiring careful viewing.

It is important to note (although not a definite rule) that these energy levels, in particular the S_1 level, can be correlated with the β-diketonate conjugation length as easily can be seen from some of them.

Table 4.3 First singlet (S_1) and triplet (T_1) energy levels of the most know and used β-diketonates organic ligands [37]

β-diketonate	**S_1 (eV)**	**T_1 (eV)**
ACAC	4.31	2.95
BTA	4.01	2.67
DBM	3.51	2.53
NTA	3.92	2.25
TTA	3.64	2.53

For instance, ACAC has the higher S_1 level and DBM the lower. Locking at Fig. 4.9 we see that the conjugation level of DBM is, without a doubt, higher that that observed for the ACAC; in TTA the S_1 level lies in the middle because its conjugation length is lower than DBM but higher than ACAC. Similar conclusions can be drawn for other β-diketonate ligands. So, as a general rule, we can state that the longer the conjugation length, the higher the excited S_1 level. In spite of that simple rule based on the more physical behavior of the conjugated polymers but clearly understandable for small molecules, some attention can be pay to the organic ligands electron behavior. For instance, let take the TFB β-diketonate ligand (with a similar structure of TTA but where the S—sulphur atom is substituted by an O—oxygen atom). We now expect the S_1 level of TFB to be at least near the S_1

level for TTA. But this is not the case, as the TTA S_1 level is of about 3.64 eV and the TFB S_1 level is of about 3.7 eV. Although not big, the difference exists. The question is that the TTA has two electron withdrawing functional groups (usually the CN and CF_3 groups), which implies a low S_1 level compared to the S_1 level of TFB having one electron donating group (usually the OC and OCH_3 groups). When an electron withdrawing group is replaced by an electron donating group the excited levels becomes higher due to the large resonance effect created by the electron donating groups, that tends to shift the π^* state and increases the energy difference. These simple questions must not be overlooked when designing a molecule and they are essential for successful outcome.

To provide a final idea about the required energy level match between the organic ligand and the RE(III) ion, the Fig. 4.12 shows a simple and concise energy diagram, allowing a fast overview.

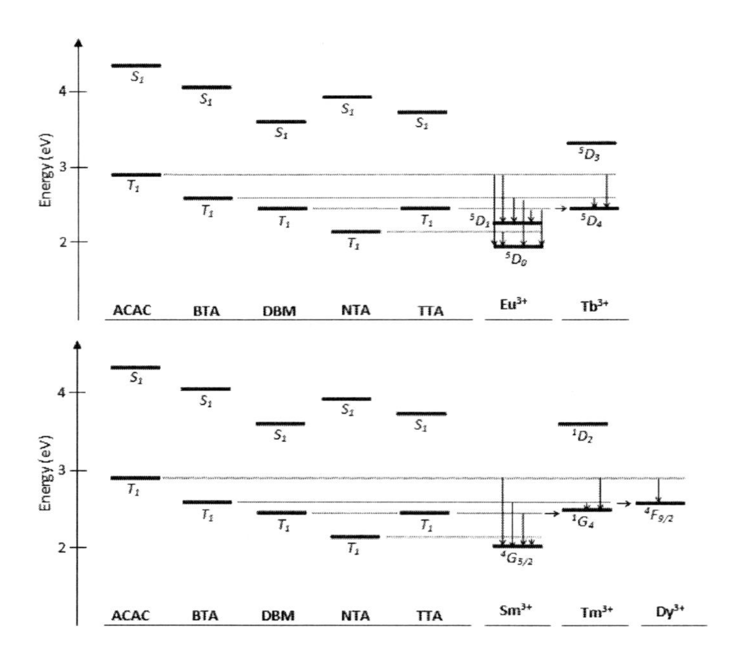

Figure 4.12 Energy level diagram of the most known β-diketonate ligands and the usual RE(III) ions with visible luminescence. Only the most important excited levels are shown. The "energy" match between

β-diketonate ligands T_1 level and lowest exited energy level of the RE(III) ion is given by the arrows.

Regarding Fig. 4.12, a few simple remarks must be made. For instance, it must be pointed that not all the possible β-diketonate/RE(III) complexes can be synthesized. Although some of them really exist, others are very rare mostly due to this complexity and high cost of the synthetic process. So, the shown diagram must be interpreted as a pedagogical "map" for the choice of the organic ligand.

From the diagram, one can see that all organic ligands are able to transfer energy from their T_1 energy states to the resonant energy levels of Eu^{3+} and Sm^{3+}. For Tb^{3+} we only expect an energy transfer from ACAC and BTA although a possible transfer from TTA an DBM cannot be excluded. For Tm^{3+} the situation is identical but the probability of an energy transference from TTA and DBM is decreased comparatively to the Tb^{3+} situation. Finally for Dy^{3+} we only expect energy transfer from ACAC, and, with some reserve, from BTA (in a similar way of TTA and DBM to Tb^{3+} and Tm^{3+} where the similar energy levels can leads to a quenching of the RE(III) luminescence [38]). These simple ideas set the theoretical bases of the molecule design. These suppositions can be override by a few different mechanisms. For instance, practically all S_1 levels can transfer energy directly to the RE(III) ions. If the probability of the respective transition is higher than the intercrossing system inside the organic part of the molecule ($S_1 \rightarrow T_1$) a new approach can be used. Also, if new organic ligands are incorporated into the β-diketonate/RE(III) system (e.g., a neutral organic ligand), there is a fair probability of having a different location for the T_1 energy state of the whole molecule and, therefore, some of the very simple previously considerations must be applied with caution.

RE(III) complexes may also be prepared using organic ligands other than the β-diketonates. Although less common employed, these are important asset for a few particular RE(III) ion for which the traditional pathway fails in efficiency and/or final efficiency vs. cost. Additionally, some diketonate complexes cannot be easily processed in to a thin film by the usual methods described in the Chapter 2, and still have a faithful reproduction of its powder luminescent behavior. Without this, their employment on an active OLED layer will be practically impossible. Finally, it is worth noting that the energy transfers mechanisms of all organic ligands, regardless of the group they belong to, run on the traditional Jablonski diagram, i.e., they depend on their singlet and triplet energy states for the electrical carrier transport and recombination. Such π–π^* and σ system due to the sp^3 carbon hybrid orbitals cannot be overcome when we wish to organically coordinate RE(III) ions to create a luminescent molecule. Further discussion will be done in next sections.

4.3.2 The Photophysics of Rare Earth Organic Complexes

4.3.2.1 The general concept

After the successful molecular design and synthesis much work must be done to study the optical behavior of a new complex. The most important data are the analysis of the emission band, which provides information on the “purity” of the RE(III) ion emission, and the estimate of the internal efficient of the energy transfer from the ligand to the resonant ion levels.

Recalling that, the molecule must absorbs energy trough its organic ligands and efficiently transfer it to the ion, we expect large absorption bands, typical of organic molecules, and a “sharp” emission spectra that, ideally, must only corresponds feature the RE(III) transitions.

The typical organic ligands used in RE(III) complexes absorbs in the spectral region ranging from the UV to the near visible (blue), with a stronger intensity in the UV region, due to the low conjugation chains involved. Curiously this feature can be very useful as base idea for a simple white-OLED, entirely based on RE(III)’s emitting (each of them) in one of the RGB color system. The main difference from the small dyes molecules or semiconducting polymers is that in these emitting organic materials, the red emitter usually absorbs in the blue spectral region and therefore cannot be used in a simple way. In theory such problem does not exists in RE(III) organic complexes, which have the further advantage of featuring a high color purity and quasi-monochromatic emission. Although we cannot state that the organic RE(III) complexes really have a single emission band, because their luminescence spectrum is composed by the RE(III) ion transitions itself, and therefore composed of several "very" sharp lines, the maximum of the absorption bands and the main (i.e., strongest) RE(III) ion transition line is, in most situations, very far. In Eu^{3+} complexes, for instance, an absorption maximum is usually observed near 350–400 nm, and the main emission is located near 612 nm. For Tb^{3+}, the absorption maxima are between 300 nm and 350 nm and the main emission is near 547 nm. As a general rule, the absorption–main emission separation is always near 200 nm (with some rare exceptions). These differences are not surprising due to the necessary energy difference between the organic T_1 state and the highest resonant RE(III) excited level as described before.

4.3.2.2 Absorption and emission spectra

In this section, we present the typical absorption (or photoluminescence excitation) and photoluminescence spectra of the most relevant RE(III) organic coordination complexes emitting in the visible region and which are, therefore, attractive for use in OLED active layers.

The photophysical spectra of Europium and Terbium (the most common) organic complexes are shown in Figs. 4.13 to 4.15. The typical ${}^{(2S+1)}D_J \rightarrow {}^{(2S+1)}F_J$ transitions are indicated, following the previously discussed electronic transition levels of such RE(III) ions emission.

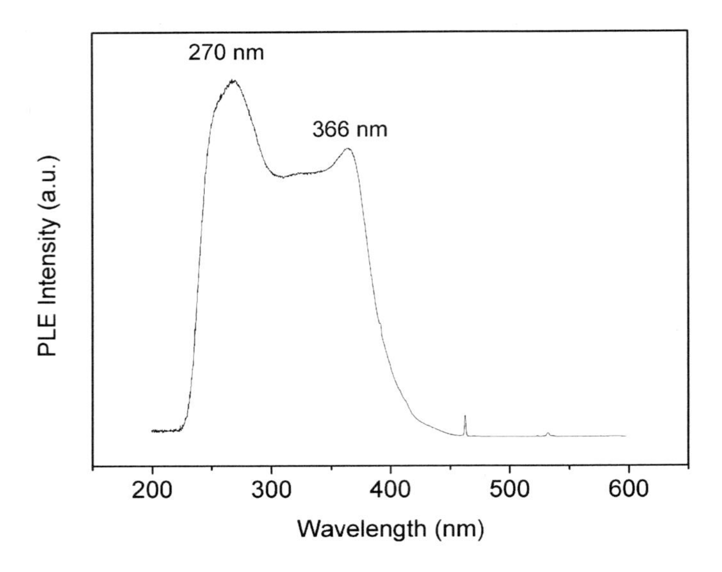

Figure 4.13 Photoluminescence excitation spectrum of a $Eu(BTA)_3$bipy complex showing the large organic ligand energy absorption. Room temperature.

The PLE spectrum on Fig. 4.13 was collected under an emission wavelength of 612 nm, the main emission of that Eu^{3+} molecule, as shown in Fig. 4.14. Some of the features are clearly related with absorption from the organic ligands, namely a very large band (from near 225 nm to near 425 nm) and is a featureless band, associated with the singlet/triplet system mechanisms discussed before. The two observed maxima, indicated in the figure are, of course related to the different organic ligands employed in the complex. Sometimes it is simple to associate one absorption band with a particular ligand, but in most cases an unequivocal attribution cannot be made. Nevertheless, the most important feature to be noted is that an efficient organic ligand absorption is available. Next figure shows the emission spectrum for the same Eu^{3+} complex.

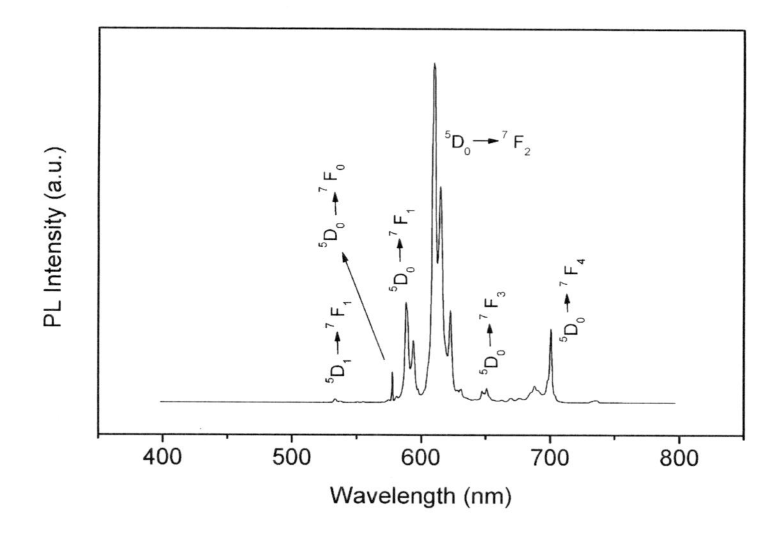

Figure 4.14 Photoluminescence spectrum of a Eu(BTA)$_3$bipy complex showing the $^{(2S+1)}D_J \rightarrow {}^{(2S+1)}F_J$ Eu^{3+}ionic well defined transitions as indicated. Excitation wavelength of 366 nm. Room temperature [17]. Reprinted from *Journal of Non-Crystalline Solids*, 354 (19–25), G. Santos, F.J. Fonseca, A.M. Andrade, V. Deichmann, L. Akcelrud, S.S. Braga, A.C. Coelho, I.S. Gonçalves, M. Peres, W. Simões,T. Monteiro, L. Pereira, Organic light emitting diodes with europium (III) emissive layers based on β-diketonate complexes: The influence of the central ligand, 2897– 2900, Copyright (2008), with permission from Elsevier.

One may note that the emission transitions are as expected very well defined and correspond to ionic emission, indicating a very good energy transference from the organic ligand to the coordinated Eu^{3+} ion, i.e., an efficient antenna process.

The transitions indicated in the figure are in agreement with the diagram shown in Fig. 4.11; thus, this Eu^{3+} complex follows the expected rules for the organic ligand (BTA) excited levels and excited resonant Eu^{3+} levels indicated in Fig. 4.12. The main emission is $^5D_0 \rightarrow {}^7F_2$ (612 nm), yielding the traditional red emission for europium complexes.

Similar considerations (of course with the necessary framing) can be made for others RE(III) coordinated organic complexes. Figure 4.15 [39] shows another instance, the PLE/PL of the Tb^{3+} complex Tb(ACAC)$_3$bipy.

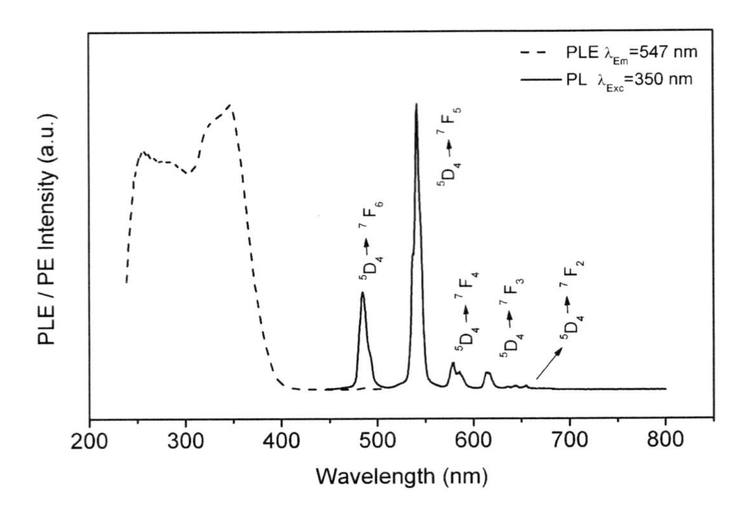

Figure 4.15 Photoluminescence excitation (dotted line) and photoluminescence spectrum (full line) of the Tb(ACAC)$_3$bipy complex. The PLE recorded wavelength and the PL excitation wavelength are shown in the figure caption. The $^{(2S+1)}D_J \rightarrow {^{(2S+1)}F_J}$ Tb^{3+} ionic emissions are represented. Reprinted from *Journal of Non-Crystalline Solids*, 354 (47–51), L. Rino, W. Simões, G. Santos, F.J. Fonseca, A.M. Andrade, V.A.F. Deichmann, L. Akcelrud, L. Pereira, Photo and electroluminescence behavior of Tb(ACAC)3phen complex used as emissive layer on organic light emitting diodes, 5326–5327, Copyright (2008), with permission from Elsevier.

We again observe a very large spectral absorption region and a relatively "sharp" emission line; the conclusions taken from the precedent Eu(BTA)$_3$bipy instance are also applicable now, i.e., there is a good organic ligand absorption with an efficient antenna effect for the energy transfer to the exited Tb^{3+} resonant energy levels. The main emission, as observed in the Fig. 4.15, is the $^5D_4 \rightarrow {^7F_5}$ (545 nm), which is responsible for the green color of Terbium based OLEDs.

The photo-physical behavior of different RE(III) complexes based on β-diketonate organic ligands, are not different from the two Europium and Terbium presented examples because they all have the same internal energy mechanisms that follows to the exact same rules. Even when different organic ligands are used, the basic principle is always the same, that is, to have an efficient antenna effect. The only differences that may be observed relate to the attempt to improve such effect by synthesizing complexes with new organic ligands aiming towards a higher internal efficiency. It is also important to stress that, in practice, the efficacy of the emitting complex is not the only factor that contributes to the OLED efficiency: we must also pay special attention to the electrical carrier transport, and to the processing of the complex into the OLED active layer. This means that, having a RE(III) coordination complex with a high internal efficiency measured by photo-physical techniques, does not automatically imply it will be a good electroluminescent material. We will develop is point further.

Other interesting RE(III) organic complexes are those based on samarium and thulium. The first one can emit strongly in the red spectral region as shown on the Fig. 4.16 [40].

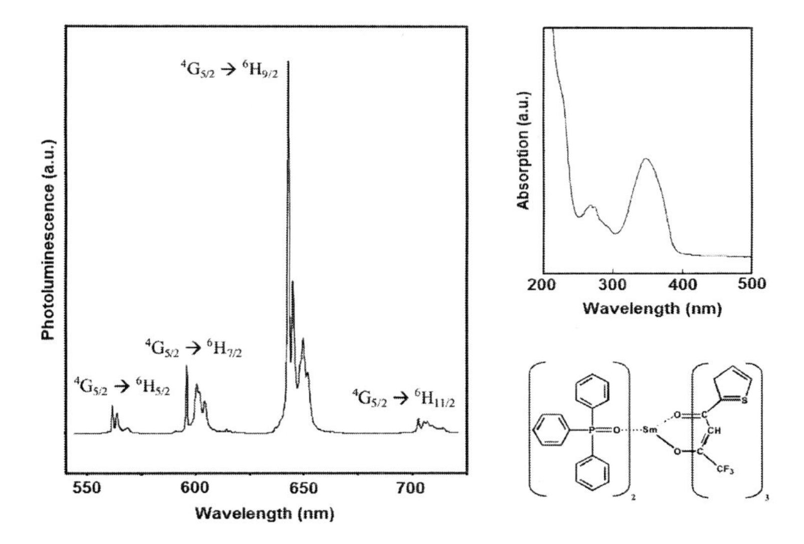

Figure 4.16 Absorption and photoluminescence spectra of a $Sm(TTA)_3(TPPO)_2$ complex at 77K. The absorption spectrum as recorded under an excitation wavelength of 348 nm. The molecular structure of the complex is shown. Reprinted from *Thin Solid Films*, 420–421, R. Reyes, E.N. Hering, M. Cremona, C.F.B. da Silva, H.F. Brito, C.A. Achete, Growth and characterization of OLED with samarium complex as emitting and electron transporting layer, 23–29, Copyright (2002), with permission from Elsevier.

In Fig. 4.16, the ionic transitions of a Sm^{3+} complex are shown, in agreement with the schematics from Figs. 4.11 and 4.12 (for Sm^{3+} levels and organic to Sm^{3+} energy transfer respectively). The used central organic ligand, TTA, only yields de-excitation from the $^4G_{5/2}$ Sm^{3+} excited level. Other authors [41] observed emissions also from the $^4G_{7/2}$ Sm^{3+} excited level using a different samarium complex, $Sm(HFA)_3(phen)_2MeOH$ that allows the energy transfer to the second Sm^{3+} excited level due to its higher organic T_1 energy level. Naturally those others organic ligands—Sm^{3+} "combinations" are possible—regarding the necessary conditions as described. The main emission is the $^4G_{5/2} \rightarrow {}^6H_{9/2}$ (645 nm) with the corresponding red emission.

Although the Sm^{3+} based complexes exhibit a strong red emission, also nearly "monochromatic" as the Eu^{3+} based complexes, the later have some advantages, in particular their simpler chemical syntheses and therefore, appear to be more attractive than the Sm^{3+}. Regardless of that, there are still some samarium complexes under development and study because using some organic ligands, the electroluminescent emissions at 563 and 598 nm can be enhanced and overcome the main emission at 645 nm, resulting in orange OLEDs.

Thulium, a typical blue emitter, is currently not very disseminated, although some work has been done with interesting results. In Fig. 4.17 the photoluminescence excitation and photoluminescence of a $Tm(PPA)_3{\cdot}2H_2O$ are shown [42].

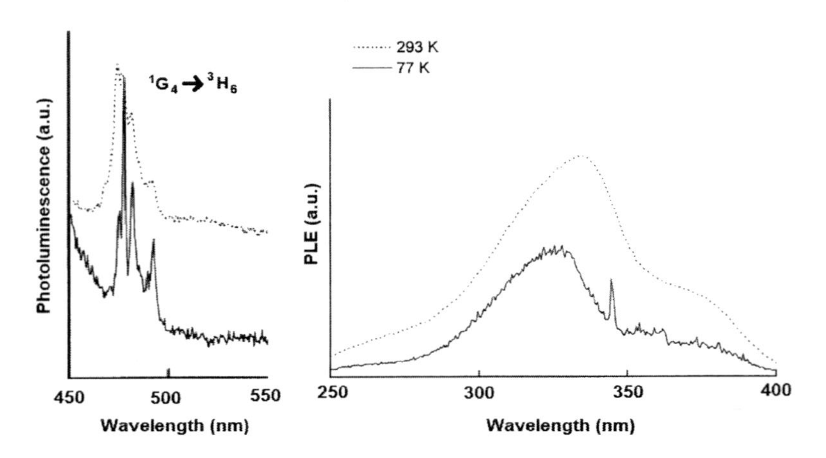

Figure 4.17 Photoluminescence and photoluminescence excitation spectra (excitation wavelength of 478 nm) of $Tm(PPA)_3{\cdot}2H_2O$ at two temperatures. The main emission transition is shown. Reprinted from *Journal of Alloys and Compounds*, 275–277, O.A. Serra, E.J. Nassar, P.S. Calefi, I.L.V. Rosa, Luminescence of a new Tmβ-diketonate compound, 838–840, Copyright (1998), with permission from Elsevier.

Using PPA as the sole organic ligand, only the main Tm^{3+} emission can be observed, i.e., the $^1G_4 \rightarrow {}^3H_6$ (478 nm) to yield the typical blue emission. Thus, and again following Figs. 4.11 and 4.12 as discussed for the other RE(III) complexes, the T_1 level of PPA must lie above (or near) the corresponding level of the BTA. The authors also observed, though quite weakly the next transitions (not shown in the figure), namely the $^1G_4 \rightarrow {}^3F_4$ (650 nm) and the $^1G_4 \rightarrow {}^3H_5$. More recently, an interesting set of results was reported for OLEDs with active layers bearing some metal oxides doped with Tm^{3+} [9].

Finally, dysprosium is also an interesting RE element for OLED active layers. As found for thulium, efficient Dy^{3+} organic complexes, are less common and relatively unknown, in comparison with the europium, terbium and even samarium complexes. Nevertheless a few interesting studies have been published. In Fig. 4.18, the absorbance and photoluminescence spectra of a Dy(BTA)$_3$phen complex is shown [43].

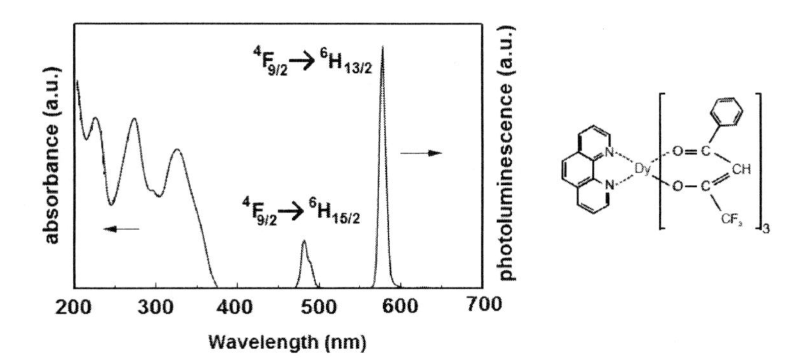

Figure 4.18 Absorbance and photoluminescence of a Dy(BTA)$_3$phen complex at room temperature. In photoluminescence, the two main Dy^{3+} transition are indicated. The molecular structure of the complex is shown. Adapted from ref. [43].

For BTA, we expect two possible transitions from the excited resonant level $^4F_{9/2}$ of the Dy^{3+}, in agreement with the ion levels shown in Fig. 4.11, namely to the $^6H_{15/2}$ (484 nm) and to the $^6H_{9/2}$

(576 nm). Again, this simple analysis follows the precedent discussion for the other RE(III). Further emissions at near 660 nm (to $^6H_{11/2}$) and 750 nm (to $^6F_{9/2}$) are very weak, thus being absent from the figure. The main emission (576 nm) yields a yellow color characteristic of Dy^{3+} complexes.

It is interesting to note that some of the photoluminescence spectra previously presented (depending naturally on the spectroscopic equipment resolution) show the splitting of the electronic dipole RE(III) ion transitions, giving other emission components, or lines, for the same transition. Such situation relates to the crystalline field effect as referred before. In fact, under small electrical (Stark) and/or magnetic (Zeeman) effects, the electronic level is separated in to $2J+1$ levels. In RE(III) complexes, such splitting is often limited to $J+1/2$ levels, which is an indication of a very low symmetry around the coordinated RE(III). Obviously, such situation is the result of incorporation of the RE(III) ion into a coordination cage of organic ligands, with consequent loss of symmetry. Nevertheless, and considering that electroluminescence intensity is lower when compared to the photoluminescence, when these RE(III) organic complexes are used in active OLEDs layers, such detailed spectral resolution is no longer observed (regardless of the spectroscopic equipment resolution) and therefore, it is not considered, giving place to a unique electronic dipole transition, usually identified by the $^{(2S+1)}J$ notation.

Much of the photoluminescence behavior observed for the RE(III) organic complexes will help plan their application in OLED development. In general, after synthesizing a complex taking into account the "molecular design" rules stated before, the photo-physical studies are the next step. Such studies, besides the simple absorption and emission spectra, which provide the expected emission CIE coordinates and show how the organic ligands exited energy levels match the RE(III) resonant excited levels, must also include the internal efficiency determination, in order to get a more precise idea on how efficient is the antenna effect. For more detailed information about these studies see [44–46].

4.4 Rare Earth Based OLED: The Devices

We have stated that RE(III) organic complexes though usually exhibiting a strong photoluminescence, may not easily translate it in high electroluminescence under electrical excitation. In this section we will recall some of the questions from Chapter 2, on how fabricate an OLED. Also knowing the process by which the organic ligand transfers energy to the RE(III) ion under electrical excitation, we see that an efficient OLED fabrication must be preceded by a carefully physical study involving the optical and electrical optimization, which will be analyzed in depth.

4.4.1 The Basic Structure

The basic structure of a RE(III) organic complex based OLED differs little from that described in Chapter 2 to illustrate the successive fabrication steps of an efficient electroluminescent device. Naturally that a slightly different approach will now be used, as the consequence of the different electro-optical nature of these complexes; but, in general, the main ideas remain valid.

Traditionally, RE(III) organic complexes, like most emitters including dyes and luminescence polymers, cannot be used in a single layer OLED. The main reasons for this are, as stated in Chapter 2, the interfaces between electrodes–organic material, the poor carrier confinement and, in some cases, the much different mobility between electron and hole in materials and the materials with more unipolar electrical characteristics. These factors must to be taken into consideration when we are setting the starting point for the development of an OLED and we must find ways/ solutions to circumvent them. Not also that, though not all good photoluminescent RE(III) organic complexes (and of course others organic emitter materials) are good electroluminescent devices, the inverse is often valid: to obtain a efficient OLED, its active layer material must be a good, and efficient photoluminescent complex.

A RE(III) organic complex based OLED is, in most and perhaps all cases, formed by a hole transport layer (HTL), the active or emissive layer (EL) and an electron transport layer (ETL) in a sandwiched structure based on the general architecture presented in Fig. 2.5. The typical electron transport materials used in OLEDs based on RE(III) organic complexes are Alq3 (*8-hydroxyquinoline aluminum*), butyl-PBD (*2-(4-biphenyl)-5-(4-tert-butylphenyl)-1,3,4-oxadiazole*), TPBI (*1,3,5-tris(2-N-phenylbenzimidazolyl) benzene*) and the OXD7 (*1,3-Bis[(p-tert-butyl)phenyl-1,3,4-oxadiazoyl]benzene*). The Fig. 4.19 shows their chemical structures.

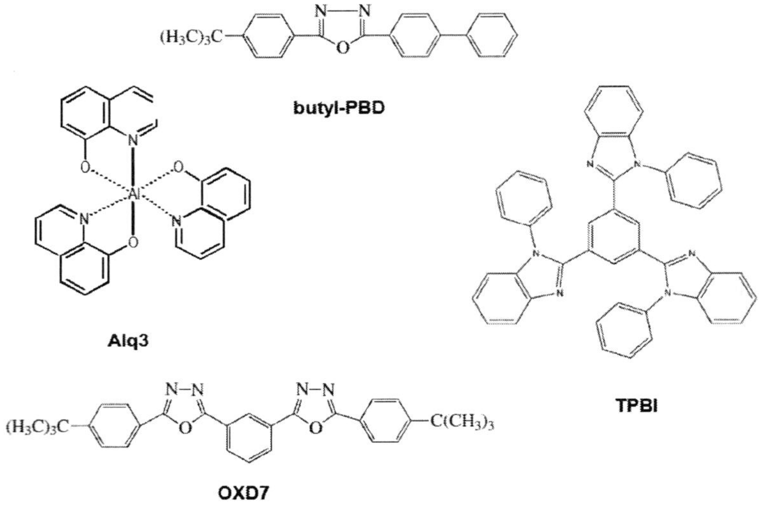

Figure 4.19 The most common electron transport materials used in rare earth organic complexes based OLEDs.

Concerning the hole transport materials the most commonly used are TPD (*N,N'-diphenyl-N,N'-bis-(3-methylphenyl-4,4'-diamine*), NPB (*N,N-bis-(1-naphthyl)-N,N'-diphenyl-1,1'-biphenyl-4, 4'-diamine*), MTCD (*1-(3-methylphenyl)-1,2,3,4-tetrahydroquinoline-6-carboxyaldehyde-1,19-diphenylhydrazone*) and CBP (*4,48-bis(Ncarbazolyl)biphenyl*). Sometimes, a non-conjugated polymer, PVK (*poly(N-vinylcarbazole)*) is also used, although in most cases it works as a host matrix for embedding the RE(III) organic complex. In Fig. 4.20, the molecular structures of these organic semiconductors are shown.

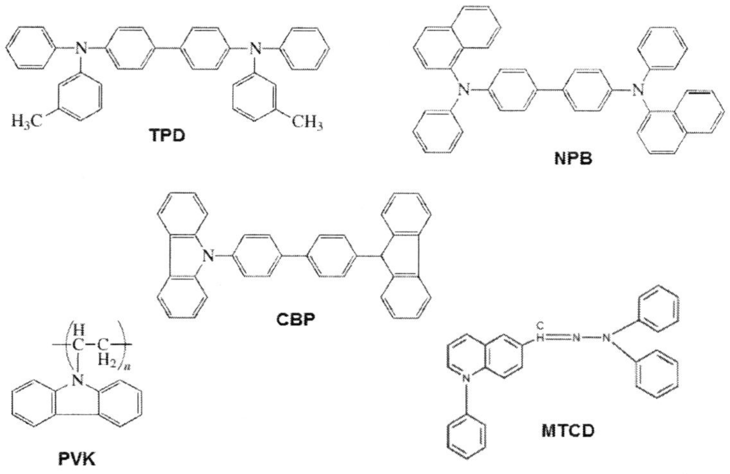

TPD NPB CBP PVK MTCD

Figure 4.20 The most common hole transport materials used in rare earth organic complexes based OLEDs.

There are other semiconducting organic materials that can be used asides from the ones herein presented, some of them for specific application. Nevertheless, the materials summarized in the last two figures represent the most employed to "sandwich" a RE(III) organic complex.

In very particular cases, a more complex OLED structure is used, with more than three layers of organic materials, bearing some layers that act as electron or hole blocking layers. The purpose is to increase the electrical carrier confinement inside the active layer (as referred before) without any interference (at least ideally) on the electron and hole injection and transport to EL. In theory such OLED design approach is an interesting solution to increase the efficiency, but sometimes in the practice the gain is too low to account for the requirements of extra layers witch imply a more complex OLED structure. But it must be noted that in the laboratory, such approach has been used with success and in some situations it is even the only efficient solution for specific RE(III) organic complexes in EL. There are two blocking materials used widely: the BCP (*Bathocuproine*) as hole blocking and a conjugated polymer, named PEDOT:PSS or *poly(4-styrenesulfonate)* as electron blocking. Their corresponding molecular structures are shown in Fig. 4.21.

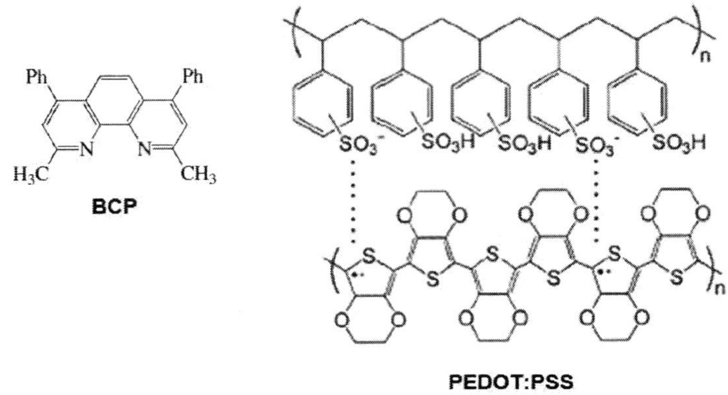

BCP

PEDOT:PSS

Figure 4.21 Structures of a hole blocking small molecule (BCP) and an electron blocking polymer (PEDOT:PSS).

Finally, a very thin layer of LiF (Lithium fluorine) may at times be used between the ETL and the cathode in order to reduce the potential barrier between the two materials. This solution is very popular because the most stable cathode metals (such as aluminum) have a work-function value above that of most common electron transport materials LUMO level. This method is quite valuable in some cases, but there are others cases for which it makes little or no difference in the OLED efficiency, thus being of no use.

The selection of electron and hole transport material as well as of an electron and/or hole blocking material, is obviously based on two main points, namely (1) the HOMO/LUMO levels of the EL, which determine the "ideal" matching levels in HTL and ETL thus conditioning the materials choice and (2) the OLED fabrication method, i.e., whether the whole device is build by thermal evaporation or it is a hybrid device, with some layers obtained by thermal evaporation and others by a wet process (spin-coating, self-assembled, ink-jet, etc.). Moreover, adding to the right choice of the materials for ETL and HTL (with or without blocking layers) the thickness of each layer may be relevant for the OLED final macroscopic behavior. In a simplified view, layers comprising materials mainly working as hole transporters should ideally be thicker than the layers of materials acting mainly as electron transporters because the electron mobility is always lower than the hole mobility; and the total thickness must have an electrical serial resistance relatively low to provide a driving voltage as low as possible. Many organic semiconducting materials have a bi-polar nature although with a prevalent electron or hole major characteristics. But concerning the RE(III) organic complexes, and except for a few very well known, the characteristic of new ones is an incognito. Consequently many times the initial development of a new OLED is a "trial and error" work, and undoubtedly, fabricating an OLED is a casuistically procedure!

In the next sections, we will follow the simpler concept adopted for RE(III) based complexes OLEDs and present some insights on the "universal" three layer structure.

4.4.2 General Cases of Rare Earth Based OLEDs: Europium, Terbium, Samarium, Thulium and Dysprosium

There are presently thousands of published articles about OLEDs based on RE(III) organic complexes, making it impossible to detail on all of them here. Moreover, much of the published work isn't strongly innovative for these OLEDs due to an almost "infinite" reproduction of the same basic solutions, with a few small changes. So, and considering the overall objectives of this book, only the fundamental matrix about OLEDs based on RE(III) organic complexes will be presented and explained. Our approach will include the basic structures and the correspondent results that elucidate this kind of devices.

4.4.2.1 Europium-based OLEDs

OLEDs based on europium(III) complexes are undeniable the most well-known among these devices [47–49]. Firstly because Eu^{3+} is a well studied element with a significant successes in the chemical synthesis; secondly, because, as can be deduced from the schemes given in Figs. 4.11 and 4.12, it is not at all difficult to find a suited organic ligand for a efficient Eu^{3+} emitter; and finally, is very attractive due to its high very pure and "monochromatic" emission.

Several approaches have been used to develop Eu^{3+} based OLEDs. A few experiments with two-layer devices have been reported, yet the most used structure comprises the traditional three-layer system, with the Eu^{3+} active layer "sandwiched" between a "p-type" and an "n-type" organic semiconductor.

Concerning the fabrication method, the Eu^{3+} based OLEDs are easily built by thermal evaporation (but not exclusively as see in next sections); all the organic materials (HTL, EL and ETL) and the metallic electrodes (traditionally aluminum as explained in Chapter 2) are suitable to thermal processing. In fact, there is no risk of material degradation (all are of "small molecules" type) and the process is simple and economic, involving small amounts of material and little loses. Moreover, the thermal evaporation process, when thickness layer is controlled, has an extremely high precision, for small device emitting areas, thus becoming very attractive. The most widely used hole and electron transport layers in this process are the TPD and Alq3 respectively, having a LUMO and HOMO levels of 2.5 eV and 5.5 eV (for TPD) and 3.1 eV and 5.6 eV (for Alq3). Considering, for instance, the much studied europium complex, $Eu(DBM)_3phen$ with, LUMO and HOMO levels of 2.5 eV and 5.5 eV, it is not difficult to observe that the interface TPD/$Eu(DBM)_3phen$ will have a short barrier of 0.1 V for electrons (and no barrier for holes) and the interface $Eu(DBM)_3phen$/ Alq3 will have also a short barrier of 0.1 V for holes (and no barrier for electrons). Based on the discussion gave in the beginning of the Chapter 2, these interfaces seems ideal. Simultaneously, the TPD helps on the "cancelation" of the hole "barrier" between the anode (ITO) and the $Eu(DBM)_3phen$ (allowing a more efficient hole injection into the Eu^{3+} complex) and by other side, the Alq3 makes the same for electrons, decreasing the initial "barrier" of 1.75 V between the cathode (aluminum) and the $Eu(DBM)_3phen$ to 0.6 V. The general device scheme, with the corresponding HOMO and LUMO energy levels is shown in Fig. 4.22.

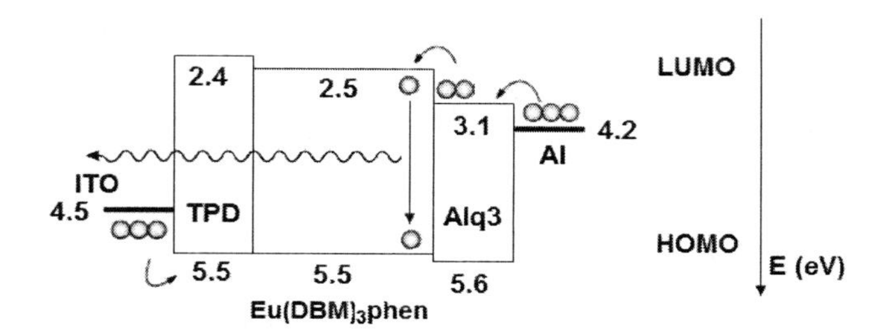

Figure 4.22 Energy levels diagram of a ITO/TPD/Eu(DBM)$_3$phen/Alq3/Al OLED. The represented thicknesses of the several organic layers are not in the same schematic scale.

A detailed OLED structure is shown in Fig. 4.23. The thickness of the OLED layers is of about 20 nm for TPD, 50–100 nm (different thickness were tested) for Eu(DBM)$_3$phen and of about 50 nm for Alq3, with an evaporation rate of about 0.2 nm/s, performed under high vacuum (better than 10^{-6} Torr) [50]. The choices of the thicknesses were mainly based on literature reports, except for the Eu^{3+} active layer, for which different thicknesses were tested in a series of experiments.

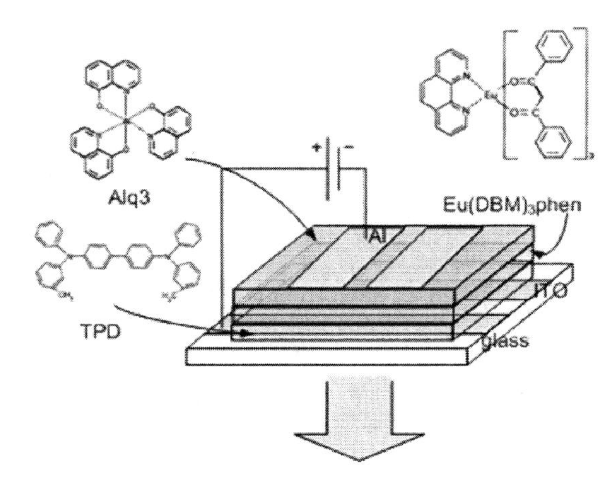

Figure 4.23 3D scheme of on ITO/TPD/Eu(DBM)$_3$phen/Alq3/Al OLED with the corresponding chemical structures of the different materials used. In this structure, each device comprises four OLEDs, indexing individually.

In cases in which there is no previous information on the usual layer thicknesses and on which material is expected to be more efficient as "n-type" or "p-type" for charge transport or injection to the active layer, the experimental method will basically be the conventional "trial and error" starting from whatever the previous knowledge there is about similar device structures. If a detailed information about the intrinsic nature of the active layer complex, in particular if it works as "p- or n-type," on the charge mobility, obtained by Hall effect and/or on the CELIV—*Charge Extraction in Linearly Increasing Voltage* techniques, and on the corresponding HOMO and LUMO levels determined by cyclic voltammeter and/or from theoretical calculations can be obtained, then the design of the device structure becomes a much easier task.

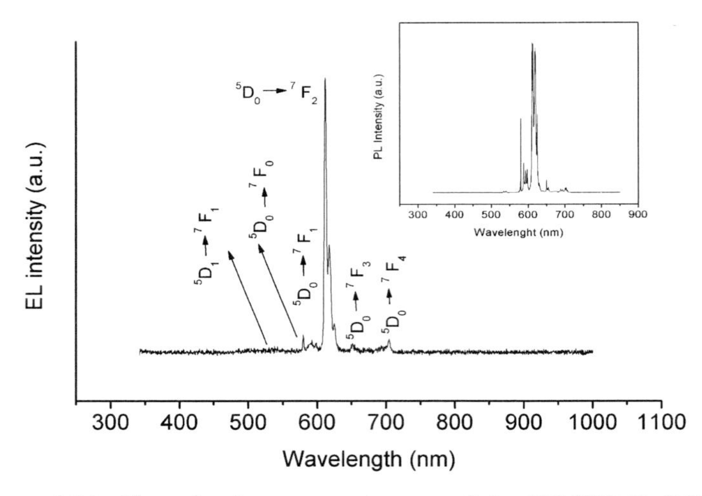

Figure 4.24 Electroluminescence spectrum of the ITO/TPD/Eu(DBM)$_3$ phen/Alq3/Al OLED under a driving voltage of 21V. Inset: the photoluminescence spectrum of the same complex, for comparison. The typical Eu^{3+} emissions are signalized.

The typical electroluminescence spectrum for the device with an EL layer thickness of 100 nm is shown in Fig. 4.24. Although the typical Eu^{3+} spectral emission can be observed, it must be noted that the intensity is much lower that observed in the correspondent photoluminescence spectrum (showed in the inset for direct comparison). Also, the electroluminescence spectrum is clearly less feature, as expected. In a first observation, one may note that the electroluminescence spectrum reproduces the photoluminescence one and, more importantly, there is no evidence of organic emission. This means that a perfect carrier confinement was achieved and, considering the resolution of the electroluminescence spectrum, an efficient antenna effect is presumed.

A plot of the electroluminescence spectra under different applied voltages is presented in Fig. 4.25. This analysis evidence the influence of the driving voltage on the CIE color coordinates with further implications for the color purity. Additionally, it may be helpful for the device technological development and application to establishment the correct working voltage, aiming at emission on a specific final color.

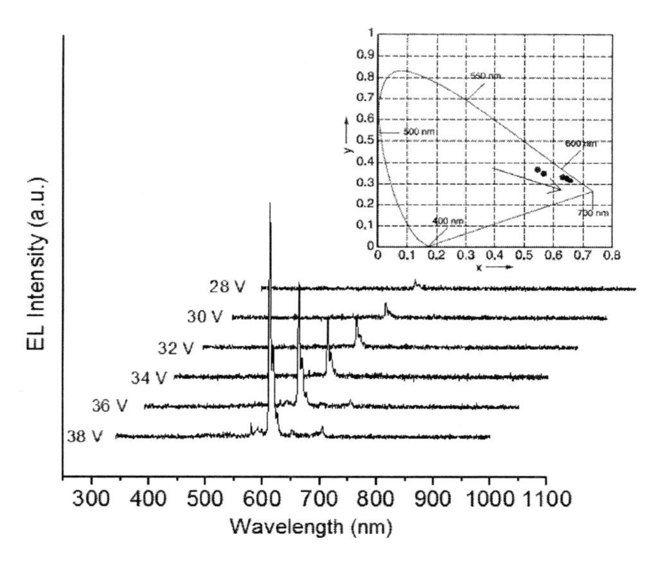

Figure 4.25 Electroluminescence spectra for the ITO/TPD/Eu(DBM)$_3$ phen (100 nm)/Alq3/Al OLED observed under different applied voltage. Inset: the influence on the CIE color coordinates (the arrow indicates the changes upon increasing the applied voltage).

We must note that from the Fig. 4.25, for higher applied voltages the color purity increases as the (x,y) CIE color coordinates approach to the color border line (in the spectral colors). These data were collected for an OLED with an active layer of 100 nm, thus requiring the high applied voltages for a correct driving.

An interesting point, we can see that when the PL spectrum in the inset of the Fig. 4.24 is compared to the PL spectrum of $Eu(BTA)_3$bipy given in Fig. 4.14 there are obvious differences in the transitions relative intensities, keeping however the dominant emission of the ${}^5D_0 \rightarrow {}^7F_2$ ($\approx$ 612 nm). Such variations are very usual and not only with europium but involves all kind of RE(III) organic complexes emitters.

The color on and Eu^{3+} based OLED, as observed for most of these devices bearing RE(III) emitting complexes, is very pure, with an emission deemed "quasi-monochromatic" due to the dominant main band at $\approx$ 612 nm corresponding to the ${}^5D_0 \rightarrow {}^7F_2$ transition. In fact, the other less significant transition emission bands may vary slightly from one europium complex to another due to some differences in the coordination sphere geometry, but this brings no influence to the overall emission.

Concerning the main transition, a simple radiometric efficiency evaluation can be estimated from ratio between the emitting optical power and the electrical power. This calculation can be obtained for the main emission only (and not for the whole electroluminescence band) Optical power results recorded for an emitting wavelength of 612 nm $\pm$ 0.5 nm are shown in Fig. 4.26.

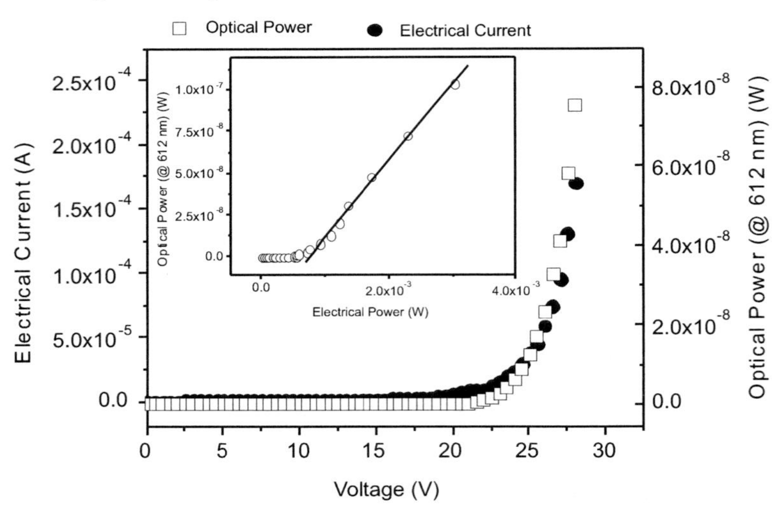

Figure 4.26 Electrical current and optical power (@ 612 nm) for the ITO/TPD/ $Eu(DBM)_3$phen (100 nm)/Alq3/Al OLED with the applied voltage. Inset: the optical power vs. electrical power. The straight line is the linear fit in the dynamic OLED region, allowing the calculation of the radiometric wall-plug efficiency.

The radiometric wall-plug efficiency η_{WW} (as described in Chapter 3) can be calculated from the linear fit of the optical power vs. electrical power plot in the OLED dynamic region. For a single wavelength calculus (the main emission at 612 nm) an efficiency of about 4×10^{-3} % is obtained, which would appear to be a very low (only one electron–hole excitation recombines for a emission at 612 nm for 25000 electron–holes pairs injected by the electrodes), we are running on still, some experimental and practical assumptions must be taken into account before a more critical conclusion is drawn. In fact, we are recording only the optical power at 612 nm; for the whole electroluminescence spectra, an increase of about 30% in the efficiency will be obtained. Secondly, we are only taking the emitting photons from the normal direction from the OLED surface (considering a Lambertian emitter) and ignoring all contributions from emitting angles different than 90°; the simple fact of not using an integrating sphere contributes to decrease in the collected data (and therefore in the final efficiency calculation) by over ten times (sometimes more). One may conclude that the cumulative effects of all these conditions will gives a value for η_{WW} clearly much more high (even less than 1% but higher than an inorganic "equivalent" emitter).

This discussion about efficiency strongly evidences the experimental difficulties in collecting and calculating figures of merit in standard conditions; this way, a comparison of data obtained by different working groups in different institutions is strongly compromised. Standardization is possible—it was done in the European project OLLA[a]—but the research groups and the industry still, work, in many cases, in opposite directions. Restricting the analysis to the physics and materials science, and revising the thousands of scientific articles published, we conclude that is very difficult to establish direct comparisons. So, we keep the discussion more focused into the research issues.

Many other Eu^{3+} organic complexes have also been extensively studied. As referred before, the thermal evaporation technique

[a] Reports and information at http://www.hitech-projects.com/euprojects/olla/

is the most widely used. Recently a set of four different Eu^{3+} organic complexes based on the same neutral ligand (*bipyridine*) but changing the central ligand was studied with interesting results [17]. We kept the OLED structure similar to that presented in the last example, i.e., ITO/TPD (20 nm)/Eu(L)$_3$bipy (50 nm)/Alq3 (50 nm)/Al (where L denotes the different ligands) fabricated by thermal evaporation. The central ligands employed were the BTA (*1-benzoyl-3,3,3-trifluoroacetonate*), NTA (*tris(1-(2-naphthoyl)-3,3,3-trifluoroacetone)*), TTA (*tris(1-(2-thieneyl)-4,4,4-trifluoro-1,3-butanedione)*) and DBM (*tris(dibenzoylmethane)*). The work compares the influence of the central ligand on the OLED figures of merit. Figure 4.27 shows the general OLED structure with the molecular structure for each complex used.

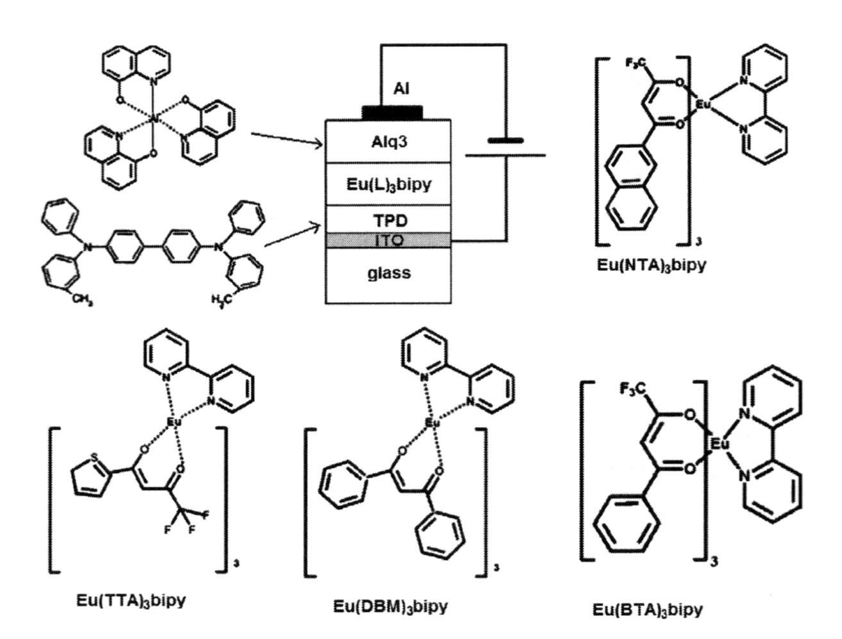

Figure 4.27 Molecular structure of the complexes used in the OLEDs and the general configuration for the devices.

The best wall-plug efficiency reaches over 1×10^{-3}% obtained again for the main emission the $^5D_0 \rightarrow {}^7F_2$, which, depending on the complex, appears between 611 and 612 nm due to the local molecular conformation and the surrounding coordination sphere influence, as reported. This value is obtained for the $Eu(DBM)_3$bipy that is lower than the reported for the $Eu(DBM)_3$phen; nevertheless we must note that the active layer thickness is different (50 nm instead 100 nm in previous result) and a much lower driving voltage is measured (about 16–17 V against more than 23–24 V) due to the low serial resistance. Using the same active layer thickness, the OLED based on $Eu(DBM)_3$phen exhibits a similar wall-plug efficiency when excluding the possible organic ligands emissions. Unfortunately, this is not the most usual case, and will be discuss latter. In advance, we can point the apparent advantage of the $Eu(DBM)_3$bipy although the $Eu(TTA)_3$bipy also reveals a good efficiency and behavior.

Another interesting example of Eu^{3+} OLEDs prepared by thermal evaporation was reported for the complex $Eu(DBM)_3$bath or (*Europium-(trisdibenzoylmethanato) bathophenanthroline*) [51]. Devices featured two (ITO/TPD/$Eu(DBM)_3$bath/Mg:Ag) or three (ITO/TPD/$Eu(DBM)_3$bath:TPD/$Eu(DBM)_3$bath/Mg:Ag) organic layers and were assembled, with a new approach: the Eu^{3+} emitting complex was used itself in the active layer or was used both in the active layer (doped with TPD) and in the electron transport layer. The authors observed photometric efficiencies of 0.02 lm/W and 0.4 lm/W for the two and three device layers respectively. Much of this work is aims to avoid the quenching effect of the cathode on the Eu^{3+} emission. This was such better achieved in the three layer device because the OLED excitation recombination region was relatively far from the cathode. This is a common issue for OLEDs and actually the solution was to abandon the two layer structure and to adopt the three layers, which the electron transport layer having a significant thickness (around 50 nm), as mentioned.

If the three layer RE(III) complex based OLEDs appear to be preferable, there are many attempts to increase the OLED performance by doping the active layer and keeping the device structure with two layers. One interesting doping material is the DCJTB (*4-(dicyanomethylene)-2-t-butyl-6(1,1,7,7-tetramethyljulolidyl-9-enyl)-4H-pyran*), a red dye; in an OLED with the configuration ITO/TPD/$Eu(DBM)_3$bath:DCJTB/Mg:Ag, the Eu^{3+}

complex acts as a host matrix; the Eu^{3+} acts as an energy acceptor for the organic ligand triplet (DBM) as usual and the DCJTB acts as an energy donor for the singlet excited state of the ligand, enhancing emission by the suppression of the exciplex emission, resulting of the others emission processes from the host or excited complex. In some cases this is not completely avoided, reducing the pure Eu^{3+} emission. The data presented in reference [52], shows that, though the photometric efficiency increases up to 5.7 cd/A, the main emission of the Eu^{3+} is broadened by emission of the organic materials, regardless of the reduction of the exciplex emission.

A very interesting result was obtained using CBP as host material and BCP as ETL/hole-blocking material [53]. The europium complex was $Eu(TTA)_3$tmphen where TTA represents *thenoyltrifluoroacetone* and tmphen is *3,4,7,8-tetramethyl-1,10-phenanthroline*. The device structure was quite complex with ITO/TPD/CBP:$Eu(TTA)_3$tmphen/BCP/Alq3/LiF/Al but a brightness over 800 cd/m^2 was achieved with an photometric efficiency of 4.7 cd/A.

Other authors presented a more simple device structure comprising a newly developed Eu^{3+} emitter, the $Eu(DBM)_3$pyzphen where the main difference is in the pyzphen neutral ligand [54], representing (*tris (dibenzoylmethane)-pyrazino[2,3-f][1,10] phenanthroline-EuropiumIII*). This complex was used as doping material in an emissive layer with *dphen* (*4,7-diphenyl-1,10-phenanthroline*) as host, known to be an electron transporting and hole blocking material. The device structure was ITO/TPD/$Eu(DBM)_3$pyzphen:dphen/dphen/Al. The photometric efficiency reaches a maximum of about 5.1 cd/A and the electroluminescence featured no emission from the organic materials.

Another report studied the influence of the neutral ligand (*NL*) on the electroluminescence of $Eu(TTA)_3$*NL* complexes [55]; several 1,10-phenanthroline derivatives were tested as *NL*, namely the tmphen (*3,4,7,8-tetramethyl-1,10-phenanthroline*), dmphen, clphen (*5-chlorine-1,10-phenanthroline*) and nphen (*5-nitryl-1,10-phenanthroline*). Using an OLED structure of ITO/TPD/CPB:$Eu(TTA)_3$*NL*/BCP/Alq3/LiF/Al, the authors concluded that the different phen substituents, methyl, chlorine and nitryl, had a pronounced effect on the device's performances: methylation yields higher efficiencies in the low and mid scale current densities; the chlorine-substituted *NL* also increased the electroluminescence performance but the nitryl-substituent group reduced the emission. Unfortunately there are no studies using only the Eu^{3+} complex in the active layer, due to problems at high electrical current density, caused not by the Eu^{3+} complex but by the CBP triplet–triplet annihilation with subsequent TTA back energy transfer associated with the near-resonance triplet energy levels of TTA and CBP [56]. These issues must be carefully evaluated when using a host matrix for the RE(III) complexes.

In a different approach, some authors [57] report an OLED with an HTL different from the traditional TPD and a new Eu^{3+} complex in the active layer. The Eu^{3+} complex was $Eu(TTA)_3(TPPO)_2$, with TTA the *α-thenoyltriuoroacetonate* and TPPO the *triphenylphosphine oxide* and the HTL was MTCD (*1-(3-methylphenyl)- 1,2,3,4 tetrahydroquinoline-6-carboxyaldehyde-1,1'-diphenylhydrazone*). The OLED structure used was ITO/MTCD/$Eu(TTA)_3(TPPO)_2$/Alq3/Al. Though the electroluminescence spectrum was free from organic material emission, i.e., only the Eu^{3+} emissions were observed, the photometric efficiency was relatively low (1.7×10^{-2} cd/A), which was attributed to a possible loss of energy transfer from the organic ligand to Eu^{3+} under electrical excitation.

Finally we must report on some of the less used tetrakis β-diketonate complexes. An interesting result was obtained with complexes of type $C[Eu(DBM)_4]$ where C is an alkali cation (Li^+, Na^+ or K^+) [58]. The employed device structure was ITO/NPB or MTCD/$C[Eu(DBM)_4]$/l. In spite of the expectations for these tetrakis β-diketonate complexes arising from their excellent, high internal efficiency photoluminescence, the OLED performance did not surpass the average typical values, although a pure Eu^{3+} emission is observed.

The several approaches to maximize the efficiency extraction from an RE(III) organic complex based OLED have been exhaustively tested with Europium for the reasons presented before. Further reading and others results besides those discussed here (and only for devices fabricated by thermal evaporation) an extensive bibliography on Eu^{3+} organic complexes and their application in OLEDs active layers is recommended (See [59, 60]).

4.4.2.2 Terbium-based OLEDs

Terbium complexes are clearly the second most studied materials in RE(III) based OLEDs [61, 62]. In spite of having less color purity than europium complexes, their strong green color, with relatively sharp emission lines quite distinct $^5D_4 \rightarrow {}^7F_{6,5,4,3,2}$ transitions, allow them to be included in the "quasi-monochromatic" emitter category. And should the main transition ($^5D_4 \rightarrow {}^7F_5$) be enhanced by a suitable organic coordination sphere, the result, concerning color purity, will be surprising. The main disadvantage of the Tb^{3+} complexes is the high location energetic location of the $^5D_4$ ion resonant level, compared to the usual excited triplet state (T_1) of the most known organic ligands. This implies a more demanding design of the organic ligand, as those typically used for Eu^{3+}, for instance, have a low or null chance to work efficiently as antennas for Tb^{3+}. Nevertheless, some efficient Tb^{3+} based OLEDs have been fabricated.

It will not be much surprising to find that the main central ligand for Tb^{3+} is ACAC (Figs. 4.11 and 4.12). The basic approach for assembling the OLED structure is not different from that described for the Eu^{3+} based ones. Such structure comprises one HTL, one ETL and the emissive layer, now comprising the Tb^{3+} complex. In a first approach, TPD and Alq3 are usually employed as HTL and ETL respectively. The best known Tb^{3+} complex is Tb(ACAC)$_3$bipy, with LUMO and HOMO energy levels of 1.8 eV and 5.9 eV, respectively. The device scheme and its corresponding energy levels (as found on the Fig. 4.22) are shown in Fig. 4.28 [39].

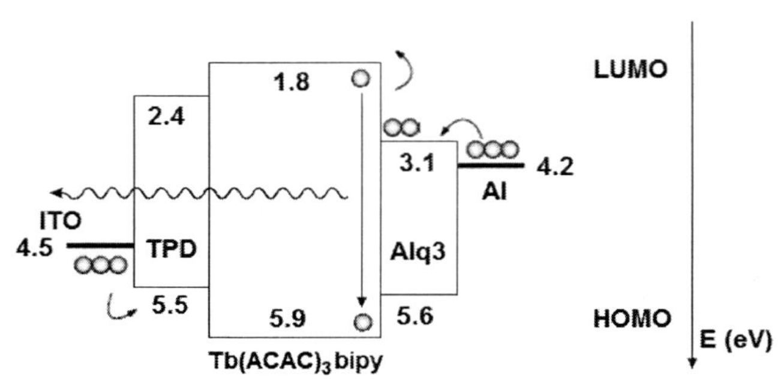

Figure 4.28 Energy levels diagram of a ITO/TPD/Tb(ACAC)$_3$bipy/Alq3/Al OLED. The thicknesses of the several organic layers herein represented are not in the same schematic scale.

A careful look at the figure reveals that the situation has changed from that observed for the Eu^{3+} complexes. The TPD/Tb(ACAC)$_3$bipy interface has no barrier neither for electrons, nor for holes and the Tb(ACAC)$_3$bipy/Alq3 interface has a barrier of 0.3 V for holes and none for electrons. In such a configuration, the TPD cannot acts as an electron blocking material and therefore there is the probability of a less efficient electrical carrier confinement. On the other hand, both TPD and Alq3 help annul the hole and electron "barriers" injection into active layer from the anode and cathode respectively.

The thickness of the OLED layers were the usual, i.e., about 20 nm for TPD, 50 nm for Tb(ACAC)$_3$bipy and 50 nm for Alq3 with an evaporation rate about 0.2 nm/s (under high vacuum and better than 10^{-6} Torr). Again, the choice of these thicknesses is mainly based on literature reports.

The electroluminescence spectrum of this OLED structure is shown in Fig. 4.29.

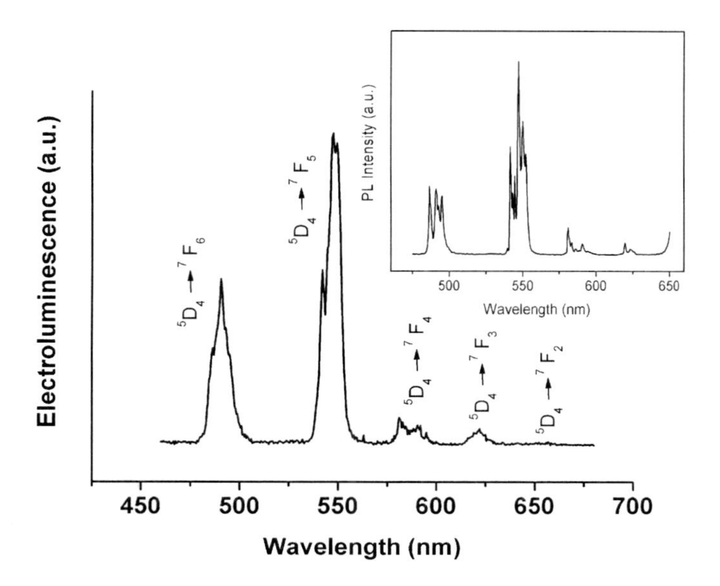

Figure 4.29 Electroluminescence spectrum of the ITO/TPD/Tb(ACAC)$_3$ bipy/Alq3/Al OLED under a driving voltage of 22V. Inset: the photoluminescence spectrum of the same complex, for comparison. The typical Tb^{3+} emissions are signalized.

As typical, the electroluminescence spectrum intensity is much lower than the measured photoluminescence, which is, as already mentioned, a consequence of the excitation mechanisms. Note also that the electroluminescence spectrum has no emissions other than the Tb^{3+} ones. This means that, although the absence of an "effective" electron blocking interface, the carrier confinement is perfect, and a "pure" Tb^{3+} emission is observed, reproducing the material photoluminescence except for a small loss of structure in each dipole transition. Due to the main emission at near 545 nm ($^5D_4 \rightarrow {}^7F_5$) the final color is green; nevertheless, there is also a relatively strong emission at near 485 nm, which in some cases cannot be "compensated" by the orange and red emissions. The CIE color coordinates thus usually fall the green–lemon region. In Fig. 4.30, the electroluminescence spectra under three different applied voltages (and their corresponding CIE color coordinates) are shown.

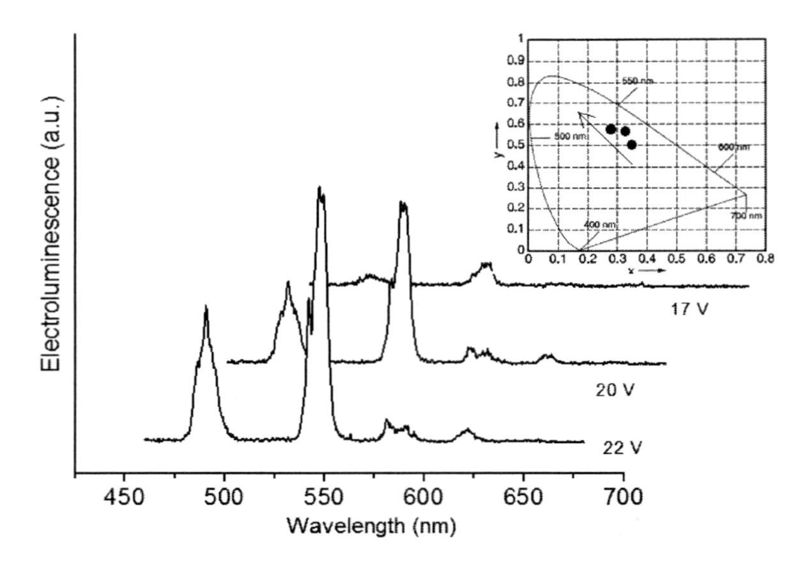

Figure 4.30 Electroluminescence spectra of the ITO/TPD/Tb(ACAC)$_3$ bipy/Alq3/Al OLED for three different applied voltage. Inset: the influence on the CIE color coordinates (the arrow indicates the changes when increasing the applied voltage).

The figure evidences that the Tb^{3+} based OLED does not have such a high color purity as that observed for the Eu^{3+} based devices. In fact, the blue–green contribution to the spectral emission is strong enough to increase the distance from the pure color border in the CIE diagram of the (x,y) coordinates. A curious aspect is that the effect of different applied voltages on coordinates do not change much the color purity although are clearly spaced in the green region of the diagram. Still, the Tb^{3+}-based organic complexes undoubtedly rise origin to excellent green RE(III) based OLEDs.

The radiometric efficiency for the main emission (at 545 nm) can be obtained as previously explained. Figure 4.31 shows the electrical current/optical power (@ 545 nm) vs. applied voltage as well the optical vs. electrical power with the corresponding linear fit for the wall-plug efficiency determination.

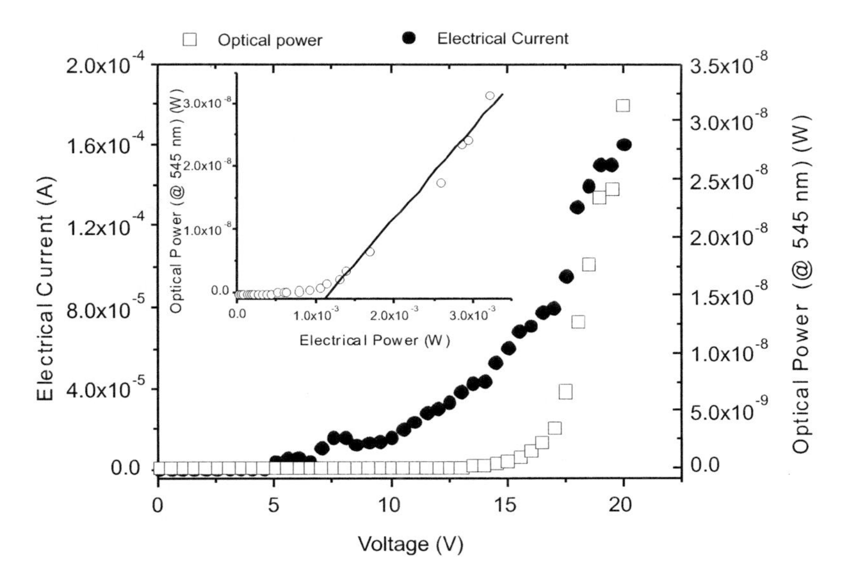

Figure 4.31 Electrical current and optical power (@ 545 nm) for the ITO/TPD/Tb(ACAC)$_3$bipy/Alq3/Al OLED with the applied voltage. Inset: the optical power vs. electrical power. The straight line is the linear fit in the dynamic OLED region allowing the calculation of the radiometric wall-plug efficiency.

The wall-plug efficiency η_{WW} (for the main emission @ 545 nm) calculated from the data on the last figure is about 6×10^{-3}%, a value higher than that obtained for the similar OLED structure with a Eu^{3+} complex. Extending it to the whole emission band, this value increases by about 40%. It must be noted that these values are obtained with the same procedure and experimental conditions as for the $Eu(DMB)_3phen$ in the last section and therefore they can be directly compared. The higher efficiencies obtained with Tb^{3+} complexes are probably related with the OLED structure and its energy levels diagram (which may favor the electrical carrier injection and transport) and perhaps with a more efficient antenna effect.

As found for Eu^{3+} complexes, Tb^{3+} OLEDs can be easily built by thermal evaporation process, which opens large possibilities for research due to the simplicity allied to the high precision of the technique. The central organic ligands differ from those of Eu^{3+} due to the needed "match" between the energy levels but in general, the neutral ligands are more or less the same. This is not surprising because the use of β-diketonates ligands with the typical and well known neutral ligands gives a solution very simple (due to its very long studied and established chemical synthesis routes) and efficient. In that way, much RE(III) organic complexes are very similar in molecular structure (a big help for chemistries) with also a much similar processing methods (a surplus for physicists and material science researches). It is evident that each complex have its own characteristics, in special related with the energy levels location (and all electrical and optical behaviors) that makes the difference.

One interesting class of organic ligands for Tb^{3+} complexes (besides the β-diketonate ACAC) is the PMIP *(1-phenyl-3-methyl-4-isobutyl-5-pyrazolone)* and its derivatives giving mixed-ligand terbium pyrazolonate complexes. For instance, the complex $Tb(PMIP)_3(TPPO)_2$ with TPPO = *tris bis(triphenyl phosphine oxide* exhibits excellent electron transport properties [63] and can be used efficiently as ETL [64]. Also, when incorporated in the OLED structure as the emissive layer, a photometric efficiency of 6.5 cd/A was achieved under low electrical current density (*J*) [65], decreasing when the *J* was increased (attributed to the exciton quenching). The device had a structure whit ITO/TPD/$Tb(PMIP)_3(TPPO)_2$/TPBI/LiF/

Al. Another OLED with very good results was obtained using a PMIP derivatives [66] with the structure ITO/TPD/Tb(eb-PMP)$_3$TPPO/ Alq/Mg:Ag, where Tb(eb-PMP)$_3$TPPO is (*tris[1-phenyl-3-methyl-4-(2-ethylbutyryl)-5-pyrazolone] (triphenyl phosphine oxide) terbium*). The device featured a photometric power efficiency of about 3.24 lm/W, be further increased to 9.4 lm/W when a BCP layer was introduced between the Terbium complex and the ETL, the Alq. Once again, the role of the BCP as a hole blocking material is evident although its use implies a more complex device. The same authors have also replaced TPD with NPB as HTL in this last described OLED structure, having found an efficiency of 11.4 lm/W. The explanation resides in the NPB HOMO level, which lies at 5.2 eV while the other HOMO levels are at 5.4 eV for TPD and 6.4 eV for Tb(eb-PMP)$_3$TPPO. Nevertheless, when the simpler device structure was used, lower color purity was observed due to contributions of emissions from the organic, terbium-free layers, arising from exciton recombination on these other regions.

The effect of the neutral ligand on the electroluminescence characteristics of a Tb^{3+} complex was also studied, in conjugation with different central ligands [67]. Four neutral ligands were considered, namely TPPO, *bipy*, *phen* and H_2O, and ten central ligands deriving from the PMIP, "combined" and tested in the conventional structure of ITO/ TPD/Tb(*CL*)$_3$*NL*/Alq/Al where *CL* stands for the central ligand and *NL* stands for neutral ligand. The authors of this study concluded that the OLED behavior was influenced by the *pyrazolone* derivative central ligand and the nitrogen or oxygen content on the neutral ligand. In many of these cases, a loss of color purity was observed due to the superimposition of the organic layers emission, broadening the spectrum.

Modifications in the traditional neutral ligands were also studied, particularly the chlorine addition to *phenanthroline* [68] having ACAC as the central ligand. Two OLEDs with structure ITO/ TPD/ Tb(ACAC)$_3$*NL*/Alq3/LiF/Al were assembled, one using plain *phenanthroline* as *NL* and the other using *Cl-phen*, and comparison of their external (non-integrated) photometric efficiencies showed that the OLED bearing Tb(ACAC)$_3$(Cl-phen) had higher values, by around 0.018 lm/W) and about 0.03 lm/W if Alq3 as ETL was replaced with Bebq2; however, associated with this higher efficiency, a high driving voltage was also reported. The results are, once again, explained by the location of the Tb^{3+} energy levels: $Tb(ACAC)_3$(Cl-phen) has a relatively low HOMO level and a high LUMO level when compared to those of $Tb(ACAC)_3$phen). Differences in the Alq3 and Bebq2 levels when the ETL is substituted are also favorable, since the LUMO level increases from 3.1 to 3.5 eV, respectively.

In the search for a more efficient (and bright) Tb^{3+} based OLED, several "non-conventional" approaches have been tested, following lines of thought similar to those observed for Eu^{3+} complexes based OLEDs. The synthesis and incorporation into an OLED of a new Tb^{3+} PMIP complex was recently reported [69]. In $Tb(PMIP)_3$DPPOC (DPPOC = *(9-(4-tert-butylphenyl) -3,6-bis (diphenylphosphineoxide)- carbazole*) the neutral ligand, has an excited level T_1 of 3.06 eV energy high enough to perform an efficient antenna effect to the $^5D_4$ resonant Tb^{3+} level. The complex was incorporated into a device with a complex structure of ITO/NPB/$Tb(PMIP)_3$/co-deposited $Tb(PMIP)_3$DPPOC/BCP /Alq/Mg:Ag/Ag, showing a pure Tb^{3+} emission (no emissions from organic compounds was detected) and a photometric efficiency of 16.1 lm/W (4.5 lm/W @ 11 V with high brightness). Though such values definitely represent an improvement, the problem of increased OLED structure complexity remains.

In all kinds of OLEDs, it is possible to occur an exciplex formation at the interface of an electric charge transporting layer (ETL or HTL) with the emissive layer [70]. The mechanism for the exciplex formation is quite simple: donor species (D) becomes excited after absorbing a photon (D*) and conjugates with an acceptor species (A) in the fundamental state, to form the exciplex (DA)*. This is one of the possible pathways, the being via the acceptor species A. The exciplex (DA)* may then decay, giving the fundamental states of the both species (D and A) and an emitted photon. Some authors point to the possibility of exciplex influence on the electroluminescence of an OLED with a basic structure of ITO/TPD/$Tb(PMIP)_3$phen/Mg:Ag [71]. According to the observed electroluminescence spectrum, the authors suggest that the TPD is first excited to its T_1 state and then forms the exciplex with the Tb(PMIP)$_3$phen in the ground state S_0 at the interface. Adding to the exciplex emission, a possible contribution of the TPD emission (blue) is also considered. The result (in extreme situation) is a pale yellowish-green color. These cases, although very interesting for color modulation (depending on the final purpose of the device) represent examples of relatively poor color purity. Concomitant with the organic molecule part emission such situations also reveal a low electrical carrier confinement.

Finally, the less common tetrakis β-diketonate complexes were also extended to Tb^{3+}. The study was from the same authors as for Eu^{3+} and comprised also ACAC, to form complexes of the general formula C[Tb(ACAC)$_4$] with C being a first group cation (Li^+, Na^+ and K^+) [58]; the device structure was kept the same, that is ITO/NPB or MTCD/ C[Tb(ACAC)$_4$] /Al allowing to observe a pure Tb^{3+} emission.

Terbium is the second most common rare earth element used as emissive ion in active layers of RE(III) complexes based OLEDs (europium being the first). Complexes with these two ions have also been reported in devices developed by others deposition techniques, particularly by spin-coating; we will focus on these hybrid structures later.

4.4.2.3 Samarium-, thulium-, and dysprosium-based OLEDs

The incorporation of complexes with these three RE(III) ions into the active OLEDs layers has been less studied. This situation is partially explained by their higher cost, but is mostly associated with difficulties in their practical application. Particularly, in the cases of thulium and dysprosium, an efficient antenna effect requires a very high excited energy level T_1 on the organic ligands, and for samarium the color is very close to the europium, in the red spectral region, which leads to lesser interest for marked applications, thought its value for fundamental research remains. In spite of that, there is much potential interests in these emitters.

Samarium is the most studied of these three RE elements. One Sm^{3+} complex, Sm(TTA)$_3$(TPPO)$_2$, was successful employed used as the active layer of an OLED with a structure ITO/MTCD/ Sm(TTA)$_3$(TPPO)$_2$/Alq3/Al [40]. The authors found a somewhat low emission, attributed to the weak luminescence of the Sm^{3+} complex and involving a possible low energy transfer from the organic ligands to the $^4G_{5/2}$ Sm^{3+} resonant level, i.e., a poor antenna effect. Although TTA appears to be a reasonable choice for working as antenna to the excited level of Sm^{3+} (Fig. 4.12), the global complex cannot achieve a good performance when incorporated on the OLED active layer. Besides the hypotheses suggested by the authors for this low performance, poor electrical carrier confinement cannot be excluded, which would require the use of further blocking materials to increase emission. In spite of that, this work present a possible solution for a Sm^{3+} based OLED with emission in the orange–red spectral region. Figure 4.32 shows the main results obtained by these authors.

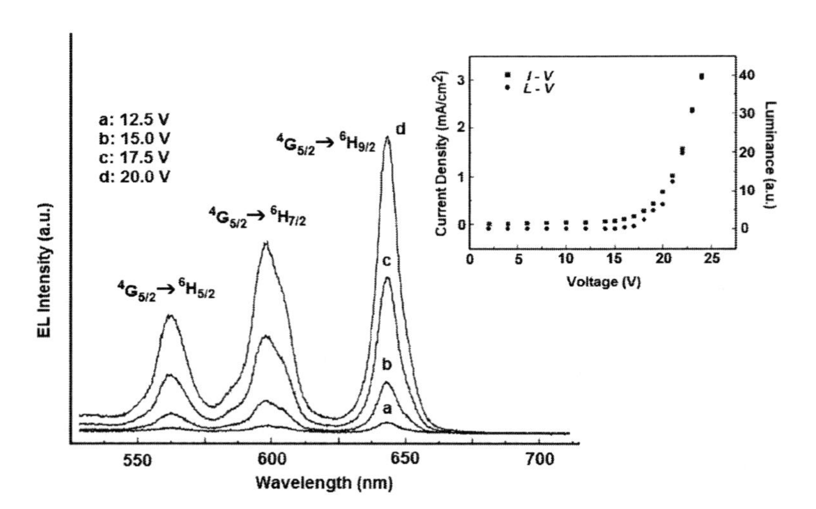

Figure 4.32 Electroluminescence spectra at different applied voltages of an ITO/MTCD/$Sm(TTA)_3(TPPO)_2$/Alq3/Al OLED. The main Sm^{3+} emission are signaled. Inset: the electrical current density and luminance vs. applied voltage. Data obtained from [40].

Other promising results on OLEDs based on Sm^{3+} have been reported. An OLED structure of ITO/TPD/$Sm(HFNH)_3$phen/BCP/LiF /Al [72], where HFNH is (*4,4,5,5,6,6,6-heptafluoro-1-(2-naphthyl) hexane-1,3-dione*), exhibited a maximum brightness of 42 cd/m^2 and a (photometric) current efficiency of 0.18 cd/A. The OLED emission was clearly orange–red and the emission increased when compared to that at the OLED with $Sm(TTA)_3(TPPO)_2$, can be associated with a more efficient antenna effect, and/or with a better electrical carrier confinement due to the BCP layer and perhaps helped by the TPD; in addition, the cathode of LiF:Al will definitely improve the electron injection, in this case.

OLEDs with different Sm^{3+} complexes and new structures have also been developed. An attempt to use the simplest and best known organic ligands with Sm^{3+} was successfully made with $Sm(BFA)_3$phen [73]. In contrast with the previous examples, these authors used a hybrid structure in which the HTL was a polymer deposited by spin coating, and the two subsequent layers—the emissive one with the Sm^{3+} complex and the HTL one—were deposited by thermal evaporation. The OLED structure was ITO/PVK/$Sm(BFA)_3$phen/PBD/Al. Although PVK may exhibit a strong blue emission, the authors didn't found any other electroluminescence than the reddish-orange characteristic emission from the Sm^{3+} energy levels. One justification in that any exciton generated inside the PVK layer was transferred to the $Sm(BFA)_3$phen active layer. And they have also considered the possibility that, when the Sm^{3+} complex was evaporated on top of the PVK, its "small molecules" could penetrate the small cavities (voids) in the PVK layer created upon deposition by spin coating, thus increasing, the excitation cross-section of the $Sm(BFA)_3$phen layer. This approach is very interesting and can be further explored in RE(III) complexes based OLEDs. Simultaneously, the PBD is known to have a "deep" HOMO level (6.2 eV) acting as a hole blocking layer; and the PVK has a "high" LUMO level (2.3 eV) acting as an electron blocking layer, which guarantees the electrical carrier confinement. As the final result, this Sm^{3+} based OLED structure presents a luminance of 135 cd/m^2 for an applied voltage of 21 V, an excellent result, when compared, for instance, with Eu^{3+} based OLEDs.

A recent [74] work reports on an OLED with an active layer comprising the complex $Sm(HFA)_3(phen)_2$MeOH (*tris (hexafluoroacethylacetonato) (phenanthroline) samarium(III) monomethanol*) and with a OLED structure of ITO/PEDOT:PSS/ PVK: $Sm(HFA)_3(phen)_2$MeOH:PBD/CsF:(Mg:Ag):Ag where the

PEDOT:PSS acts as HTL and the PKV is the active layer host matrix. The PBD works as the electron transport material and the main difference of this structure regarding the previous ones, is that this electron transport material is embedded into the active layer. The cathode is particularly complex, as an attempt to guarantee the low (and energetically compatible) work-function needed for the direct and efficient injection of electrons into the active layer. This OLED structure was totally build by spin coating (except, naturally, for the cathode) and the electroluminescence spectrum is strongly depended on the $Sm(HFA)_3(phen)_2MeOH$ concentration in the active layer. The purest Sm^{3+} electroluminescence spectrum was obtained with a mass proportion of PVK:$Sm(hfa)_3(phen)_2$MeOH:PBD 100:8:40 %wt. Decreasing the percentage of Sm^{3+} complex, the PVK emission started to appear in the blue spectral region, leading to a strong loss of color purity. The effect of the PBD concentration is correlated with the driving voltage; for a concentration of PBD of 40 %wt., the authors found the lowest driving voltage, and the highest luminance and current efficiency (135 cd/m^2 @ current density of 0.16 A/cm^2, and 0.1 cd/A @ current density of 0.08 A/cm^2). Increasing the percentage of PBD, the luminance decreased (as well as the photometric current efficiency) and the driving voltage increased. Such behavior can be related to the charge balance of holes and electrons injected in the emitting layer, since from a merely physical point of view, the device can be considered as a "single layer" OLED. Realistically speaking, the PEDOT:PSS is a high hole conductive polymer, usually viewed not as a "conductive anode" rather than as semiconducting material, thus helping on the hole injection (but not necessarily on its transport) into the active semiconducting device layer. These "one layer" devices are strongly dependent on the correct balance of the electrical charge (remember the simple explanation and discussion presented on Chapter 2), which, in turn, depends on the electrical carrier transport characteristics of each organic material/complex used. In the OLED structure presently under discussion, the PBD, as an electron transport material is needed for balancing the charges but its concentration further influences other properties: on excess may cause a quenching (too much unbalance charge) and the OLED performance would decreases. The CIE color coordinates obtained for the optimal active layer materials concentration were (0.60, 0.36), the Sm^{3+} typical reddish-orange in a good color purity. It must be noted that the OLED structure approach herein described, though not new, introduces a completely innovative path that may lead to further improvements, providing that the electrical properties of the RE(III) organic complex are "compatible" with the embedded hole (and host matrix) and the electron transport materials.

Leaving the samarium element, the following RE(III) organic complex with interest is thulium. Unfortunately, it is less studied as active OLED layer than the previous described RE(III) complexes though it exhibits a very interesting blue emission. With a high first resonant energy level—$^1G_4$ at 2.55 eV—the Tm^{3+} complexes resemble the Tb^{3+} ones. This implies that an organic ligand with a high first excited triplet T_1 state must be employed in the chemical synthesis of Tm^{3+} organic complexes for efficient antenna effect. The first and most immediate choice is ACAC. Such complex [8] was developed using *phenanthroline* as neutral ligand. The complex, $Tm(ACAC)_3$phen, was tested into an OLED with a structure of ITO/PVK/$Tm(ACAC)_3$phen/Al, but a poor photometric efficiency was obtained (0.0074 lm/W @ 16 V). Figure 4.33 shows two electroluminescence spectra for this OLED at different applied voltages.

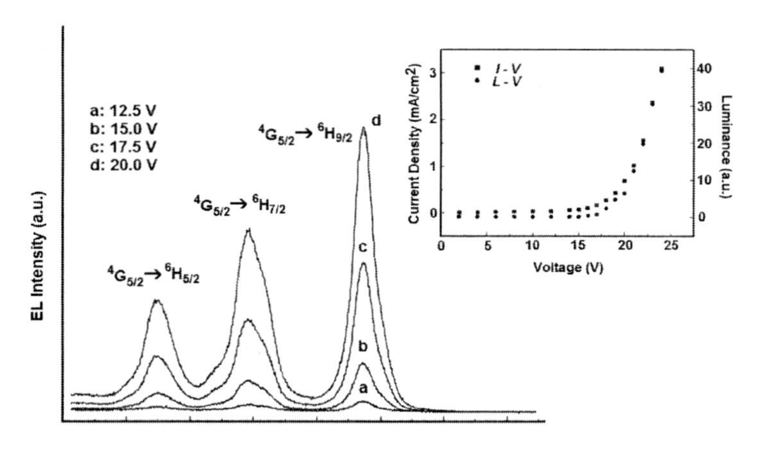

Figure 4.33 Electroluminescence spectra at applied voltages of 10 V (a) and 18 V (b) of an ITO/PKK/$Tm(ACAC)_3$phen/Al OLED. The main Tm^{3+} emission are signaled. Data obtained from [8].

Although the authors account for their results by the large energy difference between the ACAC T_1 level and the Tm^{3+} $^1G_4$ level, the absence of an ETL, which would also act as hole blocking material, has certainly contributed (and with a strong influence) to the low efficiency obtained. Again, a failure in the electrical carrier confinement is most likely. Furthermore, the symmetry of Tm^{3+} complex molecule may have an important influence, because the transition $^1G_4 \rightarrow {}^3H_6$ involves no change in parity and is strictly forbidden. This means that a choice of more "compatible" organic ligands and some changes in the symmetry around the Tm^{3+} complex may enhances the OLED performance. Adding to these, the authors also reported an overlapping of the Tm^{3+} emissions by a broad band, which may result in an emission of the organic ligands. This emission is characterized by a strong red band, recently attributed to an irreversible electrochemical reaction of the *phenanthroline* with PVK, therefore enhancing the organic contributions). Such result (also observed, for example, with Tb^{3+} in the same structure) points to the importance of the electrochemical stability of the RE(III) organic complexes on the employed OLED materials. The maximum emission obtained by the authors was 6 cd/m^2 (@ 16 V) and the CIE color coordinates were about (0.21, 0.22), i.e., a clearly blue emission (besides the low color purity).

A different target was focused when the some Tm^{3+} complex was used in an attempt to a WOLED (white-OLED) using the exciplex formation process [75]. A completely distinct OLED structure was employed, comprising ITO/m-MTDATA/Tm(ACAC)$_3$phen/LiF/Al in a bi-layer device. Following our previous reference about exciplex, it can be viewed as a charge transfer complex that is only stable in the excited state and is composed by an acceptor with high electron affinity and a donor with low ionization potential. As a result, the electroluminescence emission is red-shifted (and broadened) when compared to the own emissions of the acceptor and donor completes. As consequence, the electroluminescence efficiency is reduced, but in some cases this physical phenomena can be used as an interesting "color tuning" feature. In the cited work, the expected result was to obtain white color by overlapping of the typical blue emission of the Tm(ACAC)$_3$phen complex with the yellow emission from the exciplex between the Tm^{3+} complex and the m-MTDATA. The support for this assumption is the knowledge that the typical electroluminescence emission originated at interfacial exciplex between RE(III) organic complexes and the m-MTDATA is in the yellow spectral range [76]. To improve the "whiteness" of the white emission, the authors added another hole transport material to the m-MTDATA layer, the 2-TNATA (*4,4',4"-tris[2-naphthyl (phenyl) amino] triphenylamine*). Using these two diamine derivatives, two independent exciplex emissions would be obtained with the $Tm(ACAC)_3phen$ complex, modulating the main emission from blue too green, which is closer to the white. The best result was obtained with a m-MTDATA:2-TNATA proportion of 1:2, featuring a luminance of 47 cd/m^2, a current efficiency of 0.32 cd/A (@ 0.25 mA/cm^2) and with a CIE color coordinates of about (0.28, 0.32).

Finally, the much unexplored dysprosium, with a typical emission located at the yellow spectral region, but a very high energy for the first resonant level (the $^4F_{9/2}$ at 2.63 eV), thus drastically reducing the available organic ligands that can useful and efficiently promote the antenna effect (Fig. 4.12). Experimental data on Dy^{3+} based OLEDs are practically inexistent but an interesting result was obtained with an OLED structure of $ITO/CuPc/Dy(PM)_3(TPPO)_2/BCP/Alq3/LiF/Al$ where CuPc is (*copper phthalocyanine*) and PM is (*1-phenyl-3-methyl-4-isobutyryl-5-pyrazolone*) [77]. The authors tried several device structures but little or no Dy^{3+} electroluminescence was observed. TPD, tough used as HTL poses a huge problem because its electroluminescence bands in the blue spectral region, appeared with a relatively strong intensity and therefore overlapped the Dy^{3+} transitions, dramatically decreasing the color purity. Such behavior is not difficult to understand: the electrons, injected by the cathode (LiF:Al) directly into the $Dy(PM)_3(TPPO)_2$, were transported into the $TPD/Dy(PM)_3(TPPO)_2$ interface and then recombined to yield the two observed emissions. As no particular material was used as ETL, these results evidence a good electron transport in the Dy^{3+} complex. A second attempt (following the "trial and error" rule for materials of unknown physical properties), the authors used the Alq3 for the ETL but removed the HTL of TPD. This new approach resulted in green electroluminescence from Alq3, again superimposed with the Dy^{3+} emissions. In this device, holes were injected by the anode (ITO) directly into the $Dy(PM)_3(TPPO)_2$ layer recombining at the $Dy(PM)_3(TPPO)_2/Alq3$ interface; again, the Dy^{3+} complex proved to be a good electron transport material. A well conduced study such as this is absolutely vital for understanding the electrical nature of a new complex being the only way to develop an efficient OLED structure. The next logical step, as performed by the authors of this study, is to introduce both the HTL of TPD and ETL of Alq3. The result again evidenced a poor electrical carrier confinement, with the observation of the Alq3 band besides the Dy^{3+} emissions. This means that the $Dy(PM)_3(TPPO)_2$/Alq3 interface has no (or very little) hole blocking properties, so a layer of BCP had to be introduced between $Dy(PM)_3(TPPO)_2$ and Alq3, to afford an OLED mostly showing Dy^{3+} emissions, but with a small contribution from the TPD emission. The conclusion was that the TPD/$Dy(PM)_3(TPPO)_2$ has no much good electron blocking properties, with two possible solutions: to introduce a new layer between TPD and $Dy(PM)_3(TPPO)_2$ to act as electron blocking material or to substitute the TPD for another HTL with a higher HOMO level. The second approach was followed by the authors, replacing TPD (HOMO level of 5.6 eV) by CuPc (HOMO level of 5.3 eV). Finally, and with this configuration, both electrons and holes were successfully confined into the Dy^{3+} active layer. The best photometric power efficiency was of about 0.16 lm/W and the emission showed the blue and yellow bands attributed to the Dy^{3+} main transitions. Figure 4.34 showed these results.

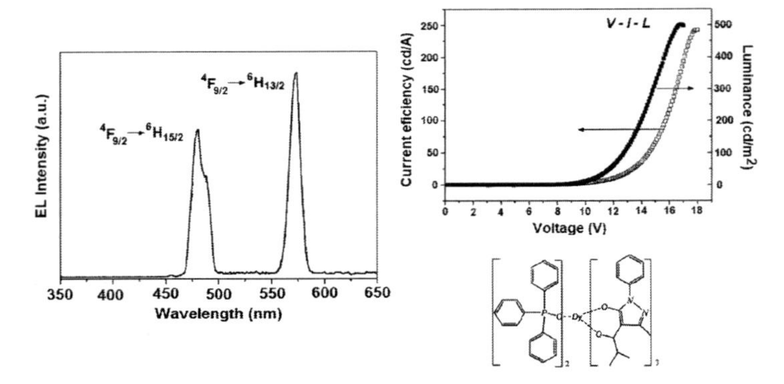

Figure 4.34 Electroluminescence spectra of an ITO/CuPc/$Dy(PM)_3(TPPO)_2$ /BCP/Alq3/LiF/Al OLED. The main Dy^{3+} emission are signaled, and the molecular structure of the emitter is show at the bottom. The Voltage–Current Efficiency–Luminance is also shown. Reprinted from *Synthetic Metals*, 104(3), Z. Hong, W. Li, D. Zhao, C. Liang, X. Liu, J. Peng, D. Zhao, Spectrally-narrow blue light-emitting organic electroluminescent devices utilizing thulium complexes, 165–168, Copyright (1999), with permission from Elsevier.

More work on devices with Dy^{3+} complexes is available, focusing on the achievement of near WOLED emission and not specifically on the Dy^{3+} emission itself. A simple discussion about such OLED structures and results will be presented further ahead.

4.4.2.4 Special cases of electrical carrier confinement in rare earth based OLEDs

In the last section, a detailed discussion was presented on carrier confinement in a Dy^{3+} based OLED. Although the potential barrier at the organic interfaces can play an important role in the process, there are others possible causes involved. In this section we will discuss an interesting study made with $Eu(DBM)_3phen$, in a previously presented OLED structure with TPD and Alq3 as HTL and ETL respectively. This example will illustrate different origin for loss of electrical carrier confinement.

In Fig. 4.22 we may see that the OLED structure ITO/TPD/$Eu(DBM)_3phen$/Alq3/Al has an energy diagram that, at least, in theory, should allow an efficient electrical carrier confinement, though with low potential barriers at the organic interfaces, with the TPD and Alq3 perform the double functions of HTL and electron blocking material and ETL and hole blocking material, respectively. In Fig. 4.24 a perfect Eu^{3+} emission was thus observed. Such behavior occurs (even under relatively high driving voltages) when the organic layers, obtained by thermal evaporation, are deposited with a rate within 0.2–0.4 nm/s. Nevertheless, when the $Eu(DBM)_3phen$ layer is deposited at rates under 0.1 nm/s, in principle, favoring a better molecular structure deposition in the film, the OLED electroluminescence spectra exhibits a large band in the green spectral region that emits cumulatively with the Eu^{3+} transitions. Such band is independent from the applied voltage although (as expected) it increases with the voltage increase [50]. It is interesting to note that, such behavior is absolutely dependent on the RE(III) organic complex evaporation rate. As a consequence, the OLED radiometric efficiency increases over two times (when compared with the "pure" Eu^{3+} emissions for similar OLEDs prepared at higher deposition rates) but the color purity decreases dramatically, leading to an almost yellow emission (CIE color coordinates of (0.45, 0.43)). Considering that no other parameters of the device fabrication were changed, the answer for this less efficient electrical carrier confinement must reside in the local molecular structure formation inside the film, which is very sensitive to the growth rate.

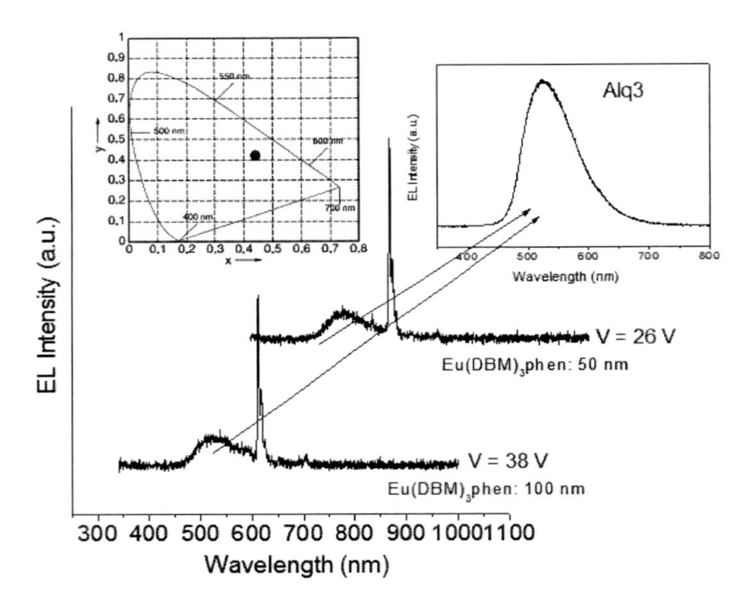

Figure 4.35 Typical EL spectra with the green emission (centered at 530 nm) for two OLEDs with a structure ITO/TPD/Eu(DBM)$_3$phen/Alq3/Al with different emissive layers thickness, growth with very low evaporation rate. Inset: the EL spectrum of a single layer ITO/Alq3/Al OLED, and the CIE color coordinates (cf. with Fig. 4.25). Reprinted with permission from Z.-F. Li (2007) Synthesis, structure, photoluminescence, and electroluminescence properties of a new dysprosium complex, *J. Phys. Chem. C*, **111**, 2295–2300. Copyright 2007American Chemical Society.

Result for the low rate evaporated OLED are shown in Fig. 4.35. The same behavior was also observed for an OLED with the same structure and the complex Eu(DBM)$_3$bipy, but interesting not for complexes with other central ligands, as TTA, NTA or BTA. This is an indication that the β-diketonate DBM is sensitive to thermal evaporation procedures. In order to explain these results, we will assume that the hole barrier at the $Eu(DBM)_3phen$ and Alq3 interface is slightly modified when the evaporation rate of the Eu^{3+} complex layer changes. At low evaporation rates we presume that additional energy levels are formed, promoting a hole density injection to the Alq3 from the $Eu(DBM)_3phen$ at the materials interface. Further electron from Alq3 and with such holes can recombine at the interface. This means that intrinsic defects that are electrically active near the HOMO level are created. Thus, the final OLED EL band, will present two contributions one from the Eu^{3+} emissions and other due to molecular orbitals recombination from the Alq3.

A schematic representation is shown in Fig. 4.36.

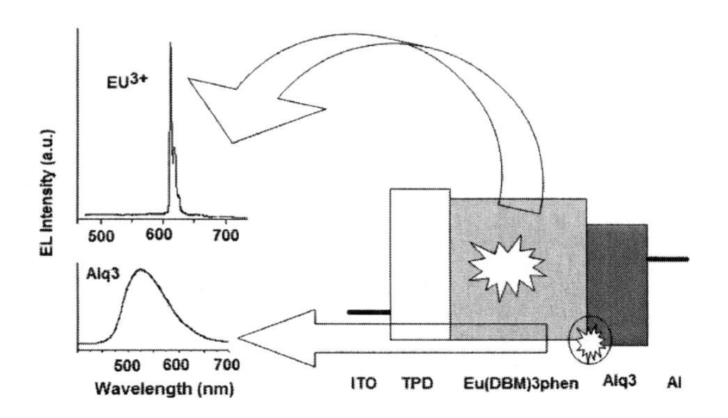

Figure 4.36 Schematic representation of the OLED structure origin for the two emissions observed in the Fig. 4.35 under the OLED structure and emissive layer growth parameters described for such results.

This result reveals an additional problem that needs to be considered whenever one is designing a novel OLED. As a final observation on the physical parameters of OLED fabrication, it must be noted that all organic materials (polymers, small molecules, dyes, metallo-organic complexes, etc.) can exhibit different molecular orientations when deposited into a very thin film, which for major cases, are not really problematic as the macroscopic optical and electrical data reveals no significant effects, but may pose some problems in particular cases. The different local molecular conformation allows the π–π^* orbitals to overlap, cross-link, attract, repulse, or form others geometrical effects due to the orientation of the molecules inside the solid thin film, inevitably creating several intrinsic defects (due to the nature of the "amorphous" film) that may be electrically active by promoting the carrier injection and/or transport, or by blocking the electrical flow. As a result, the device behavior may change very easily.

Naturally, the problem can be overcome by using one of the most mentioned solutions, inserting a hole blocking material between the Eu^{3+} complex active layer and the ETL or changing the ETL itself. In the first case, a more complex OLED structure is obtained; in the second case, we must assume that the new ETL, when acting as hole blocking material at the emissive layer, does not increase the potential barrier to the point of rendering the device unusable. Our early experiments with Eu^{3+} complexes comprising with the best known central ligands (i.e., DBM, TTA, NTA and BTA) showed that replacing Alq3 with butyl-PBD to change the HOMO level from 5.6 eV to 6.2 eV, lead to an interface disruption when the applied voltage reached a value slightly above the driving voltage, without any increase of the efficiency.

4.4.3 Special Cases of Rare Earth Based OLEDs: Energy Transfer from a Host Matrix

Besides the conventional RE(III) complexes based OLEDs, where the active layer is often composed by the complex itself, there are innumerous possibilities of using different semiconducting organic materials, with own optical and electrical properties (covering in a first approximation all the "desired" characteristics, depending on the specific choice), opens, naturally, new pathways for OLEDs development. For the RE(III) organic complexes, in particular, and similarly as will be preferential for TM complexes, modification in the physical (optical and electrical) mechanisms associated with the active layer is expected. The most common and, perhaps, efficient method is to disperse the RE(III) complex into a host matrix, typically made of a semiconducting polymer, to form a RE(III) complex-polymer blend. This has two main purposes: first, the matrix serves as support for the thin film deposition by a wet process, usually spin coating, as otherwise an uniform and homogeneous active layer could not be obtained; secondly, and more importantly the optical and electrical features, it serves as an energy pump to the organic ligands of the emitting complex, further increasing its electroluminescent emission.

For a successful energy pumping from an host polymer to a RE(III) complex a basic requirement must be fulfilled: the lowest excited triplet state T_1 of the polymer must be located at a higher energy than the (at a minimum exigency) RE(III) complex organic excited triplet state T_1, to allow the energy transfer from the polymer to the RE(III) complex. Naturally that all the non-radiative process cannot efficiently competes to this energy internal conversion. In a simple description, the process can be successively established by the following steps:

1. Due to the higher electrical conductivity of the host polymer, the excitons are preferentially (but not exclusively) formed in this material;
2. Further electron–hole recombination yields polymer emission;
3. If the polymer emission band overlaps with the RE(III) organic complex absorption band, then this significant amount energy arising from the polymer emission can be absorbed by the organic ligands at the RE(III) complex;
4. Further antenna effect occurs "inside" the RE(III) organic complex as described before;
5. Final touch: using a high conductive polymer to match increase in the energetic density of the excited T_1 state of the RE(III) organic ligands, which results in a much higher efficiency of the emission of the RE(III) complex.

This theoretical schematic appears to be very simple, but in practice there are several issues to be considered before successfully applying this approach. First, we must perform a complete photo-physical study of the materials involved (polymer and RE(III) organic complex) in to learn their relaxation mechanisms, important for the polymer–RE(III) complex energy transfer, and the precise spectral location of the polymer emission band and of the RE(III) organic ligands absorption band. The polymer emission is traditionally a phosphorescence emission (excited triplet to fundamental singlet state) with high probability of internal conversion between energy states of similar spin multiplicity, whereas in the case of the organic ligands one more troublesome because they absorb in the UV to near blue spectral region and the known polymers don't efficiently emit in this spectral region. Obviously, a more complex emitting layer blend can be make using, for instance, a non-conductive polymer to guarantee the perfect thin film formation and embedding the RE(III) organic complex into another organic small molecule that must: (1) be more conductive than the organic ligands of the RE(III) complex and (2) emits by a phosphorescence mechanism in the spectral region of the RE(III) complex. There are already a few small molecules adapted for mixing with specific RE(III) complexes, but still a lot of hard work will have to be done before achieving an OLED more efficient that obtained with an active layer composed solely by the RE(III) organic complex. In a first approach we will focus on the polymer + RE(III) organic complex blended active layer. Figure 4.37 shows a simplified scheme for this process.

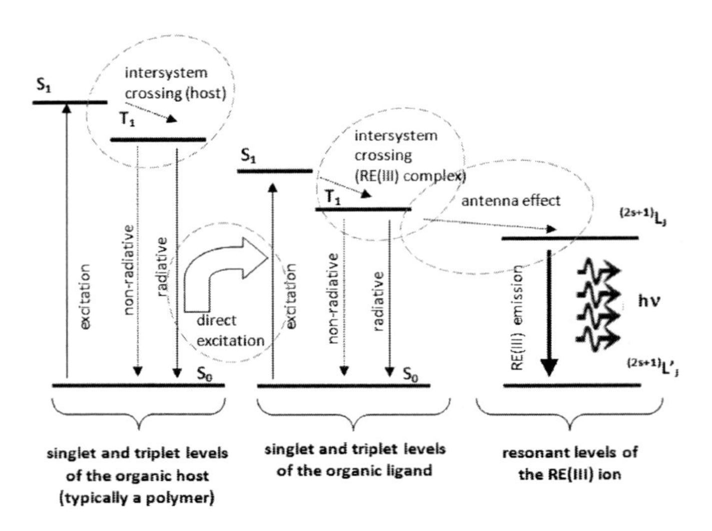

Figure 4.37 Diagram for energy transfer between a host material (typically a polymer) to the RE(III) organic ligands (with further antenna effect). $^{(2s+1)}L_J$ and $^{(2s+1)}L'_J$ are respectively the excited resonant and the fundamental levels of the RE(III) as previously discussed.

This figure is a more complete diagram than the simple idea concept shown in Fig. 4.7, depicting the overall framework. Usually the energy transfer process from the organic host to the RE(III) organic complexes follows the already discussed Förster mechanism ("long-range" dependence and represented in the last figure as "direct excitation" singlet-singlet). In the Förster mechanism, the speed rate of the energy transfer can be described by [78]:

$$X = \int |\langle 1{,}2^*|H'|1^*{,}2\rangle^2 g_1(E) g_2(E) dE \tag{4.3}$$

where $g_1(E)$ and $g_2(E)$ are the normalized spectral type functions and 1 and 2 are the functions of the host and the guest (in our cases the RE(III) organic complexes) species respectively. Equation 4.3 shows that the speed rate of the energy transfer mechanism between the host and the guest is controlled by the overlapping of the host emission spectrum and the guest absorption spectra, as referred before. A strong overlap implies a faster energy transfer process. Nevertheless, we cannot exclude the direct electrical carrier trapping by the RE(III) organic complex when the host material is used as hole transfer material and the complex works as an electron trap. In such a mechanism, the electrical carriers trapped into the RE(III) organic ligands, induce ligand excitation to the excited levels S_1 and T_1, subsequently promoting the antenna effect to the RE(III) resonant excited levels.

Before the energy can successful be transferred to the RE(III) excited resonant level, we must pay attention to the following questions: (1) Which are the non-radiative energy loss mechanisms inside the organic host and/or RE(III) complex organic ligand, and (2) what is the importance of the radiative emission from the same both organic materials? We have already seen emission from organic materials in ETLs and HTLs in the previous sections. So we cannot exclude this issue now.

The most widely used polymer to acts simultaneously as host matrix and "energy pump" in OLED active layers is the PKV.

It dissolves easily into the most solvents used for RE(III) organic complexes, it is cheap and has good electrical conductivity properties. Unfortunately, its main emission is located at the blue spectral region and therefore not all interesting RE(III) organic complexes can be efficiently pumped by this polymer. As a first study, it is very important to known how the bands (emission from polymer and absorption from RE(III) organic complex) overlaps. The result will give, in a first approximation an idea about how efficient can be the process.

Several attempts have been made in order to extract a quantitative information on the effectiveness of the energy transfer from the host material to a RE(III) organic complex (or another final energy acceptor) considering the spectral bands involved. One of the simplest methods, yielding good results can be expressed by [79]:

$$\phi = \frac{I_{\text{norm}}\left(\%\text{RE(III)}\right) - I_{\text{norm}}\left(0\right)}{I_{\text{norm}}\left(\%\text{RE(III)}\right)} \tag{4.4}$$

where φ is the transfer rate parameter, $I_{\text{norm}}(\%\text{RE(III)})$ and $I_{\text{norm}}(0)$ are the intensity ratios of the RE(III) emission peak to the host material intensity, for films containing %RE(III) and 0% of the RE(III) organic complex respectively. To determine the intensities, the emission spectra of the different RE(III) concentrations on the host material are separated into two compounds, one for the host material and another for the RE(III) organic complex, and further integrated. Equation 4.4 measures the energy transfer process and isn't a direct result of band overlapping, but rather results from the blend emission band. As the φ result approaches to the unity, the more efficient is the energy transfer.

It must be also noted that building an OLED based on this energy transfer process must takes into account two important points, the thin film layer formation and the physical limits for the energy transfer process. The first involves the concentration of the RE(III) organic complex employed in the host blend, as higher concentrations lower the polymer matrix concentration and cause a bad film formation. Remember that this layer must be deposited by spin coating or another wet technique due to the presence of a polymer and therefore no suitable emissive layer can be formed it the polymer is too diluted. Regarding the second issue, a particular concentration of the RE(III) organic complex may promote the non-radiative emissions and causes a luminescence quenching. If this "quenching concentration" exists, the RE(III) organic complex concentration limit is ruled by whichever limitation happens first, bad film formation or luminescence quenching; if not, the structural film formation is the only constraint. Naturally that when a material other than a polymer is employed as "energy pump," e.g., the well known PBD, which is a small molecule, these referred problems become even more complex.

The most extensively studied host for RE(III) complex energy transfer is PVK associated with europium complexes. This does not by chance: the PVK emission is a broad band located, at the near-UV - blue spectral region and several absorption bands of the organic ligands in Eu^{3+} complexes are also typically broad bands extended from the UV to blue. A relative overlap occurs with good probability of energy transfer.

One of the first Eu^{3+} complex used for this "energy pump" mechanism was $Eu(TTA)_3bipy$. The active layer employed a blend of PVK:$Eu(TTA)_3bipy$ of 95:5 in mass. The absorption band of the Eu^{3+} complex organic ligands and the emission band of PVK are shown in Fig. 4.38, evidencing the region where the two spectra overlap. Although the overlapped area appears seems to be relatively small, we must remember that it is a mere prediction for the probability of energy transfer from PKV to $Eu(TTA)_3bipy$ and no quantitative information on the possible energy transfer can be extracted.

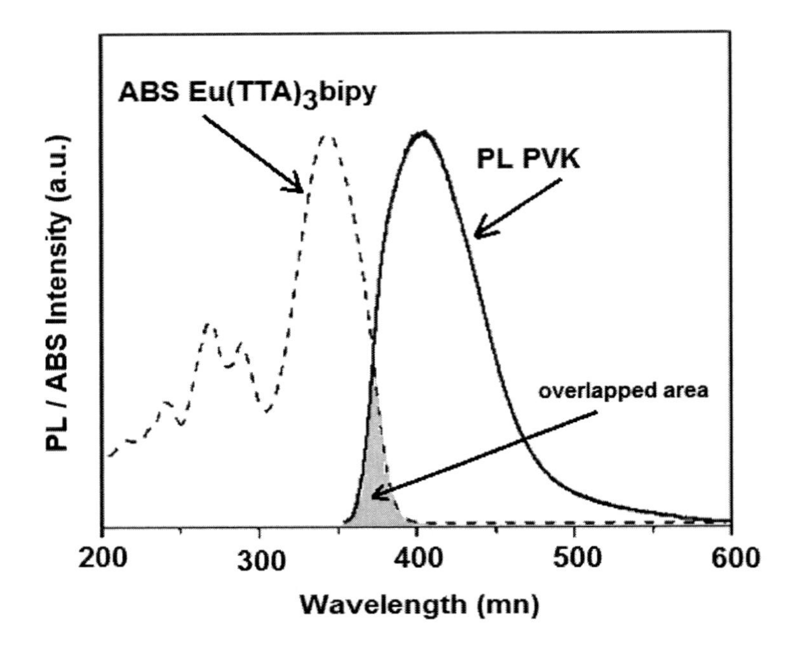

Figure 4.38 Absorption band of $Eu(TTA)_3bipy$ and emission band ok PKV. The shadow area gives the overlapped region of the two spectra.

In order to fabricate the OLED the energy levels must be considered. The LUMO and HOMO levels of PVK are respectively 2.3 eV and 5.8 eV; and for the $Eu(TTA)_3bipy$ the LUMO and HOMO levels are located at 3.2 eV and 6.2 eV respectively. Considering that the PKV is almost a hole transport material, theoretically a good electrical carrier confinement would be achieved with a HTL having a HOMO level between the ITO work-function (4.5 eV) and 5.8 eV: the PEDOT:PSS with a HOMO level of 5.1 eV, is a good candidate. The ETL, a material with a LUMO level near 3.2 eV and a HOMO level below 5.8 eV (for hole blocking) is desired. The butyl-PBD is the best choice (LUMO and HOMO levels of 3.1 eV and 6.2 eV). The energy diagram is shown in Fig. 4.39.

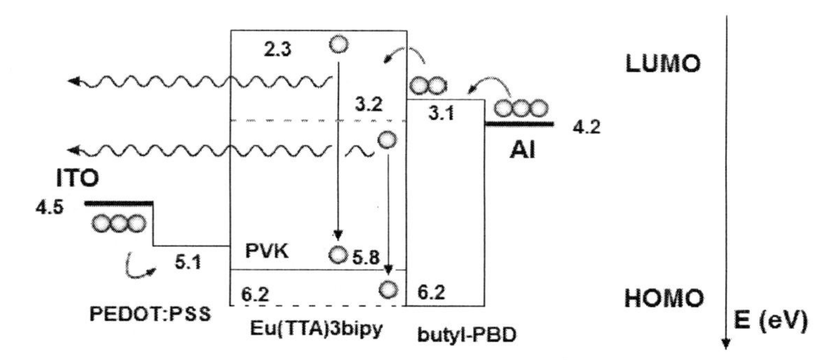

Figure 4.39 Energy levels diagram of a ITO/PEDO:PSS/PVK:Eu(TTA)$_3$bipy/butyl-PBD/Al OLED. The represented thicknesses of the several organic layers are not in the same scale. The two possible electroluminescence emissions (PVK and Eu^{3+} complex) are shown.

The PEDOT:PSS is usually considered as a "pure" hole transport material with very high conductivity, therefore its LUMO level is omitted from the energy levels diagrams. Several authors also consider that the PEDOT:PSS layer behaves much more as hole injecting than as HTL.

In the OLED structure [80], the PEDOT:PSS and the PVK:Eu(TTA)$_3$bipy layers were deposited by spin coating with a thickness of about 80 nm and 150 nm, determined by profilometry, and the butyl-PBD was deposited by thermal evaporation (20 nm thick). The electroluminescence spectra at different applied voltages are shown in Fig. 4.41. The first observation is the extreme high applied voltage regime required for device operation although with low electrical current indicating a high resistance. And a strong electrical stability, as no disruption was observed.

The second observation is the appearance of two broad emission bands, one in the blue spectral region and another in the orange–red region. The first is typically associated with the PVK emission and the second can be ascribed to the organic ligand emission due to the specific OLED structure and the high applied voltages (i.e., electrical fields). Consequently, the CIE color coordinates are of (0.49, 0.37) for an applied voltage of 40 V while at 35 V there are (0.5, 0.4)). These values remain within the red color region, but they represent lower color purity than that observed in pure Eu^{3+} emission.

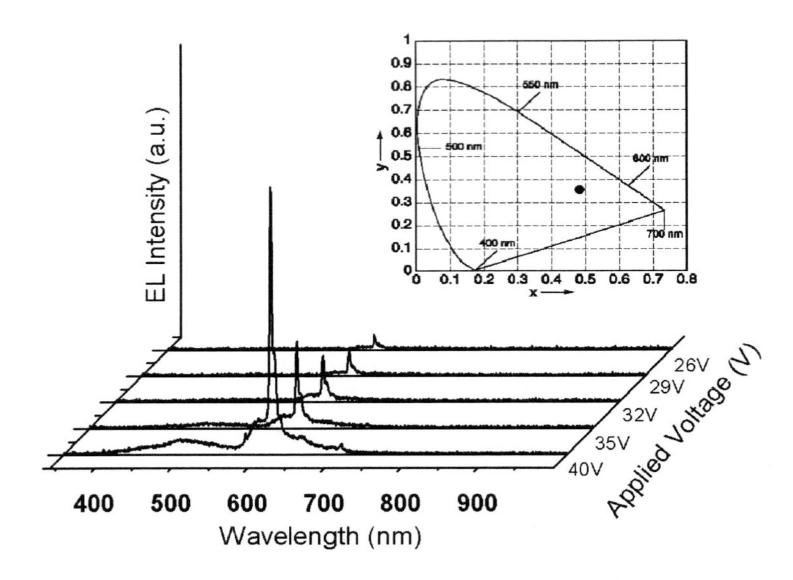

Figure 4.40 Electroluminescence spectra of the ITO/PEDOT:PSS/PVK(95%):Eu(TTA)$_3$bipy(5%)/butyl-PBD/Al OLED at different applied voltages. Inset: the CIE color coordinates (@ 40 V) that departs from the Eu^{3+} emission typical ones (cf. Fig. 4.25) due to the broad emission bands.

A simple φ calculation from Eq. 4.4 gives a value about 0.81 (applied voltage of 35 V) and 0.77 (applied voltage of 40 V) indicating a good energy transfer process from PVK to Eu(TTA)$_3$bipy. The radiometric efficiency for the single emission at 612 nm ($^5D_0 \rightarrow {}^7F_2$) under an applied voltage of 40 V is near 2×10^{-3}%, value higher than that obtained for the ITO/TPD/Eu(TTA)$_3$bipy/Alq3/Al thermally evaporated OLED.

The high electrical resistance observed is clearly a problem. Nevertheless, a different structure employing the same Eu^{3+} complex was tested with much interesting results [81]. The OLED structure was ITO/PEDOT:PSS/PVK:Eu(TTA)$_3$bipy/Ca/Al leading to a much lower driving voltage, which also helped reduce the organic emission. The calcium cathode (with a low work-function) makes all the difference, and the ETL can be omitted. These results are quite promising, opening a new range of applications for the "energy pump" Eu^{3+} organic complexes based OLEDs.

Recent work we have obtained, an excellent result employing $Eu(NTA)_3bipy$ in an OLED of the ITO/PEDOT:PSS/PVK(100-*x*):$Eu(NTA)_3bipy(x)$/butyl-PBD/Al. The best result was obtained for $x = 10\%$. Figure 4.41 shows the electroluminescence spectrum for an applied voltage of 25 V.

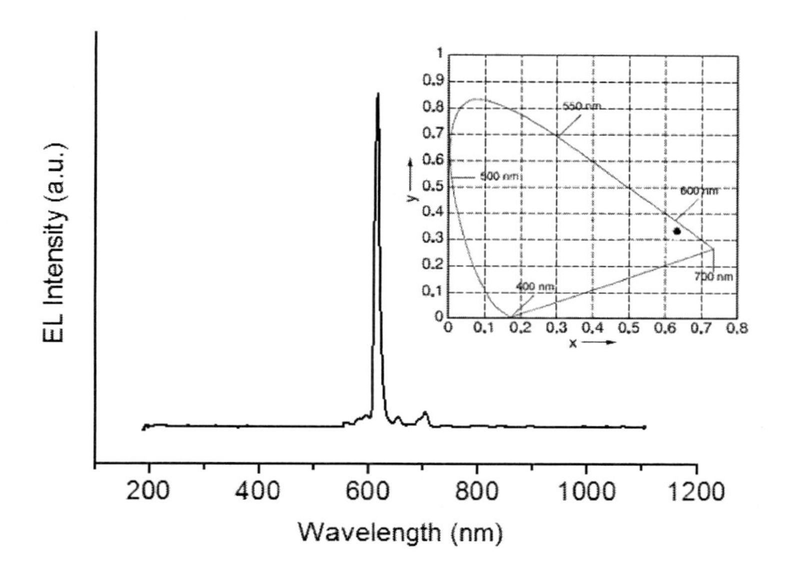

Figure 4.41 Electroluminescence spectra of the ITO/PEDOT:PSS/PVK(90%):$Eu(NTA)_3bipy(10\%)$/butyl-PBD/Al OLED for an applied voltage of 25 V. Inset: the CIE color coordinates that reproduces the "pure" typical Eu^{3+} emission (cf. Fig. 4.25).

As expected for NTA, no emission from the $^5D_1$ Eu^{3+} excited level was observed, and, most importantly, a "pure" Eu^{3+} emission was achieved, with no organic contribution (from PVK and/or the organic ligands). The HOMO and LUMO levels, lower than those of other known central ligands may contribute to this good behavior. In any case, we demonstrated that the "energy pump" principle is applicable to Eu^{3+} organic complexes using the PKV as host material.

Another work has been reported an attempt to use a host polymer for energy transfer to Eu^{3+} organic complexes, bearing

DBM as central ligand. Two distinct OLEDs structures were made [82]: a two layer device (ITO/PVK:Eu(DBM)$_3$*NL*/pEt-NTAZ/Mg:Ag) and a five layer device (ITO/PEDOT:PSS/PVK:PBD:Eu(DBM)$_3$*NL*/BCP/Alq/Cs-doped TMCD/Al) where the *NL* can be the *phenanthroline* (*phen*) or *4,7-diphenylphenanthroline* (*p-phen*) and pEt-NTAZ (*3-(4-biphenylyl)-4-(4-ethylphenyl)-5-(1-naphtyl)-1,2,4-triazole*) is a hole blocking material and the Cs-doped TMCD (*Cs-doped 6,12,19,25-tetramethyl-7,11 : 20,24-dinitrilo-dibenzo[b,m]1,4,12,15 tetra-azacyclo-docosine*) the electron-injecting layer. Both OLEDs have shown an electroluminescence strongly correlated with the concentration of Eu^{3+} complex. The best result having been obtained for the five layer structure with a proportion of 7:3 of PVK:PBD and 5% of Eu^{3+} complex in the emitting layer. In this structure, the interfaces between electrodes and organic layers probably exhibit a more "ohmic" behavior, which promotes the injection of the electrical carrier; also and the energy transfer from the host polymer to the Eu^{3+} organic complex is more efficient due to the presence of the PBD.

Another interesting work used the Eu(DBPM)$_3$phen (*tris(dibiphenoylmethane)-mono (phenanthroline)-europium*) embedded into a host polymeric matrix constituted by PVK and the polymer *poly(2-methoxy-5-(2-ethyl-hexyloxy)-1,4-phenylene)* in an OLED structure of ITO / PEDOT:PSS/Eu(DBPM)$_3$phen:PVK:polymer/LiF/Ca/Ag [83]. The result shows a "pure" Eu^{3+} emission with no emissions from any organic ligand and/or polymer. The Eu(DBPM)$_3$phen:PVK:polymer mass was 2:6:1. The low driving voltage (5.2 V), good luminance (maximum luminance of 150 cd/m^2 for a electrical current density of 95 mA/cm^2) and relatively high efficiency (0.47 cd/A at a luminance of 50 cd/m^2) revealed an efficient OLED structure, most likely due to the large overlap between the emission band of the polymer *poly(2-methoxy-5-(2-ethyl-hexyloxy)-1,4-phenylene)* and the absorption band of Eu(DBPM)$_3$phen.

Terbium(III) organic complexes have also been tested in the "energy pump" concept of OLED active layers due to great interest on their green emission. Although these will require the use of an organic ligand with a triplet excited state T_1 of higher energy than for Eu^{3+} based complexes, some results have shown that it is possible to follow similar approach to that already discussed for europium.

An interesting result was obtained with a few unusual Tb^{3+} complexes, the $TbY(o\text{-MOBA})_6(\text{phen})_2{\cdot}2H_2O$ and the $Tb(o\text{-MOBA})_6(\text{phen})_2{\cdot}2H_2O$, where the organic ligands are *o*-MOBA (*o-methoxybenzoic acid*) with *phenanthroline* and water [84]. The authors tested two different OLED structures, namely ITO/PVK: Tb^{3+} complex/LiF/Al and ITO/PVK: Tb^{3+} complex/BCP/Alq/LiF/Al with the difference obviously lying in the ETL and the hole blocking material of the second device structure. When using the first device, structure both OLEDs clearly exhibits the Tb^{3+} emissions without any presence of organic ligands and/or PVK emissions; but emission of the $TbY(o\text{-MOBA})_6(\text{phen})_2{\cdot}2H_2O$ was more intense. In the second structure with this complex into the active layer, the emission showed a small contribution of the PVK for applied voltages above 16 V whereas using the $Tb(o\text{-MOBA})_6(\text{phen})_2{\cdot}2H_2O$, such PVK emission only appeared for applied over 22 V; for both complexes the overall emission was clearly higher than that observed with the first OLED structure (nevertheless with a decrease in the color purity).

The same authors have also used the above mentioned OLED configurations to test another unusual Tb^{3+} complex with ligands from the same family [85]. The complex is $TbY(\text{p-MBA})_6(\text{phen})_2$wherep-MBAis(*p-methylphenylformicacid*).The two OLEDs with the configurations ITO/PVK: $TbY(\text{p-MBA})_6(\text{phen})_2$/LiF/Al and the ITO/PVK: $TbY(\text{p-MBA})_6(\text{phen})_2$/Alq/LiF/Al, have both shown a pure Tb^{3+} emission, although the second OLED structure exhibited a much high brightness (333.7 cd/cm^2 at an applied voltage of 23 V), most certainly due to the better electron injection/transport due to the presence of an ETL. Also, the authors report that the basic complex, $Tb(\text{p-MBA})_3\text{phen}$ (i.e., without the Y^{3+}) has a much lower performance, concluding that the presence of the Y^{3+} increases the host $\rightarrow$ guest energy transfer. Figure 4.42 shows the result.

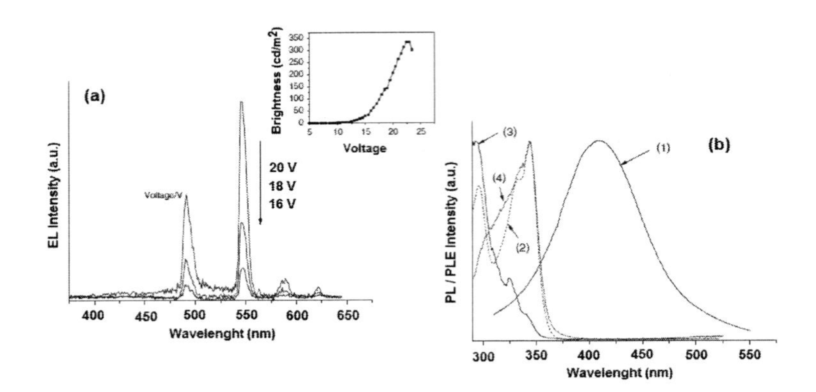

Figure 4.42 (a) Electroluminescence spectra of the ITO/PVK: $TbY(p\text{-}MBA)_6(phen)_2$/Alq/LiF/Al OLED at different applied voltages. Inset: the brightness vs. applied voltages. (b) Photoluminescence spectrum of PVK (1) the photoluminescence excitation spectra of PVK; (2) $TbY(p\text{-}MBA)_6(phen)_2$; (3) the blend. Reprinted from *Journal of Luminescence*, 122–123, Z. Chen, Z. Deng, Y. Shi, Y. Xu, J. Xiao, Y. Zhang, R. Wang, Electroluminescent devices based on rare-earth complex TbY(p-MBA)6(phen)2, 671–673, Copyright (2007), with permission from Elsevier.

A different Tb^{3+} complex, without neutral ligand, which would, in principle make it less efficient, was also tested [86]. The complex, $Tb(DPAB)_3$ where DPAB is (*4-diphenylamino-benzoic acid*) was used as the active layer in an OLED structure of ITO/ PVK:$Tb(DPAB)_3$ (50 wt%)/TPBI/Alq3/LiF/Al, and exhibited a maximum luminance of 230 cd/m^2 (@ 20 V) and a maximum efficiency of 0.62 cd/A. The electroluminescence spectrum comprised only typical "pure" Tb^{3+} emissions even under higher applied voltages.

Several other experiments with different Tb^{3+} complexes were performed (see for instance [87, 88]) but for most cases the electroluminescence spectra of the OLEDs fabricated with these complexes revealed emissions from the organic ligands and/or PVK indicative of the difficulty in obtaining the perfect energy transfer to improve the Tb^{3+} emission by this "energy pump" mechanism. Nevertheless, we expect further improvements as new Tb^{3+} complexes and/or host materials are under development.

For other interesting RE(III) organic complexes there is not much work done using the host-to-guest energy approach. The most relevant work was studied for a Sm^{3+} complex and was already discussed in the last section. Naturally, the physical mechanisms involved (not discussed before) are those we have now introduced.

4.4.4 Unusual Rare Earth Organic Complexes with Applications of Interest

Besides the well-known interest of the RE(III) organic complexes for OLEDs active layers (a "quasi-monochromatic" emission with high color purity), there are other applications of great interest, not directly related with the pure and narrow near line emissions. Without modifying the molecular mechanisms intrinsic to the RE(III) organic complexes, these applications seek to employ the known individual RE(III) emissions in a different structure in order to achieve novel goals. The most attempted work involves, not surprisingly, the WOLED (white-OLED), with a few molecular modifications to enhance the individual emissions.

The first simple concept that is applied, takes into account two important features of the RE(III) organic complex emitters: (1) much of them can easily be coordinated with the same (or very similar) organic ligands; and (2) each of them has its own characteristic emission. Based on this points, some authors have tried to use a blend of two complexes with distinct RE(III) ions in order to achieve to so-desired color tuning. Still much of the research done so far yielding no effective results or has only (for the moment) an academic interest, but the idea may still be further developed to give very important practical results in a near future.

One of such blend was the $Sm(TTA)_3(TPPO)_2$:$Eu(TTA)_3(TPPO)_2$ (0.7:0.3) where a Sm^{3+} and a Eu^{3+} were coordinated with the well known central and neutral ligands [89]. The authors used an thermally evaporated OLED structure of ITO/MTCD/$Sm_{(0.7)}Eu_{(0.3)}$$(TTA)_3(TPPO)_2$/Alq3/Al. The results show the contributions of the emissions of the two RE(III) complexes in superimposition with a very broad emission band, tentatively attributed to the molecular electro-phosphorescence of TTA. The CIE color coordinates varied with the applied voltages: at 10 V the color coordinates were (0.622, 0.372), typically red and attributed to the main Eu^{3+} emission; increasing the applied voltage, the color gradually changed to the coordinates of (0.493, 0.465), typically yellowish, due to the increase of the molecular electro-phosphorescence contribution from TTA. The authors found that, although the emission depends on the contribution from Sm^{3+} and Eu^{3+}, is also governed by the intramolecular energy transfer and by the electro-phosphorescence contribution from the TTA ligand of the Sm^{3+} complex.

Concerning the WOLED based on RE(III) organic complexes, none of the reported solutions afforded the exactly white color coordinates although, as first approach are good solutions. One of these first experiments was carried with an $Eu_xTb_{1-x}(ACAC)_3$phen dinuclear organic complex [90] and used the traditional RGB the approach: Eu^{3+} for red, Tb^{3+} for green and a small molecule, TPB or NPB for blue. The device structure was also conventional of ITO/PVK:NPB or PVK:TPD/$Eu_xTb_{1-x}(ACAC)_3$phen/Al. The best electroluminescence spectrum was obtained with the PVK:NPB (instead of TPD) and $Eu_{0.2}Tb_{0.8}(ACAC)_3$phen and had a CIE color coordinates of (0.328, 0.335) for an applied voltage of 6.5 V (we note that the "pure" white CIE coordinates are (0.33, 0.33)). It is interesting to note that, in a direct relation with the antenna effect both for Eu^{3+} and for Tb^{3+}, when the molar ratio of the Tb^{3+} ion was increased, the applied voltage required for obtaining the white emission decreased. A possible explanation is that although the antenna effect is expected in the binuclear complex, the probability of energy transfer to Tb^{3+} increased as the Tb^{3+} molar ratio increased, displacing the color coordinates to the green region, passing through the white, which, in the CIE diagram, is more or less in the middle. The closer energetic proximity of the ACAC triplet excited state T_1 to the $^5D_4$ Tb^{3+} excited resonant level to that of the $^5D_0$ Eu^{3+} excited resonant level may account for this higher probability. Simultaneously the green light is enhanced in the electroluminescence spectra at low applied voltages.

A similar approach was used for another dinuclear complex, the $Eu(BTFA)_3$phenterpy$Tb(ACAC)_3$, where the difference lies in the use of specific central organic ligands for Eu^{3+} (BTFA) an Tb^{3+} (ACAC) [91]. The neutral ligand, also working as a bridging ligand for the two RE(III) ions, is the *phenterpy (4',4''''-(1,4-Phenylene)-bis-(2,2':6',2''-terpyridine))*. The blue emission is still assured by NPB (as in the previous case). The most successful OLED structure was ITO/NPB/$Eu(BTFA)_3$phenterpy$Tb(ACAC)_3$/Alq3/Al, featuring the color coordinates of (0.28, 0.35). This result, although interesting, does not represent such a pure white as the previous one. Again the OLED color coordinates were found to depend on the Eu^{3+} and Tb^{3+} emissions (antenna effect) and on the organic ligands electro-phosphorescence as well as the contribution from NPB. As a result, the color can be controlled by the bias voltage and the thickness of the different OLED layers.

A different approach was recently used by synthesizing a new $Eu(TTA)_3$*NL* complex with a double emission (bluish-green and red), which combined to give a pure white emission [92]. Detailed information on the *NL* can be obtained in the article's supporting information. The interest of this work is that the complex comprises only Eu^{3+} without any other RE(III) element. The blue emission is obtained from the organic ligands and the device structure employed was ITO/NPB/$Eu(TTA)_3$*NL*/Alq3/Al and have a CIE color coordinates of (0.34, 0.35). As expected, the color emitted by the device is dependent on the applied voltage.

4.5 Final Considerations about Rare Earth Based OLEDs

Considering all the topics discussed in this chapter on the use of rare earth elements in active (emissive) layers of OLEDs, some final remarks can now be given. In fact, the research and development around the RE elements for OLED applications is a fascinating world that started more than thirty years ago and seems to keep growing. The main reason lies obviously in the new organic complexes with these ions that everyday add new results to the efficiency and color purity improvement.

This chapter focused on the fundamental concepts of the RE(III) ions when inserted on an organic coordination sphere, from the internal physical mechanisms related to light emission to OLED development with an active (emissive) layer based on such complexes. The most widely employed structures and organic ligands were discussed. Also, some special application of the RE(III) complexes were presented.

It must be also noted that the organic–RE complexes used for light emission are all based on 3^+ ions, because under such oxidization state they emit in the visible spectral region, opening new and unexplored scientific and market areas. The "quasi-monochromatic" emission typically of RE complexes, has caused a "revolution" in the world of OLEDs where the usual framework were the large emission bands from the π–π^* organic systems. Naturally that this new approach will have its own market and applications niche and will not need to compete with the widely used conjugated polymers or small dye molecules.

Several advantages make RE(III) complexes very attractive, many of which are completely related to their intrinsic nature, which makes them unique. We can stress the following points:

- The RE(III) ion "blindage" effect, by which the *5* (*s*, *p*, *d*) and *6s* electrons acts as shields for *4f* electrons, thus allowing their electronic transitions to take place without any influence from the surrounding media;
- The immediate consequence of the precedent point is that these RE(III) ions have sharp, strong spectral lines, at very specific wavelengths and therefore exhibit "quasi-monochromatic" emission, in contrast with those usually observed for organic light emitters;
- On the other hand, rare earth ions are lithe or not capable of absorbing energy by themselves, requiring the creation of an organic coordination sphere around them that will be fulfilling the role of an energy absorption;
- The organic coordination sphere must be able to efficiently transfer the absorbed energy of its triplet excited state T_1 to the resonant excited levels of the RE(III); this implies a careful choice of organic ligands to achieve the best energy

level compliance, i.e., the T_1 level must be slightly above the RE(III) resonant excited levels;

- The organic ligand must not allow an efficient electron–hole recombination itself either by non-radiative mechanism (energy completely lost in the molecular structure vibration) either by radiative ones (the emission will overlaps the RE(III) emissions); all the energy lost by the recombination inside the organic matrix itself, represents a decrease in the energy transferred to the RE(III) ions.
- Taking into account the two precedent points, it is currently possible to synthesize RE(III) organic complexes with a organic to RE(III) energy transfer near 100%, thus exhibiting a very intense emission;
- The energy transfer from organic ligands to the RE(III) ion is called "antenna effect" as the organic ligands act like an antenna for energy absorption and delivery; such effect can be explained by the Dexter or Förster mechanisms;
- With a correct OLED structure, a strong electroluminescence with only the RE(III) emissions can be obtained. Such a result is the final target of the work in this field.

Besides the RE(III) ions focused in this chapter, there are others with interesting and promising applications, not referred here because their usual emissions are in the near-IR region. In such classes, a special attention must be given to Ytterbium (Y) (main emission at 978 nm), Neodymium (Nd) (main emission at 1065 nm) and Erbium (Er) (main emission at 1509 nm). This last one is currently under strong research for applications in optical communications. Interestingly, its emission can be strongly enhanced by the "energy pump" concept discussed in the section 4.4.3 when NPB is used as host material.

The RE(III) ions can easily adapt to several different inorganic and organic ligands, thus bringing an enormous flexibility for novel synthesis and a wide range of complexes to explore. For application in OLED active layers, the most common RE(III) ions complexes are tris β-diketonates, although tetrakis β-diketonates and *pyrazolonate* complexes have also been successful used.

Finally, the several spectral visible colors obtained from the different RE(III) organic complexes with high purity bring added interest for market technological applications. In fact "quasi-monochromatic" emissions are much desired, from RGB displays to luminescence probes for biomedical analysis and solid state lasers. Also due to the general simple device structure that can be successfully employed, the use of RE(III) complexes into flexible emitters is a real possibility.

References

1. Kalinowski, J. (2005). High-electric-field quantum yield roll-off in efficient europium chelates-based light-emitting diodes, *Appl. Phys. Lett.*, **86**, 241106/1–3.
2. Miyamoto, W. (1999). Electroluminescent properties of a Eu-complex doped in phosphorescent materials, *J. Luminescence*, **81**, 159–164.
3. Kido, J. (1994). Bright red light-emitting organic electroluminescent devices having a europium complex as an emitter, *Appl. Phys. Lett.*, **65**, 2124–2126.
4. Liang, F. (2003). Oxadiazole-functionalized europium(III) b-diketonate complex for efficient red electroluminescence, *Chem. Mater.*, **15**, 1935–1937.
5. Ma, D. (1999). Bright green organic electroluminescent devices based on a novel thermally stable terbium complex, *Synth. Metals*, **102**, 1136–1137.
6. Wang, J. (2001). First oxadiazole-functionalized terbium(III) b-diketonate for organic electroluminescence, *J. Am. Chem. Soc.*, **123**, 6179–6180.
7. Kido, J. (1990). Electroluminescence in a Terbium Complex, *Chem. Lett.*, **19**, 657–660.
8. Hong, Z. (1999). Spectrally-narrow blue light-emitting organic electroluminescent devices utilizing thulium complexes, *Synth. Metals*, **104**, 165–168.
9. Lima, S. A. M. (2006). Low voltage electroluminescence of terbium- and thulium-doped zinc oxide films, *J. Alloys Compounds*, **418**, 35–38.

10. Hong, Z. (2000). White light emission from OEL devices based on organic dysprosium-complex, *Synth. Metals*, **111–112**, 43–54.

11. Katkova, M. A. (2005). Coordination compounds of rare-earth metals with organic ligands for electroluminescent diodes, *Russ. Chem. Rev.*, **74**, 1089–1110.

12. Kuz'mina, N. P. (2006). Photo and Electroluminescence of Lanthanide(III) Complexes, *Russian Journal of Inorganic Chemistry*, **51**, 73–88.

13. Jones, C. J. (2002). *d- and f- Block Chemistry*, Wiley-Interscience, UK.

14. Bruce, J. L., Lowe, M. P. and Parker, D. (2001). Photophysical aspects of Lanthanide(III) Complexes, in *The Chemistry of Contrast Agents in Medical Magnetic Resonance Imaging* (ed. Merbach, A. E. and Toth, E.), John Wiley & Sons, UK, pp. 437–460.

15. Guillet, J. (1987). *Polymer Photophysics and Photochemistry*, Cambridge University Press, UK.

16. Bunzli, J. C. and Eliseeva, S. V. (2009). Basics of lanthanide photophysics, in *Lanthanide Spectroscopy, Materials and Bio-Applications* (eds. Hänninen, P. and Härmä, H.), Springer Verlag, Germany.

17. Santos, G. (2008). Organic light emitting diodes with europium (III) emissive layers based on b-diketonate complexes: The influence of the central ligand, *J. Non Cryst. Sol.*, **354**, 2897–2900.

18. Weissman, S. I. (1942). Intramolecular energy transfer in the fluorescence of complexes of europium, *J. Chem. Phys.*, **10**, 241–217.

19. Gonçalves e Silva, F. R. (2002). Visible and near-infrared luminescence of lanthanide-containing dimetallic triple-stranded helicates: energy transfer mechanisms in the Sm(III) and Yb(III) molecular edifices, *J. Chem. Phys. A*, **9**, 1670–1677.

20. Latva, M. (1997). Correlation between the lowest triplet state energy level of the ligand and lanthanide(III) luminescence quantum yield, *J. Luminescence*, **75**, 149–169.

21. Sato, S. (1970). Relations between intermolecular energy transfer efficiencies and triplet state energies in rare earth beta-diketone chelates, *B. Chem. Soc. Jpn.*, **43**, 1955–1962.

22. Andreiadis, E. S. (2009). *Edifices Luminescents a base de Lanthanides pour l'Opto-electronique*, PhD Thesis, Universit Joseph Fourier, Grenoble, France.

23. Bindhu, C. V., Harilal, S. S., Varier, G. K., Issac, R. C., Nampoori, V. P. N. and Vallabhan, C. P. G. (1996). Measurement of the absolute fluorescence quantum yield of rhodamine B solution using a dual-beam thermal lens technique, *J. Phys. D: Appl. Phys.*, **29**, 1074–1079.

24. Lakowicz, J. R. (1999). *Principles of Fluorescence Spectroscopy*, 2nd edn, Kluwer Academic/Plenum Press, USA.

25. Bunzli, J.-C. (2000). Trivalent lanthanide ions: versatile coordination centers with unique spectroscopic and magnetic properties, *J. Alloys Comp.*, **303–304**, 66–74.

26. Eliseeva, S. V. (2005). Heteroleptic complexes of terbium(III) phenylanthranilate ($Tb(PA)_3$) with triphenylphosphine oxide (TPPO): a $Tb(PA)_3(TPPO)_2$-based electroluminescent device, *Russ. J. Inorg. Chem.*, **50**, 534–540.

27. Sinha, A. P. (1971). Fluorescence and laser action in rare earth chelates, *Spectrosc. Inorg. Chem.*, **2**, 255–265.

28. Yan, B. (1998). Intramolecular energy transfer mechanism between ligands in ternary rare earth complexes with aromatic carboxylic acids and 1,10-phenanthroline, *Photochem. Photobiol. A: Chem.*, **116**, 209–215.

29. de Sá, G. F. (2000). Spectroscopic properties and design oh highly luminescent lanthanide coordination complexes, *Coordination Chem. Rev.*, **196**, 165–196.

30. Phillips, D. (1985). *Polymer Photophysics*, Chapman & Hall, UK.

31. Roundhill, D. M. (1994). *Photochemistry and Photophysics of Metal Complexes*, Plenum Press, USA.

32. Yu, G. (2000). Soluble europium complexes for light-emitting diodes, *Chem. Mater.*, **12**, 2537–2541.

33. Binnemans, K. (2005). Rare-earth beta-dikeonates, in *Handbook on the Physics and Chemistry of Rare Earths* (eds. Gschneidner, K. A., Bunzli, J.-C. G and Pecharsky, V. K.), Elsevier, The Netherlands.

34. Robinson, M. R. (2000). Synthesis, morphology and optoelectronic properties of tris[(*N*-ethylcarbazolyl)(3',5'-hexyloxybenzoyl)methane](phenanthroline)-europium, *Chemical Comm.*, **17**, 1645–1646.

35. Yu, J. B. (2005). Efficient electroluminescence from new lanthanides (Eu^{3+}, Sm^{3+}) complexes, *Inorganic Chem.*, **44**, 1611–1618.

36. Bunzli, J.-C. G (2006). Benefiting from the unique properties of lanthanide ions, *Acc. Chem. Res.*, **39**, 53–61.

37. Kim, H. J. (2002). Ligand effect on the electroluminescence mechanism in lanthanide(III) complexes, *Op. Mater.*, **21**, 181–186.

38. Zheng, Y. (2001). A comparative study on the electroluminescence properties of some terbium β-diketonate complexes, *J. Mater. Chem.*, **11**, 2615–2619.

39. Rino, L. (2008). Photo and electroluminescence behavior of $Tb(ACAC)_3$phen complex used as emissive layer on organic light emitting diodes, *J. Non Cryst. Sol.*, **354**, 5326–5327.

40. Reyes, R. (2002). Growth and characterization of OLED with samarium complex as emitting and electron transporting layer, *Thin Solid Films*, **420–421**, 23–29.

41. Hin, Z. (2008). Optical and electroluminescent properties of samarium complex-based organic light-emitting diodes, *Thin Solid Films*, **516**, 2735–2738.

42. Serra, O.A. (1998). Luminescence of a new Tm^{3+} β-diketonate compound, *J. Alloys Compounds*, **275–277**, 838–840.

43. Johannes, H.-H. (1996). Vacuum deposited thin films of lanthanide complexes: spectral properties and application in organic light emitting diodes, Institut für Hochfrequenztechnik, TU Braunschweig, Annual report, pp. 29–35.

44. Binnemans, K. (2009). Lanthanide-based luminescent hybrid materials, *Chem. Rev.*, **109**, 4283–4374.

45. Turro, N. J. (1991). *Modern Molecular Photochemistry*, University Science Book, USA.

46. Ronda, C. (2008). *Luminescence: From Theory to Applications* (ed. Ronda, C.), Wiley-VCH, Germany.

47. Chen, F. (2009). Progresses in electroluminescence based on europium(III) complexes, *J. Rare Earths*, **27**, 345–355.

48. Xu. H. (2010). Novel light-emitting ternary Eu^{3+} complexes based on multifunctional bidentate aryl phosphine oxide derivatives: Tuning photophysical and electrochemical properties toward bright electroluminescence, *J. Phys. Chem. C*, **114**, 1674–1683.

49. Tang, H. (2009). Synthesis, thermal, photoluminescent, and electroluminescent properties of a novel quaternary Eu(III) complex containing a carbazole hole-transporting functional group, *J. Mater. Sci.: Mater. Electron.*, **20**, 597–603.

50. Santos, G. (2007). Light emission optimization of europium based complex in multilayer organic light emitting diode, *Proc. SPIE* **6655**, *Organic Light Emitting Materials and Devices XI*, pp. 66551U/1–7.

51. Liang, C. J. (2000). Improved performance of electroluminescent devices based on a europium complex, *Appl. Phys. Lett.*, **76**, 67–69.

52. Hong, Z. R. (2003). Efficient red electroluminescence from organic devices using dye-doped rare earth complexes, *Appl. Phys. Lett.*, **82**, 2218–2220.

53. Fang, J. (2003). Efficient red organic light-emitting devices based on a europium complex, *Appl. Phys. Lett.*, **83**, 4041–4043.

54. Xin, Q. (2006). Improved electroluminescent performances of europium-complex based devices by doping into electron-transporting/hole-blocking host, *Appl. Phys. Lett.*, **89**, 223–524.

55. Fang, J. (2007). Ligand effect on the performance of organic light-emitting diodes based on europium complexes, *J. Luminescence*, **124**, 157–161.

56. Adachi, C. (2000). Electroluminescence mechanisms in organic light emitting devices employing a europium chelate doped in a wide energy gap bipolar conducting host, *J. Appl. Phys.*, **87**, 804–805.

57. Reyes, R. (2002). Growth and characterization of OLEDs with europium complex as emission layer, *Braz. J. Phys.*, **32**, 535–539.

58. Quirino, W. G. (2008). Electroluminescent devices based on rare-earth tetrakis β-diketonate complexes, *Thin Solid Films*, **517**, 1096–1100.

59. Takada, N. (1994). Strongly directed emission from controlled-spontaneous-emission electroluminescent diodes with europium complex as an emitter, *Jpn. J. Appl. Phys.*, **33**, L863–L866.

60. Wang, Y.-Y. (2007). Efficient electroluminescent tertiary europium(III) b-diketonate complex with functional 2,2'-bipyridine ligand, *Synth. Metals*, **157**, 165–169.

61. Xin. H. (2003). Carrier-transport, photoluminescence, and electroluminescence properties comparison of a series of terbium complexes with different structures, *Chem. Mater.*, **15**, 3728–3733.

62. Xu. H.-B. (2008). Crystal structure and photo- and electroluminescent properties of a "sandglass" terbium cluster, *Inorg. Chem. Commun.*, **11**, 1187–1189.

63. Gao, X. (1998). Electroluminescence of a novel terbium complex, *Appl. Phys. Lett.*, **72**, 2217–2219.

64. Huang, L. (2002). Blue organic electroluminescent devices based on a distyrylarylene derivative as emitting layer and a terbium complex as electron-transporting layer, *J. Luminescence*, **97**, 55–59.

65. Liang, C. J. (2007). The electroluminescent decay mechanism of rare-earth ions in OLEDs based on a terbium complex, *IEEE Photonics Technol. Lett.*, **19**, 1178–1180.

66. Xin, H. (2003). Efficient electroluminescence from a new terbium complex, *J. Am. Chem. Soc.*, **125**, 7166–7167.

67. Gao, X. (1999). Photoluminescence and electroluminescence of a series of terbium complexes, *Synth. Metals*, **99**, 127–132.

68. Kim, Y.K. (2000). Chlorine effect on electroluminescence of Tb complexes, *Synth. Metals*, **111–112**, 113–117.

69. Chen, Z. (2009). A highly efficient OLED based on terbium complexes, *Org. Electron.*, **10**, 939–947.

70. Cocchi, M. (2002). Efficient exciplex emitting organic electroluminescent devices, *Appl. Phys. Lett.*, **80**, 2401–2403.

71. Xin, H. (2002). Photoluminescence and electroluminescence of the exciplex formed between a terbium ternary complex and *N,N'*-diphenyl-*N,N'*-bis(3-methylphenyl)-1,1'-diphenyl-4,4'-diamine, *Phys. Chem. Chem. Phys.*, **4**, 5895–5898.

72. Yu, J. (2005). Efficient electroluminescence from new lanthanide (Eu^{3+}, Sm^{3+}) complexes, *Inorg. Chem.*, **44**, 1611–1618.

73. Stathatos, E. (2003). Electroluminescence by a Sm^{3+}-diketonate-phenanthroline complex, *Synth. Metals*, **139**, 433–437.

74. Kim, Z. (2008). Optical and electroluminescent properties of samarium complex-based organic light-emitting diodes, *Thin Solid Films*, **516**, 2735–2738.

75. He, H. (2009). Interfacial exciplex electroluminescence between diamine derivatives with starburst molecular structure and tris(acetylacetonato)-(mono-phenothroline) thulium, *J. Alloys Compounds*, **470**, 448–451.

76. Li, M. T. (2006). Tuning emission color of electroluminescence from two organic interfacial exciplexes by modulating the thickness of middle gadolinium complex layer, *Appl. Phys. Lett.*, **88**, 091108/1–3.

77. Li, Z.-F. (2007) Synthesis, structure, photoluminescence, and electroluminescence properties of a new dysprosium complex, *J. Phys. Chem. C*, **111**, 2295–2300.

78. Huang, S. H. (2001). *The Principles and Methods of Laser Spectroscopy*, Jilin University Press, China.

79. Diaz-Garcia, M. (2002). Energy transfer from organics to rare-earth complexes, *Appl. Phys. Lett.*, **81**, 3924–3926.

80. Santos, G. (2008). Improvement of europium based organic light emitting diode structure using a polymeric conductive host, Proc. 24th Polymer Processing Society, 540–543.

81. Deichmann, V. A. F. (2007). Photo- and electroluminescent behavior of Eu^{3+} ions in blends with poly(vinyl-carbazole), *J. Braz. Chem. Soc.*, **18**, 330–336.

82. Li, T. (2004). Organic electroluminescent devices using europium complex-doped poly(N-vinylcarbazole), *Polym. Adv. Technol.*, **15**, 302–305.

83. Fang, J. (2006). High-efficiency spin-coated organic light-emitting diodes based on a europium complex, *Thin Solid Films*, **515**, 2419–2422.

84. Shi, Y. (2007). Electroluminescence characteristics of a new kind of rare-earth complex: $TbY(o\text{-}MOBA)_6(phen)_2 \cdot 2H_2O$, *J. Luminescence*, **122–123**, 272–274.

85. Chen, Z. (2007). Electroluminescent devices based on rare-earth complex $TbY(p\text{-}MBA)_6(phen)_2$, *J. Luminescence*, **122–123**, 671–673.

86. Zhang, L. (2008). A terbium (III) complex with triphenylamine-

functionalized ligand for organic electroluminescent device, *J. Luminescence*, **128**, 620–624.

87. Lin, Q. (2000). Green electroluminescence generated from the thin film based on a soluble lanthanide complex, *Synth. Metals*, **114**, 373–375.
88. Lepneva, L. (2009). OLEDs based on some mixed-ligand terbium carboxylates and zinc complexes with tetradentate Schiff bases: Mechanisms of electroluminescence degradation, *Synth. Metals*, **159**, 625–631.
89. Reyes, R. (2004). Voltage color tunable OLED with (Sm,Eu)-β-diketonate complex blend, *Chem. Phys. Lett.*, **396**, 54–58.
90. Zhao, D. (1999). White light emitting organic electroluminescent devices using lanthanide dinuclear complexes, *J. Luminescence*, **82**, 105–110.
91. Quirino, W. G. (2006). White OLED using β-diketones rare earth binuclear complex as emitting layer, *Thin Solid Films*, **494**, 23–27.
92. Law, G.-L. (2009). White OLED with a single-component europium complex, *Inorg. Chem.*, **48**, 10492–10494.

Chapter 5

Transition Metal Complexes: The Path to High-Brightness OLEDs

Transition metals ions coordinated to organic ligands give rise to the most intense light emitters known in the field of OLEDs. These devices also have a few drawbacks: they present a relatively complex internal physical mechanism for electrical carrier recombination, have a large spectral emission bands, and some of them are based on expensive transition metals; however, and in spite of all the final result is a really surprising brightness. Naturally, these transition metals complexes are serious candidates for general lighting and even for decorative and architectural environment illumination, as, besides the high brightness, widely spectral colors are also obtained. This chapter addresses the fundamental notions about transition metal-based OLEDs, starting, as expected, with a discussion about the physical mechanisms of their radiative decay, setting ground for the understanding of the second part of the chapter, dedicated to OLEDs development and device structures employed. Some aspects of the chemical structure nature and the corresponding device working mechanisms are also focused, including a special discussion on light electrochemical cells (LECs) based on the observed redox behavior (although it must be noted that for the scope of this book those devices will be generically designated as OLEDs). The final

Organic Light-Emitting Diodes: The Use of Rare-Earth and Transition Metals
Luiz Pereira

www.panstanford.com

goal is to establish knowledge for further (and improvements) developments in this field.

5.1 Introduction—The Choice of the Transition Metal Complexes

If, in Chapter 4, the choice of rare earth organic complexes for OLEDs active layers was clearly established, it is now time to see what transition metal complexes have to offer. As stated, the usual strong luminescence obtained with these complexes anticipates a very bright OLED, providing there's a correct structure to take advantage of these luminescent properties. In this chapter, we discuss such issues.

Emitting complexes based on transition metals (TMs) are commonly included in the "triplet emitters" group [1] due to their known "triplet paths" for exciton recombination with further increase of the internal efficiency (based on the phosphorescence emission). Though a few particular merely organic emitters also present this behavior, the strong (and efficient) emission from the TMs clearly opens new possibilities.

The most interesting TMs complexes emit from the lowest excited triplet states with a high-emission quantum yield and, as the ground energy state is always a singlet, a phosphorescence mechanism is observed. As mentioned, their lifetime (which may viewed, in a first approximation as the reverse of the transition probability) is usually several orders of magnitude higher than the fluorescence, though for practical applications such question is completely irrelevant. In fact even when compared to the "fast phosphorescence" typical of rare earth organic complexes (in lasting micro-seconds), the typical TMs phosphorescence of TM comprises lasting up to few milliseconds (typically much less). This way, the very bright response of TM organic complexes-based OLEDs (covering also the visible spectrum—and some IR emission) makes all the difference.

Additionally, almost every interesting transition metals organic complexes exhibits a redox behavior [2] (depending on the triplet states), affording special properties to the devices based on such materials and constituting an extra advantage of TM complexes over the RE ones. This redox properties responsible for the denomination of these devices as LECs allow, for example, the development of a single layer OLED based on a ruthenium organic complex, without any special considerations about on work-functions values of the cathode and the anode electrodes [3, 4]. This brings an enormous potential of application with several points of interests for the industry.

The basic internal mechanisms of energy transfer and radiative emission depend on the organic ligands, as naturally expected, but the molecular behavior under excitation is quite distinct from that of the RE complexes from the last chapter. In fact, on TM organic complexes there is no energy transfer from the organic ligands to the metal ion resonant levels, but a *metal to ligand charge transfer* [5] (often called as MLCT). The excited electrical carriers interact with the organic ligands to produce a broad-band "main" emission resembling that of the merely "pure" organic emitters, but which depends on the TM ion. This specific behavior results of the formation of molecular energy levels which in turn are dependent on the TM ion and organic energy levels. Naturally that the choice of ligands for TM-emitting complexes assumes an even more critical importance than for REs, as it will not only influence light emission profile but will also modulate the redox potentials. Besides the MLCT mechanism, others can be applied to the TM complexes and will be discussed latter.

With the excited electronic levels formed from both metallic ion and organic ligands combined, it is not so difficult to understand the formation of a highly effective internal triplet state in TM-emitting complexes. In fact, the emitting excited energy states are always triplets and the TM organic complex must have an efficient intercrossing system from the excited singlet state to these excited triplet states. Such mechanism is usually called *triplet harvesting* and, in a simple description, corresponds to an efficient energy accumulation in the lowest excited triplet state. Further details on this are presented in later sections.

Although there are several TMs with potential interest, for emitting complexes for OLEDs, a relatively restricted group is considered the most interesting for combining high brightness and high efficiency with a simplest device structure and emission in a dominant color. In this group, the most important TM ions are ruthenium in a 2^+ oxidation state, i.e., Ru(II), featuring an orange emission [6–8], rhenium, in a 1^+ oxidation state, i.e., Re(I), with a yellow emission [9–11], iridium, in a 3^+ oxidation state, i.e., Ir(III), with a broad color range emission [12–15], from blue-green to red, osmium, in a 2^+ oxidation state, i.e., Os(II), typically emitting in the red [16, 17] and the platinum in a 2^+ oxidation state, i.e., Pt(II), featuring an red emission [18]. Complexes of others TM elements have also been studied namely, chromium [19] and iron [20] (for instance) but unfortunately with low scientific interest for their use in OLEDs active layer.

It is also important to note that some of the most representative TM ions used in OLED active layers, present different oxidation states available for coordination with organic ligands, such as Os(II) or Os(III) allowing to further broaden the range of the molecular design of specific complexes of interest [21].

All these features make TM organic complexes one of the most interesting and important pathways for development of high efficacy OLEDs, in a completely different physical approach to that of the previous chapter for rare earth-based devices. The new molecular internal mechanisms, associated with the efficient results and vast possibilities for chemical synthesis, make TM emitting complexes the best choice for a new scientific "highway" towards high-brightness OLEDs.

5.2 The Basic Transition Metal Organic Complex Systems and Application to Light Emission

Transition metals are classified, from the chemical point of view, as *d-block* elements of the, i.e., elements with an electronic configuration characterized by partially filled *d* or *f* orbitals, with a few exceptions, as the groups 11 and 12 of the IUPAC Periodic Table including copper, silver, gold, zinc, cadmium, and mercury, and others like ytterbium and nobelium. Nevertheless, in this book such elements will be not explored.

The main configuration group does not naturally include the elements with the d^0 or d^{10} and f^0 or f^{14} electronic configuration (and considering the exceptions cited above). The chemical nature of the transition metals is clearly different from that of the rare earth elements in particular in what concerns the oxidative state, probably one of the most important properties for the synthesis of complexes, to be further used as OLED active layers. As viewed in last chapter, the RE elements and their coordination complexes of the interest for OLEDs, are predominantly in a 3^+ valence state, whereas the TM elements can easily occur in more than one valence state, particularly from 1^+ to 3^+ giving rise of different complexes with naturally different luminescence characteristics.

In spite of this, is not difficult to observe that the scientific interests are focused in a few TM coordination complexes with a specific TM oxidative state; the one that will maximizes emission when used in an OLED active layer. The exceptional properties of TM complexes, including the powerful color emission, are dominated by the partially filled *d* or *f* orbitals and such characteristic opened a new opportunity for the high-brightness OLEDs.

5.2.1 Physical Structure of Different Transition Metals Coordinated Ions

The transition metal elements comprising the three rows of the Periodic Table, 4, 5 and 6, exhibit a different electronic configuration, based on a distinct core shell. Row 4 has a typical electronic configuration based on argon, row 5 is based on krypton, and row 6 on xenon. Their respective electronic configuration are $1s^2\ 2s^2\ 2p^6\ 3s^2\ 3p^6$, $1s^2\ 2s^2\ 2p^6\ 3s^2\ 3p^6\ 3d^{10}\ 4s^2\ 4p^6$ and $1s^2\ 2s^2\ 2p^6\ 3s^2\ 3p^6\ 3d^{10}\ 4s^2\ 4p^6\ 5s^2\ 4d^{10}\ 5p^6$ hereinafter called [Ar], [Kr] and [Xe], respectively. It must be noted that the transition metal elements of row 6, except the lanthanum, with a core electronic configuration based on [Xe], also have a full $4f$ orbital, i.e., $4f^{14}$ and sometimes their core representation is represented as [Xe]($4f^{14}$).

The traditional TM elements used for OLED active layers, all have partially occupied *d* orbitals (and some also the *s* orbital also). Table 5.1 shows the electronic configuration of the most interesting TM elements (although some of them will not be extensively considered in this book but represents some important phosphorescence emitters).

In the table, the inclusion of copper and zinc which, as mentioned before, are exceptions to the "intrinsic" nature of the TM elements, makes underlined sense. These elements exhibit the major properties recognized and found on the remaining TM elements presented in the table (and of course in others "pure" TMs) and are the base of many widely used triplet organic complexes not only in OLEDs [22, 23] but also in organic photovoltaic devices [24, 25] (as ruthenium in dye-sensitized photovoltaics [26]).

These electronic configurations with a partially filled *d* orbital, pose a different situation for the shielding effect when compared to the one described for the rare earth elements, thus leading to a distinct electronic behavior. In fact, considering the *s*, *p* and *d* orbitals (as well as the *f* orbital for the heavy TM elements), an electron *x* in an *nd* × orbital (valence electron) is not fully shielded by the electrons in the $(n+1)s^2$ sub-shell [27]. This is almost the case of the TM elements and a calls for different framework from the simple only used for the rare earth elements.

Table 5.1 Electronic configuration of the most interesting transition metals elements for use in OLEDs

Element	Chemical Symbol	Electronic Configuration
Copper	Cu	[Ar] $3d^{10}$ $4s^{1}$
Zinc	Zn	[Ar] $3d^{11}$ $4s^{2}$
Ruthenium	Ru	[Kr] $4d^{4}$ $5s^{1}$
Rhenium	Re	[Xe]($4f^{14}$) $5d^{5}$ $6s^{2}$
Osmium	Os	[Xe]($4f^{14}$) $5d^{6}$ $6s^{2}$
Iridium	Ir	[Xe]($4f^{14}$) $5d^{7}$ $6s^{2}$
Platinum	Pt	[Xe]($4f^{14}$) $5d^{9}$ $6s^{1}$

See the text for explanation about [Ar], [Kr] and [Xe] electronic configurations.

5.2.2 Emission in Transition Metal Organic Complexes

The emissions of TM organic complexes can be discussed in terms of the organic ligand orbitals (the $\pi\pi^*$ system), the metal ion orbitals (considering the TM as part of d-block we have a dd^* system) or even in terms of both organic ligand and metal ion orbitals (the $d\pi^*$ system). The first is usually called as ligand-centered, the second as metal-centered and the last the already mentioned MLCT (metal to ligand charge transfer). Is this section, a brief discussion of each system will be given, including an explanation of the energy levels of the TM complexes. This will help understand of luminescence properties and the further application of the complexes as OLEDs active layers.

5.2.2.1 Transition Metal complexes luminescence based on the organic ligand: the ligand-centered and metal-centered mechanisms

The luminescence of the TM complexes can be processed by different ways. The most fundamental are those governed by the ligand (ligand centered—LC) or by the metal (metal centered - MC). The more usual (for the aim of this chapter) is the process of the metal to ligand charge transfer (MLCT). This last mechanism will be addressed in the next section. Now, we briefly described the two first.

In the LC mechanism the luminescence system is governed by the orbitals of the organic ligands but strongly influenced by the presence of the TM ion. This means that the simple $\pi\pi^*$ system of merely organic emitters cannot be directly applicable and we must consider the triplet emitter system.

When TM coordination organic complex acts as a triplet emitter, the emission from the excited triplet energy levels to the fundamental singlet (which in a simplified way can be viewed as the phosphorescence in the Jablonski diagram, Figure 3.10) the rule for the forbidden transition due to the spin multiplicity, is naturally relaxed. This relaxation is mainly due to the spin–orbit coupling induced by the TM ion. The complex nature of the energy transfer and emission (radiative and/or non-radiative) involving the triplet harvesting concept, may be one principal distinction for the rare earth emitters described in last chapter. This triplet harvesting can be roughly described in the following steps:

(1) Upon excitation, the molecule under takes in an excited state which must, according to the spin statistics (schematically represented in the Figure 3.10), must create an excited population of 1:3 of singlets and triplets;

(2) Due to the strong spin–orbit coupling created by the presence of a heavy transition metal in the center of the coordination complex, the probability of a $S_1 \rightarrow T_1$ intercrossing system (both excited states) becomes higher than that of a fluorescent $S_1 \rightarrow S_0$ transition, thus allowing 25% of excited population to be added to the T_1 levels;

(3) Due to this highly probable intercrossing system, all the excited population becomes "available" at T_1 levels (that already have 75%);

(4) Moreover, the strictly forbidden (by spin multiplicity) $T_1 \rightarrow S_0$ transition, but now allowed by the spin–orbit coupling (due to the presence of the TM ion) leads to a phosphorescence emission;

(5) This phosphorescence (radiative emission) $T_1 \rightarrow S_0$ involves all the excited population (from S_1 and T_1 levels).

This very interesting and efficient system will obviously require a high probability transition for the phosphorescence (and also for the intersystem crossing), otherwise the desired triplet harvesting cannot be achieved. Although the fundamental promoter of this system is the coordinated TM ion, the organic ligands also play an important role in the definition (and location) of the energy levels. So, once again, the molecular "design" is as a very important, if not essential, tool for the whole efficiency.

When compared to the "traditional" organic emitters, the advantage is obvious: organic emitter's can only efficiently takes the emission from the $S_1 \rightarrow S_0$ transition (i.e., fluorescence) as the transition probability for the $T_1 \rightarrow S_0$ emission (phosphorescence) is traditionally low implying a loss of about 75% of the excited population that recombines in a non-radiative way. Figure 5.1 shows a simplified diagram of that system.

Another important feature of a TM organic complex is the splitting of the triplet excited levels T_1 into the three energy levels. This splitting is called the zero-field splitting and is controlled by the spin orbit coupling, although its magnitude depends on the MLCT nature of the emitting states. We will focus on this matter later, but for now, we should note that the efficacy of an emitting complex is strongly dependent on the local molecular configuration and the charge relationship between the TM ion and its organic ligands.

Many interactions can influence phosphorescence, in particular the spin orbit coupling and its effects on the molecule's energy level splitting. Most of these effects (with much more interest for the quantum mechanics theory) will not be discussed in the present book, as they fall beyond its scope. For more details on the singlet and triplet states, refer to the previously presented descriptions such on section 3.3.3.1 where the diagram is firstly introduced. For some further explanations see [28].

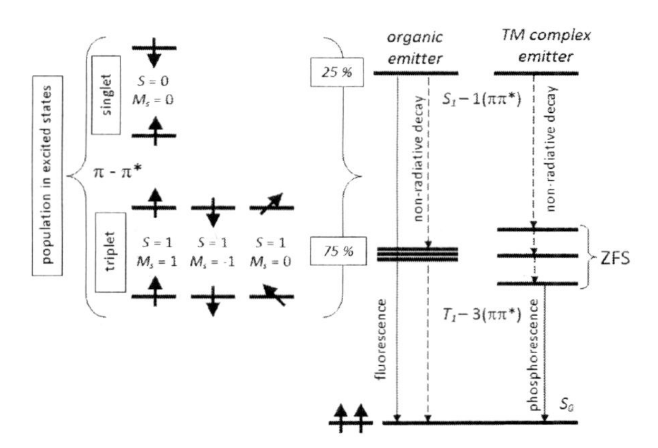

Figure 5.1 The triplet harvesting mechanism considering the distributed population in excited states T_1 and S_1. The zero field splitting (ZFS) effect is shown on the TM complex emitter pathway. For a simple comparison, the organic emitter pathway is also presented. The spin orientations for the several energy states are shown.

Contrarily to the LC mechanism introduced in the last section, the metal-centered (MC) mechanism depends on the metal ion *dd** orbitals, which is important for a series of points. In particular, and similarly to the LC mechanism, the spin–orbit coupling, originated from the *d*-orbitals can be (as reported for the π-orbitals) large enough to allow a transition probability from triplet–singlet (formally forbidden) and thus the MT complex will emit (phosphorescence). We should stress that all TM ions of much interest for OLED applications exhibit a luminescence behavior dependent on their *d*-orbitals.

One of the most explained theories for this interaction is the *crystal field theory* (CFT) [27], first proposed to explain the properties of metal ions in a crystal lattice. The CFT is essentially a model based on electrostatic interaction that takes into account the electric field arising from the electronic charge of the ligands and its influence on the electrons energies in the metal ion *d* orbitals. Following (and based on) the initial model for the CFT, a new one was developed in order to include the covalent contributions to the formation of the metal ion–ligand energy levels, which is known as *ligand field theory* (LFT) [27]. These models can, in a simple way, explain the main properties of TM ions coordinated with organic ligands, in particular their spectroscopic properties, of major interest for OLEDs. However, it must be noted that for a more precise result, the *molecular orbital theory* (MO) must be included [27], besides the LFT, (i.e., the covalent situation) a widely applicable model which fits better with the properties of TM organic complexes.

One first observation to the theories focused in the last paragraph is that they are geometry-dependent, i.e., the final energy diagram obtained by applying the theory depends on the molecular geometry. The simplest case is the octahedral geometry very common in TM organic complexes of the interest, excepting Pt(II) complexes. In this geometry, the *d* orbitals are distributed over the x, y and z Cartesian directions, the d_{z2} and d_{x2-y2} orbitals being oriented along the main axes and the d_{xy}, d_{xz} and d_{yz} oriented along directions between the axes. These orientations, resulting from an octahedral geometry, easily create an electronic density distributed into a spherical "shape" equivalent to a system composed by six ligands in an octahedral complex where the

TM ion is located at the core of the sphere. Naturally, the presence of the metallic ion in this electronic distribution will increase the energy of the electron in the referred *d* orbitals due to the repulsive force created by the surrounding cloud of charges. In the energy diagram we will obtain five discrete (but degenerated) energy states. When the spherical electronic charge is distributed over an octahedral geometry, we have six equivalent points, one at each octahedron vertex, corresponding, from the physical point of view, to the partial remove of the five *d* orbitals degeneracy. This is brought about by the electrostatic field and we are in the presence of the crystal field effect from the CFT. Further energy splitting may occur because some *d* orbital electrons (d_{z2} and d_{x2-y2}) are more often near the six octahedral points of the electronic sphere than the other *d* electrons orbitals, thus causing the d_{z2} and d_{x2-y2} energy levels to be located above the others. Figure 5.2 shows this energy diagram.

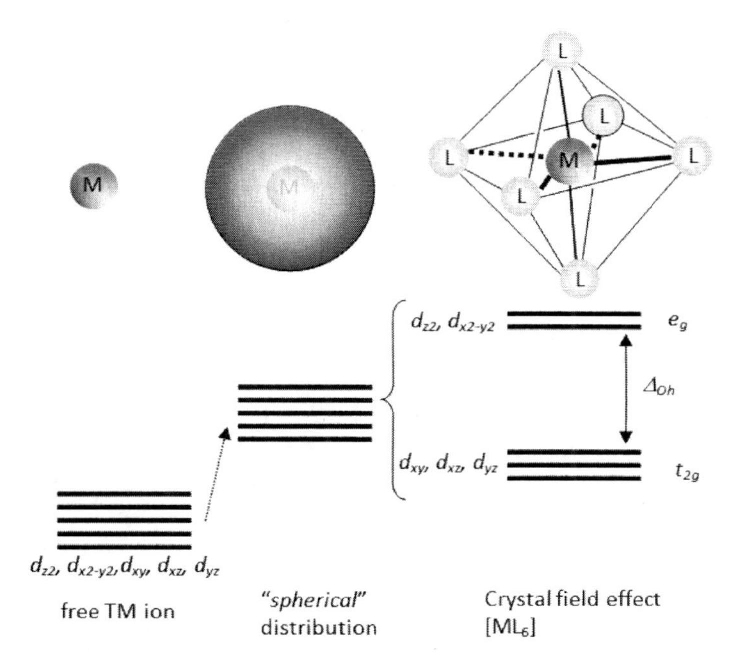

Figure 5.2 Simple energy level diagram for the *3d* orbitals of a metal ion with octahedral geometry under the CFT.

In Fig. 5.2, the final energy diagram is divided (split) into two main groups of energy levels with distinct symmetries: the e_g comprising the levels from the degenerated d_{z2} and d_{x2-y2} orbitals, and the t_{2g} comprising the levels from the remaining d_{xy}, d_{xz} and d_{yz} degenerated orbitals. These two symmetries (from the group theory [29]) will not be discussed in detail here but are quite simple to understand: the t_{2g} indicates three (t) degenerated orbitals that are labeled g (*gerade*) and are symmetric with respect to an inversion centre (in the octahedron); the e_g indicates two (e) degenerated orbitals with the same energy, undistinguished.

One important parameter for these complexes is the magnitude of the ΔO_h splitting (O_h denotes the octahedral geometry point group) between the two groups of d orbitals. Such splitting, called *crystal field splitting* (or ligand field splitting) is given by Δ and is usually denominated $10Dq$. Also, the splitting is depended on the organic ligand field and, of course on the specific TM ion within the coordination sphere. The value of the splitting Δ is particularly important in TM complexes used in OLEDs (related to the ligand field strength) as it can be correlated with the luminescence quenching [30].

Taking into account the energy diagram from Fig. 5.2 the several possible energy states arising from the excitations of an electron at the ground state t_{2g} can be obtained by determining the electron–electron interactions following the group theory mechanisms [31]. We will not discuss such mechanisms here as they are not essential for the aim of this book. Considering all the excitation possibilities, the final number and type (spin multiplicity and energy) of these levels can be obtained with the known Tanabe–Sugano diagrams [32]. In the several obtained energy states there is, due to the increase of the metal–ligand bond, a possible overlap of the potential energy surfaces of both ground and excited states at low energies [33]; consequently the non-radiative decay probability can be higher than the radiative one, leading to an emission quenching at room temperature. This may becomes a serious problem for TM complexes based OLEDs if the emitting energy level, either from the organic ligands or derived from the MLCT system, is energetically close to the excited energy levels of the *dd** orbitals.

A first solution to the problem may be to increase the *Dq* value as much as possible, which will work up to a certain limit. In fact, the efficiency of some emitting levels (in the organic ligands) is the combined result of contributions from the MLCT and the emitting triplet level, thus comprising on inevitable influence of the occupied *d* orbitals. For that, the splitting Δ cannot be exaggerated [33]. This matter points to problems related with the molecular design and choice of suitable organic ligands for an emitting TM complex, which we will briefly discuss later.

Besides the octahedral geometry (typical of d^6 TM ions, very favorable in the final energy balance and allows a relative low electrostatic repulsion of the organic ligands), there are others geometries with particular interest. Perhaps the second more interesting is the planar geometry (point group D_{4h}), which is typical of d^8 TM ions. This molecular geometry can be derived from the previous discussed octahedral geometry, where the two points on the *z*-axis (top and bottom) are displaced towards the "infinite" with the consequent formation of a single molecular local plane for the distributing points. In a planar geometry the energy of the d_{z2} orbital is lower than d_{xy} and the value of the splitting (Δ) is governed by the difference between the d_{x2-y2} and the d_{xy} orbitals.

Another possible molecular geometry for these complexes is the tetrahedral one typical of d^{10} TM ions and belonging to the point group T_d, which can be derived from the much simpler cubic geometry, less common but also found, in a few very heavy TMs. The description for the tetrahedral geometry can be made, as for the octahedral obe, by considering the coordination sphere inserted into a cube, with the TM core at the center of the cube. The tetrahedral system is then obtained by removing alternatively four regions of electron density defined in the cubic geometry. There are some interesting differences from the octahedral geometry. Firstly, the electrostatic repulsive forces between the electronic density of the organic ligands and the electrons in the d_{z2} and d_{x2-y2} orbitals are smaller than the corresponding repulsive forces between the electronic density of the organic ligands and the electrons in the d_{xy}, d_{xz} and d_{yz} orbitals; this implies that the

d_{z2} and d_{x2-y2} orbitals are located at a lower energy than the others and the fundamental state is now an *e* level (two sublevels from the d_{z2} and d_{x2-y2} orbitals) and the excited level is a *t* level (with three sublevels from the d_{xy}, d_{xz} and d_{yz} orbitals). Secondly, the splitting between the *e* and *t* groups (denoted by Δ_T) is (4/9) of the Δ_{Oh} value.

Naturally, other molecular geometries can be found in TM complexes, in particular the dodecahedron, useful for connecting eight equivalent organic ligands [27]. In this book we will keep our discussion in the most "popular" molecular geometries, like the octahedron, the tetrahedron and the plane. The influence of the molecular geometry in the phosphorescence luminescence from the triplet emitter of the TM complex energy levels will be discussed later.

5.2.2.2 Transition metal complexes luminescence based on the organic ligand: the metal to ligand charge transfer mechanism and the energy levels diagram

The metal to ligand charge transfer (MLCT) can roughly be described as the process where an electron moves from a "metallic" molecular orbital to an "organic ligand" molecular orbital, in a process that affords oxidation of the metal (and ligand reduction). In practice, this process is usual in organic complexes of metal ions with highly-filled *d* orbitals acting as electron donors to the organic ligand π orbitals located at low energies, as is very common in aromatic organic ligands. The MLCT mechanism is therefore related with both the metal *dd** and the organic ligand $\pi\pi^*$ orbitals.

The understanding of the MLCT nature is quite simple. Let us assume a simplified model considering only one HOMO level from the organic ligand and another from the TM ion. From the previously described organic ligand and metal orbitals, we can establish four distinct orbitals (two "ground state" π and *d* and two "excited state" π^* and d^*) leading to a simple model with two HOMO levels ($HOMO_\pi$ and $HOMO_d$) and two LUMO levels ($LUMO_{\pi^*}$ and $LUMO_{d^*}$). Considering also that the organic ligand, as mentioned, has low energy states and the TM ion a high energy state, we can assume that the energy levels are ordered as $HOMO_{\pi} < HOMO_{d} < LUMO_{\pi^*} < LUMO_{d^*}$.

Four excitations can arise on this system, involving the four levels. The specific levels of the excitation determine the acronym for the transition, as follows:

1. An $HOMO_{\pi} \rightarrow LUMO_{\pi^*}$ excitation. This excitation involves only the energy levels belonging to the organic ligand, i.e., is a ligand-centered excitation of type $\pi\pi^*$. It is denoted by LC;
2. An $HOMO_{d} \rightarrow LUMO_{\pi^*}$ excitation. This excitation involves a TM ion energy level and an organic energy level. In practice, it corresponds to electron (charge) transference from the TM ion to the organic ligand of type $\pi\pi^*$. We are now in the presence of the metal to ligand charge transfer. It is denoted by MLCT;
3. An $HOMO_{d} \rightarrow LUMO_{d^*}$ excitation, involving only the energy levels belonging to the TM ion, i.e., a metal-centered excitation of type dd^*. It is denoted by MC;
4. Finally, an of $HOMO_{\pi} \rightarrow LUMO_{d}$ excitation, involving an organic energy level and a TM ion energy level. In practice, it corresponds to electron (charge) transference from the organic ligand to the TM ion of type $d\pi^*$. We are now in the presence of the ligand to metal charge transfer. It is denoted by LMCT.

Some consideration can be derived from the above mentioned four excitations. First, we expected both MC and LMCT excitations not to be very probable in our currently discussed model due to the fact that if the organic ligand has a low energy states and the TM ion a high energy states, the $LUMO_{d^*}$ lies at high energies, lowering (if not annulling) the respective transition probability. Moreover, the MC transitions are forbidden by symmetry (Laporte rule [34][a]) although this rule may be relaxed by several molecular electronic effects which introduce a molecular distortion. Thus, in our hypothesis only the two first

[a] The Laporte rule stays that an electronic transition is forbidden if the molecular symmetry or asymmetry relatively to an inversion center is conserved, i.e. the parity keeps unchanged.

excitations must be taken into account. These excitations are the LC and MLCT. Is such an approach correct? If we consider almost TM organic complexes of interest for OLEDs have excitation and de-excitation processes with follow the MLCT (the main one) and the LC mechanisms, the assumption is correct. Moreover, we must remember that low $LUMO_{d^*}$ energy states for TM ions lead to a luminescence quenching at room temperature. So, for an TM organic complexes to be efficient on an OLED active layer, they must always feature high-lying energy levels, so further discussion will only be on the MLCT and LC processes, excluding the others. In Fig. 5.3 the schematic transitions are shown.

As a first approach, we then stipulated energy level diagram including a ground level and four exited levels ordered by increasing energy as MLCT < LC < MC < LMCT, the last two are not involved in the electronic transitions, is suitable; however, understanding emission mechanisms, requires further considerations on the MLCT and LC levels. As referred in the section 3.3.3.1 for the $\pi\pi^*$ orbitals of organic ligands, the electrostatic interactions given by the Coulomb Integral J and the Exchange Integral K induce splitting in those energy levels, so six levels are expected. These are listed in Table 5.2.

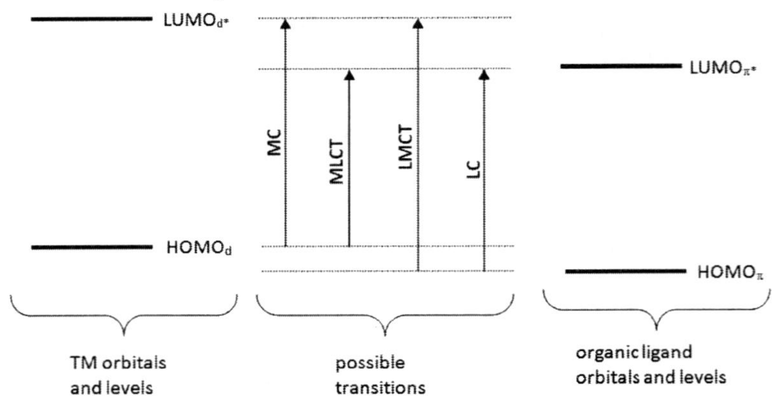

Figure 5.3 The four excitation transitions that can occur in a transition metal organic complex when the organic ligand has a low energy states and the TM ion a high energy states. See text for more explanations.

Table 5.2 Energy levels from the ligand-centered (LC) and metal to ligand charge transfer (MLCT) after the splitting due to the electrostatic interactions

Level	Split level	Configuration
LC	Singlet	^{1}LC ($^1\pi\pi^*$)
	Triplet	^{3}LC$_1$ ($^3\pi\pi^*_1$)
	Triplet	^{3}LC$_2$ ($^3\pi\pi^*_2$)
	Triplet	^{3}LC$_3$ ($^3\pi\pi^*_3$)
MLCT	Singlet	^{1}MLCT (^{1}dπ^*)
	Triplet	^{3}MLCT$_1$ (^{3}dπ^*_1)
	Triplet	^{3}MLCT$_2$ (^{3}dπ^*_2)
	Triplet	^{3}MLCT$_3$ (^{3}dπ^*_3)

A qualitative indication on the location of the energy levels can be obtained from the general equation concerning the Exchange Integral, given by [35]:

$$K \propto \left\{ \gamma(r_1)\gamma^*(r_2)\left[\frac{1}{r_{12}}\right]\gamma(r_2)\gamma^*(r_1) \right\} \tag{5.1}$$

where $\gamma\gamma^*$ represents the orbitals under evaluation (in our case these will be $\pi\pi^*$–$\pi\pi^*$ or $d\pi^*$–$\pi\pi^*$ and represent the HOMO and LUMO wavefunctions), r_1 and r_2 the electron coordinates and r_{12} the distance between two electrons.

Normally the wavefunctions associated to the MLCT energy states are more extended than those of the LC as they involve $d\pi^*$ orbitals; so, predictably the splitting of the MLCT energy states will be lower than that expected for the LC energy states. Simultaneously for the LC energy states, increasing the conjugation length (as defined in chapter 2) leads to a stronger overlap of the $\pi\pi^*$ wavefunctions, with a consequent decrease of the Exchange integral K and of the splitting value. So, when considering the splitting of the MLCT and LC levels, we must take into account that the first will be lower than the second.

Although the splitting of the MLCT and LC energy levels can be, in a qualitative way, easily addressed, the order of the final electronic levels in our simplified model is not so straightforward. This order depends on the difference between $HOMO_{\pi}$ and $HOMO_{d}$ and on the splitting introduced by the Exchange Integral K. This lowest triplet state is usually not easy to predict. Much information can be obtained either by experimentally data from spectroscopic measurements, either from theoretical calculations, or from their combinations, but in many cases we only having an approximated model to accommodate the experimental results.

The complexity of a real-case scenario involving the several intrinsic mechanisms as spin–orbit coupling, molecular geometry distortions, etc., is very difficult to quantify and often an approximation to the simple model herein discussed, is used. Moreover, the model has proved very useful to explain luminescence data of the TM complexes, allowing, interpreting results with a good approximation. Figure 5.4 shows the simplified energy diagram.

Besides the complexity, it is very important to known in particular the lowest energy. The question in not surprising: remembering the "triplet harvesting" concept, an efficient OLED, that imply that an efficient TM complex triplet emitter, must take advantage of the phosphorescence mechanism that in turns is dependent on the "lowest" excited energy state. As seems before, the obtained phosphorescence is only allowed by the spin–orbit coupling that relaxes the strictly forbidden spin parity transition observed without this coupling. In particular, the quantum mechanical interactions that mixed the LC and MLCT states (both singlet and triplet) is crucial for the properties of the lowest excited energy state.

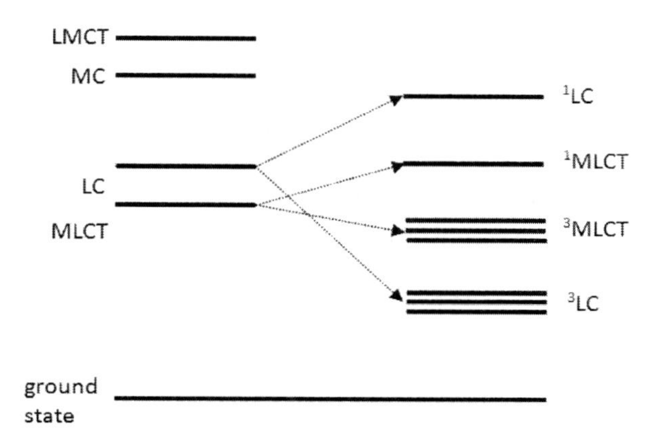

Figure 5.4 Schematic diagram of the splitting of the MLCT and LC energy states, based on the model represented in Fig. 5.3. See the text for more explanations.

While a TM complex with a pure triplet energy state will give probably a low efficient emission, another TM complex with a "mixture" between the LC and MLCT energy levels can be a good candidate as the spin–orbit coupling with the high singlet energy states will be efficient. The reason can be simply explained. The spin orbit coupling is known to be efficient [36] for energy states with MLCT character and inefficient with those of LC character. This way, two distinct situations may occur: if, the lowest excited triplet state is of MLCT character a strong spin–orbit coupling with a high-energy singlet state is expected leading to a highly probable radiative emission (phosphorescence) from that state; if, the lowest excited triplet state is of LC character the spin–orbit coupling with the high energy singlet state will have a low or null probability. The only way for this state become an "efficient" triplet emitter, is by mixing with an MLCT character state with an indirect spin–orbit. This mixture is usually made by a configuration interaction[b] resulting from electron–electron interaction.

---

[b] The configuration interaction gives an optimized many-electron state from a mixing of many-electron wavefunctions obtained from different electronic configurations.

As previous discussed, the spin–orbit coupling involving the lowest excited triplet state leads to zero field splitting (ZFS) on this state. The higher the spin–orbit coupling, the higher the ZFS and, in an extension of our discussion, higher the ZFS, the higher will be the MLCT character of that state. As the ZFS can be estimated by optical spectroscopy, its value provides a first approximation to the nature of the lowest excited triplet state. From literature data, a ZFS up to near 10^{-4} eV is attributed to a LC character state, weakly disturbed by the MLCT state; ZFS between around 5×10^{-4} eV to near 5×10^{-3} eV are typical of a LC character state with expressive MLCT perturbation; and ZFS above than 6×10^{-3} eV (up to near 3×10^{-2} eV) have a dominant MLCT character in the emitting state. Recently an interesting compilation of the ZFS and the respective lowest excited triplet state character for some TM complexes was made, providing useful reference information [34].

Further improvements on this simple model can be made, but will not be detailed here. When required, some specific notions will be presented along the text, but in general, this simple model is enough to understand the basic mechanisms governing the luminescence of the TM complexes with interest.

5.2.3 Molecular Structure of a Transition Metal Organic Complex and the Related Luminescence Mechanism

The configuration of the energy levels in a TM complex usually depends not only on the configuration of the TM ion itself, but also on the local geometry. Most TM ions (and perhaps all those of interest for OLED active layers) have their *d* orbitals partially filled and often have *f* empty orbitals, which translates into or absence or small blindage effect, in contrast to the rare earth ions. The *3d* orbitals in TM thus, have a more pronounced interaction with the organic ligands and this reflects on their energy states. Accordingly, the luminescence properties of TM complexes will not follow the simple and predictable "path" for the rare earth complexes and new models of interpretation will be required.

Luminescence of TM complexes can be viewed as a global molecular mechanism not individually assigned to the TM ion or the organic ligand itself. The emission arises from the whole molecule as the combined contribution of *both* the TM ion *and* the organic ligand. Several factors, on the molecular level, can therefore influence that luminescence. In particular, attention must be paid to the following points:

- The atomic number Z of the TM ion. This point is quite simple to understand: on increasing the Z number, we will be handling a heavy metal and therefore the magnitude of the spin–orbit coupling will increase, with strong influence on the phosphorescence emission;
- The molecular geometry. As briefly discussed, different geometries changes energy levels of the *dd** orbitals and consequently the energies states of the molecule (MC, LC, LMCT and MLCT), with implications for the final luminescence behavior;
- The choice of the organic ligand for the specific TM ion and eventually the expected molecule geometry. In fact, one way to avoid the internal quenching originated by the excited *dd** energy states and to improve the phosphorescence emission is by increasing the gap between quenching and emitting states (e.g., LC/MLCT). This can be obtained by the use of an organic ligand which promotes the shift of the *dd** states to the high energy region, in practice, by increasing the *Dq* value as much as possible. Still this shift must not be excessive because overly high *Dq* values avoid the necessary destabilization of occupied *dd** energy states needed for introducing the MLCT character on the emitting triplet sate, required for an efficient phosphorescence. A balanced situation is the best option;
- The field strength of the organic ligand. This point is related with the previous one, as the higher this field strength, the higher *Dq* value due to electrostatic interaction. But the organic field strength also depends on the TM ion. Normally the magnitude of the field strength of TM increases as we move downwards in the Periodic Table. For instance, we expected a more pronounced field strength for Re, Os, Ir and Pt (all in third row) than for Ru, from the second row.

And we must also keep in mind the Z number increases from Re to Pt and must also considered in the final molecular design. The rule of thumb is to go for organic ligands of lower field strength as the used TM are of increased row numbers of the Periodic Table row and to simultaneously play TM's Z number and the optimized Dq value;

- The use of ancillary ligands is widely disseminated. In general, ancillary ligands are defined as those ligands that provide the appropriate electronic environment around the metal core (in our case the TM ion) but do not take part in any transformation that the TM ion undergoes (some of the β-diketonates and neutral ligands given in chapter 4 fall within this classification). The field strength of these ligands will further follow the rule from the last point, i.e., to increase on associated gap of the MC transition (avoiding the phosphorescence quenching) without compromising the MCLT character on the triplet emitting state;
- As the molecular modifications on the organic ligands can "modulate" the location of the triplet emitting level (for a specific TM ion), an often explored idea is the "color tuning." The TM ion is kept the same and modifications are performed solely on the organic ligands to induce changes in the energy of the emission given by $\Delta E = (E_{\text{triplet emitting state}} - E_{\text{ground state}})$. This molecular "manipulation" is an incredible advantage of TM complexes, affording extraordinary results: for example, in Ir complexes, it is possible to have emissions from the blue to the red; with Ru, from green to yellow; with Os, blue and red; with Pt, yellow and green (and even near white) and so on. There are naturally, limits to the color tuning, much of them related with the previously mentioned points (to be further discussed later);
- Neutral charged complex are preferred over electrically charged ones. Most experimental results, point to, an average, superior efficiency for electrically neutral complexes. The intrinsic ionic property of electrically charged complexes has been pointed as the reason for this phenomenon.

We will now further develop on some of these points leaving the organic ligands for next section. For now we will see how the molecular geometry influences the phosphorescence. Particularly important are, the mentioned octahedral, planar and tetrahedral geometries, which will be compared, but bearing in mind that the formed complexes may not present perfectly well defined geometries, and usually some distortions is. In fact, the degree of flexibility of the ligands, the electrostatic interactions and even the influence of neighbor molecules, often cause geometry distortion. In that case, the general analyses become very complex but some indications can be obtained from the experimental determination of the ZFS, where the expected values range is not observed. In particular, the measured values become often lower that the expected indication a molecular symmetry decrease due to distortion in the molecular geometry. Nevertheless, we will work with the ideal geometries (octahedral, planar and tetrahedral) even if the molecules are slightly shifted from ideality. Moreover, the planar geometry frequently presents two distinct energies states corresponding to the *dd** orbitals, under weak and strong organic ligand field. We will also discuss this specific situation.

As mentioned before, the energy levels of the TM ions *dd** orbitals depend on the specific molecular geometry; a diagram of the energy levels for the cited molecular geometries is shown in Fig. 5.5 .

One important literature statement [36] is that the spin–orbit coupling in the same π^* orbital ($LUMO_{\pi^*}$), is only efficient if different *d* orbitals ($HOMO_d$) are involved in the process. In Fig. 5.5 we can see that, for the octahedral and tetrahedral geometries all the *d* orbitals of the $HOMO_d$ level are very close and thus when one experiences an excitation there the others nearby will also be involved, in agreement to the above stated. On the contrary, in the planar geometry (regardless of the organic ligand field but more pronounced for strong field) the splitting on the *d* orbitals from the $HOMO_d$ level is relatively high and an effective mix of two of them in the spin–orbit coupling is unlikely. Thus, the final triplet emitting state on this molecular geometry is predicted to have less MCLT character than the other two geometries.

This qualitative discussion helps, explain the luminescence experimental results for some TM complexes of known (or estimated) molecular geometries, and sheds some light on a phenomena exhibited by some TM complexes used in an OLED active layer, with a different behavior from what was predicted using theoretical calculations and from the luminescence data obtained for the complex alone. More importantly for TM complex-based OLEDs, the symmetry distortions may lead to a significant loss of the MLCT character of the triplet emitting level, with a consequent efficiency decrease.

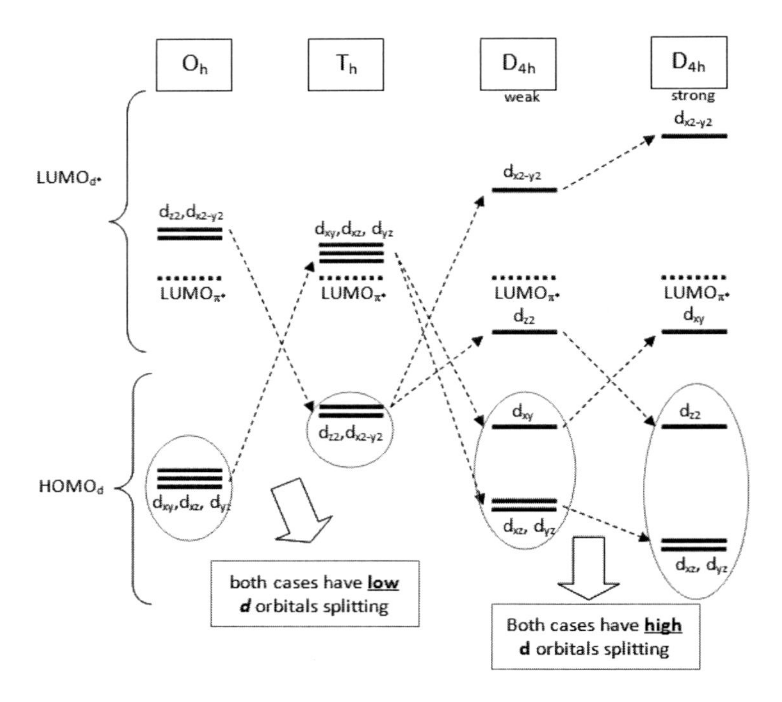

Figure 5.5 Schematic representation of the energy levels associated with dd* orbitals and their splitting under different (most common) molecular geometries (O_h—octahedral, T_h—tetrahedral, D_{4h}—planar). The $LUMO_{\pi^*}$ corresponds to the energy level of the π^* orbital from the organic ligand.

This means that, along with the molecule design, the conditions and parameters used in the OLED processing plays a very important role in the final result.

Asides from the high brightness of TM-based OLEDs another strong point of interest is the possibility of color tuning. In fact this is a very attractive experimental possibility based, once again, on the molecular design, taking into account the expected location of the *dd** orbitals (and the corresponding energy levels) of the TM ion under question. To change the emitting color, we must modify the location (in energy) of the lowest triplet state responsible for the emission. This can be done using organic ligands with the $\pi\pi^*$ orbitals with energy states at different locations. The further construction of the energy level, following the simple model of the Figs. 5.3 and 5.4, should give different locations for the triplet emitter. There are a many experimental results showing that this approach is very efficient with a wide range color tuning using the same TM ion. Another case involves the substitution of one organic ligand when the molecule has two of different type each of them handling one emission.

Perhaps, the most impressive case in color tuning of a TM complex is the Ir (typically Ir(III)) for which complexes with different ligands were synthesized and used as OLEDs active layers, with efficient emissions ranging from the blue to the red. Others examples, with a less color tune range available, will be discuss later.

The most investigated different mechanisms for the color tuning process employing a TM complex can be summarized as follows:

1. *Specific and proper functionalization of the organic ligands.* Functionalization of the organic ligands with substituent groups which can alter the molecule symmetry or the electronic density redistribution in the bonded structure (induction effect that depends on the involved atoms electronegativity). In this process, the aim is to control the metal–ligand bonding and the organic ligand orbitals energy, and, further color tuning by influence on the lowest excited triplet state is achieved. Particularly, the introduction of electron or hole donating groups will handle the organic ligand conjugation length acting on the HOMO level, donating or removing electrons to/from the metal;

2. *Playing with the conjugation length of the organic ligands.* Considering the pure conjugation length effect; an increase in this length increases the extent of overlapped orbitals on the organic ligand leading to a widely spread orbital cloud. The higher or lower conjugation lengths have strong effects on the molecule. By tuning such effect (associated with the organic ligand bandgap) we can play with the molecular structure of the ligand, increasing or decreasing the size and the degree of aromatic nature and creating localized electronic density. These effects will further change the location of the lowest excited triplet state. Naturally, the usual problems must be considered, in particular the molecular distortion (with the already known effects) and the high energy triplet emitter limit that balances the final OLED efficiency;
3. *Changing the emitting color by replacing its original organic ligand.* When possible, one ligand is changed and the others are kept the same. One common molecule structure formed by this process includes n-cyclometalated ligands (particular case of the chelate ligands[c] where at least one σ-donor atom is either sp^2 or sp^3 carbon and includes several ligands with C, N and P atoms in the bounding location) and m-ancillary ligands, where n and m determines the number of such ligands. It may occur that emission spectrum will be constituted by an independent contribution of each ligand type. The measured CIE color coordinates will thus be the sum of each contribution. The color tuning can them be achieved by handling each ligand independently, i.e., replacing one of them in the chemical synthesis process. Usually the cyclometalating emission is related with the LC level [37] and the ancillary ligand is related with the MLCT [38].

These mechanisms are the most employed, but not the only ones available. Much of them depends on the chemical procedure used for the molecular synthesis and will not be discussed here,

[c] A chelate ligand is an organic ligand where the ligating atoms and the coordination center form at least one closed loop considering the atoms connectivity.

because with more or less differences, the basic principles follow those already discussed.

Although very attractive, some limitation in the color tuning exists. Perhaps one of the most important problems is the one which derives from the molecular geometry distortion and from the interference of high energy transitions, in special those associated with the MC state. These problems are particularly important when the color is tuned towards the blue (high energy emission) where the interference of the MC transition is more likely to appear. Another usual problem derives from the solution of the previous. To compensate the interference from the MC transition, an acceptor-like organic ligand with strong field can be used but in general the risk of the large destabilization of the *d** orbitals is high with a drastic decrease of the emission efficiency (loose of the MLCT character of the triplet emitting state).

Figures 5.6 and 5.7 show the simple and qualitative diagrams for how to perform the color tuning following the discussion above.

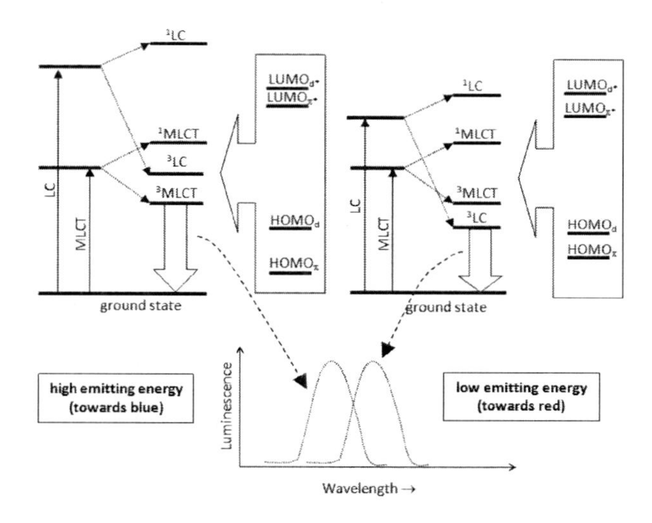

Figure 5.6 Schematic representation of the color tuning process in TM complexes by the changing only the organic ligands with different HOMO/LUMO levels. In this case, an extreme situation is shown where the emitting level changes in its mainly character. Bellow a simple representation of the qualitative spectral location of the emission. For simplification, the excitation is shown on energy levels not split.

In the case presented in Fig. 5.6, the levels of the TM ion remain unchanged, the levels of the organic ligand changed (e.g., by increasing/decreasing the conjugation length or by specific functionalization), leading to a change in the spectral region color of the emitting triplet phosphorescence. The example is an extreme situation where the triplet emitting level changes its LC/MLCT character. The diagram does not take these changes into account or other possibilities, as the molecular geometry deformation. So, the emission band shape shows no changes when the spectral peak moves towards the blue or towards the red. Obviously, in a real case scenario such situation is unlikely. Also, it must be noted that the ^{1}LC level at the left-hand scheme has a limit to be located at high energy. If relatively close to the MC (cf. Fig. 5.4) the interference of the latter becomes a problem.

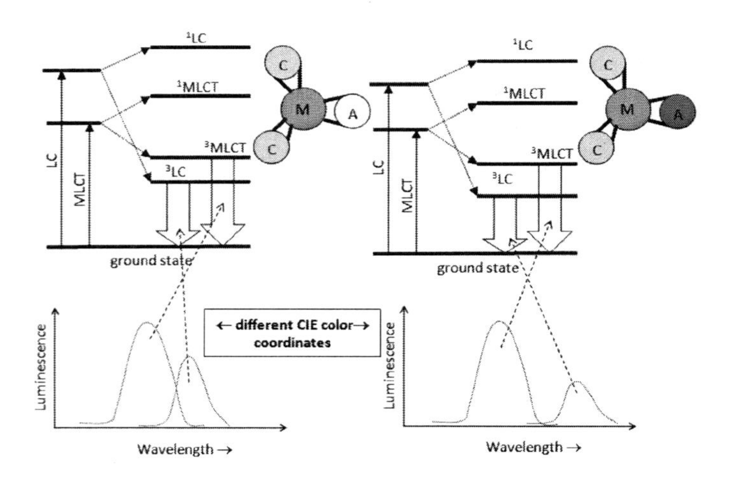

Figure 5.7 Schematic representation of the effect on the emitting color of a TM complex with two emitting levels. The molecule picture has a metal ion (M) two cyclometalated (C) ligands and one ancillary (A) ligand. Changing this ancillary ligand the LC emission changes modifying the final color. Bellow a simple representation of the qualitative spectral location of the emissions. For simplification, the excitation is shown on energy levels not split.

Figure 5.7 presents another other possibility for color tuning in a TM complex having two distinct and independent emissions, one from the LC state and another from the MLCT state. In that case by substituting the ancillary ligand that, often changes the emission attributed to the LC state while the emission belonging to the MLCT state remains unchanged, the final color is modified. Once again we must stress that the mentioned collateral effects (like the molecular geometry distortion, etc.) are not taken into account for the schematic spectral emission band shape and location. If both cyclometalated ligands are changed, the high energy peaked band will change (related with the MLCT state).

As a final consideration about the specific points focused in this section about TM complexes, it must be emphasized some important conclusions:

- The molecular geometry strongly influences the efficiency of the phosphorescence. Nevertheless, a "pure" and well defined geometry is not much expected to be found in a thin film active layer of an OLED, so we must be prepared for deviations in the predicted result;
- The color tuning, appearing as an important possibility, is relatively strong limited by the possible efficient combination of the *dd** and $\pi\pi^*$ orbitals location of the TM ion and of the organic ligand, respectively. Further "molecular engineering" to overcome some drawbacks in the most straightforward pathways, usually doesn't give the so desired efficient result;
- So, pay must be taken on the balance between the molecular structure and the final efficiency.

5.3 Transition Metal-Based OLED: The Material Chemical–Physical Properties

Absorption and emission on TM organic complexes are as pointed, attributed to the whole molecule and not to the individual organic ligand or TM ion. Even in the case of the triplet emitter being an energy state with a dominant LC character, influence of the MLCT character is needed to achieve efficient phosphorescence. And for the absorption, the $d \rightarrow \pi^*$ excitation is required for the MLCT energy state definition. In this perspective, the molecular relation between the TM ion and the organic ligands used is fundamental to the final results, and the luminescence behavior will change accordingly.

5.3.1 Typical Molecular Structure of a Transition Metal Organic Complex and Energy Absorption/Emission Schemes

5.3.1.1 General transition metal complex molecule structure

The molecular structure of the TM complex depends on the oxidation state of the metal ion. In the chemistry of TM complexes, two main types of structures can be found. These structures are the TM organic cyclometalates complexes or with cyclometalates and ancillary organic ligands. Some TM complexes does not obey to such distinct division of ligands and the organic coordination sphere is made by ligands that although can be included on both class referred, the designation is not used. The choice of the ligands to use in the complex will depend not only on the TM ion, but also and perhaps most importantly on the emission color and max efficiency for the new complex.

For now, let us consider the ligands division as referred. The generic chemical structure of a TM organic complex can thus be given as $M(n)(Cy^\wedge Cy')_n$ or $[M(n)(Cy^\wedge Cy')_{(n-1)}(L^\wedge X)]$ where M(n) is the TM ion, *n* is it oxidation number, Cy^Cy' a cyclometalate organic ligand and L^X an ancillary organic ligand where *L* denotes a neutral entity and *X* an anionic one. On other hand, the cyclometalate organic ligands can have different Cy entities and the most used for TM complexes are the C^C, C^N, N^N and C^P (C—carbon, N—nitrogen and P—phosphate). In the general molecular structures under discussion, the final electrical charge of the TM complex is zero. The positive electrical charge given by the oxidization number *n* is compensated by the negative charge of the ligands. So, the cyclometalate and the ancillary organic ligands may have, a negative electrical charge (−1). If a different final net charge balance is desired, the structure must be changed accordingly, e.g., from $M(n)(Cy\^{}Cy')_n$ to $[A][M(n)(Cy\^{}Cy')_{(n-1)}(B)]^{-1}$, with *A* and *B* being other organic ligands, generally not included in the original (cyclometalate and ancillary) class. In this example, it is supposed that both new ligands are electrically neutral; if not, a different number of them will be necessary to achieve the final desired electrical charge. We will discuss these points later with practical examples.

Besides the generic structure of the TM organic complexes, there are a few particular points of higher relevance for the final properties of the emitting complex, one of them being the geometry. An important aspect of the molecular geometry is the isomerism. In general, this transformation implies a different point group (even inside the same general point group) with a modification of the molecule symmetry and therefore its luminescence properties. Figure 5.8 shows a scheme of the two most common isomers fount in TM complexes, the *fac* (denoting *facial* species) and the *mer* (denoting *meridional* species).

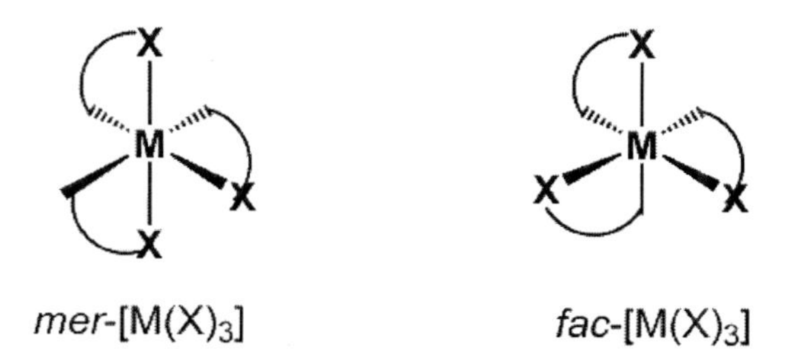

Figure 5.8 Schematic representation of the isomerism. M is a transition metal ion with 3^+ oxidization (n = 3) and X is a cyclometalate organic ligand.

These basic concepts about the molecular structure of TM complex, allow us to have an overview of the relevant phenomena for molecular design and helps further understand how to "play" with the different organic ligands.

5.3.1.2 Molecular structure of a transition metal organic complex and the energy absorption and emission

Considering the basic molecular structure given before for a TM organic complex, we can now discuss on the influence of each part of the molecule on energy absorption and emission, with a highlight on photoluminescence and electroluminescence. As stated, we regard the TM organic complex, as composed by (i) a TM ion at the core of an organic coordination sphere that can comprises two kind of organic ligands, (ii) the cyclometalate, and (iii) the ancillary ones. In the following explanation we will consider a well-known iridium-based complex.

Figure 5.9 shows the molecular scheme of the (*iridium(III) bis(2-phenylpyridinato-N,C2')acetylacetonate*) complex abbreviated, as $Ir(ppy)_2ACAC$. The TM ion is iridium with a valence of 3^+, Ir(III), the ppy (*phenylpyridinato-N,C2'*) is the cyclometalate organic ligand with a valence of 1^- and the ACAC (*acetylacetonate*) is the ancillary organic ligand with a valence 1^-. Thus the molecule is electrically neutral.

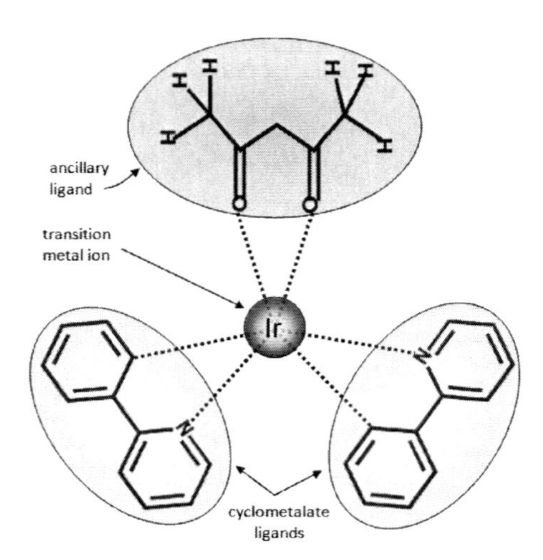

Figure 5.9 A molecular scheme of an Ir(III) complex, the iridium(III) bis(2-phenylpyridinato-N,C2')acetylacetonate, known as $Ir(III)(ppy)_2ACAC$. The shadow areas represent the different parts of the organic molecule as indicated.

The luminescence process is, as stated result of a collective molecular emission; thus, instead of an emission feature close to the organic parts or TM ion, we expect the *dd** orbitals of the Ir(III) to interact with the $\pi\pi^*$ orbitals from the ppy and ACAC to afford a novel global molecular emission. Figure 5.10 shows a scheme for the photoluminescence of the $Ir(ppy)_3ACAC$.

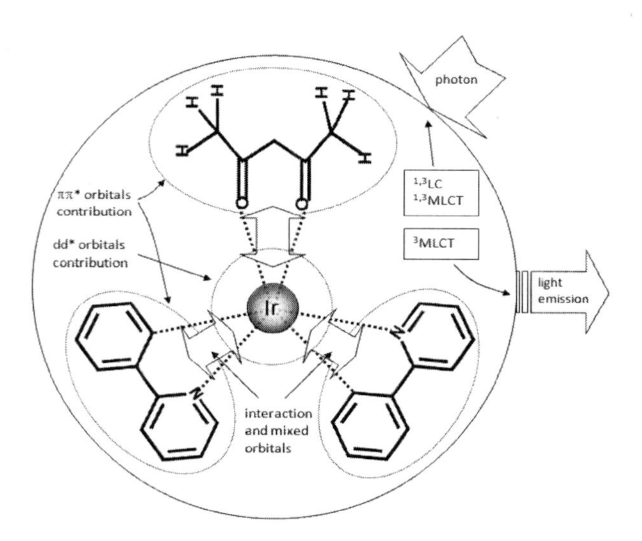

Figure 5.10 A typical photoluminescence process for the $Ir(ppy)_2ACAC$ molecule. Upon photon excitation an electron is promoted to the ^{1,3}LC and ^{1,3}MLCT levels, created by the organic $\pi\pi^*$ orbitals or organic mixed $d\pi^*$ orbitals (we exclude the MC and LMCT). By intersystem crossing, the charge relaxes predominantly to the ^{3}MLCT level with further phosphorescence emission.

Considering the nature of the energy states associated with the electronic transitions (organic and/or organic mixture character), both absorption and emission spectrum are expected to be featureless, i.e., constituted by large bands extended over a large spectral (wavelength) region, specially the emission, since it arises from a triplet emitting state. This represents a very different system from that of the RE complexes, having a featureless absorption bands for the organic ligands and sharp emission lines due to the intra-ionic recombination in the RE ion.

In TMs the organic and/or organic mixture character states are involved in both the absorption and the emission. In the case of $Ir(ppy)_2ACAC$, an electron is excited, by a photon, to the upper levels, namely the $^{1,3}LC$ and $^{1,3}MLCT$ (in UV and near the UV-blue spectral region [39]). After, a relaxation to the ^3MLCT state by intercrossing system and internal conversion will occur. In strict terms, the photoluminescence spectra also shows a component of the ^3LC state (with a MLCT character) [40], but the ^3MLCT character level is to be considered the main excited triplet emitter. Note also that this ^3MLCT level may contain some contribution of the ^3LC character. With these reserves, we keep in our discussion the ^3MLCT and ^3LC designation. The emission is a large band in the green spectral color. This process can be viewed in Fig. 5.11.

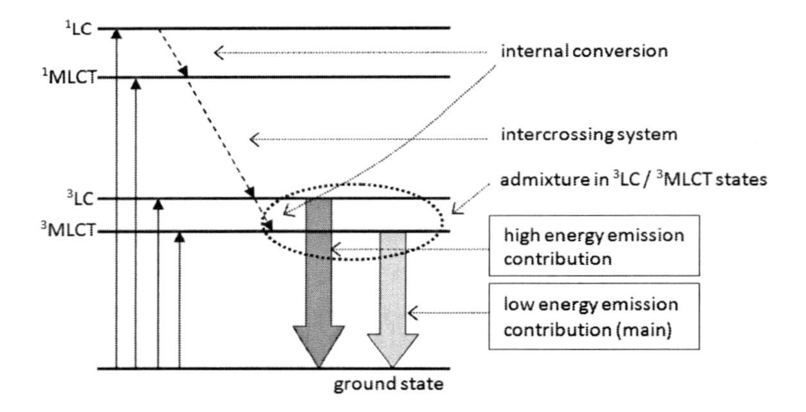

Figure 5.11 Scheme for absorption and emission of the $Ir(ppy)_2ACAC$. The main emission (phosphorescence) band is attributed to the ^3MLCT level although a less intense (but more featured) emission from the ^3LC level (with a MLCT character) appears at high energy.

We can conclude that identifying the lowest triplet emitting state is not particularly simple. Normally, the main designations are used (^3MLCT or ^3LC) but, we must always to bear in mind that both characters are involved, though one naturally predominate.

A useful tool to provide information on the nature of the lowest triplet emitting state and on the efficiency of the phosphorescence is the time-resolved spectroscopy. This technique measures the decay time of the emission (lifetime), which correlates with the inverse of the transition probability, i.e., a high lifetime corresponds to a low transition probability and vice-versa. For reference, only a simple discussion will be given. As mentioned before the triplet to singlet transition is forbidden in a pure organic emitter, and must be overcome by spin–orbit coupling. In fact, the increase of the spin–orbit coupling magnitude increases the relaxation of the forbidden transition rule. This must be accompanied by an increase of the respective transition probability, which in turn, decreases the transition lifetime. Recalling the information about the spin–orbit coupling involving the MLCT and LC states, we know that the coupling is strong when a MLCT character state is involved (as a matter of fact, the spin–orbit coupling is not expected with the "pure" LC state). So, in conclusion, as the MLCT character of the emitting energy level increases, the spin–orbit coupling and transition probability also increase, the corresponding lifetime decreases. Thus, the phosphorescence lifetime measured by time resolved spectroscopy helps assess the MLCT character of the emitting level. Fast transitions are correlated with a dominant MLCT character and slow transitions will be the contrary. However, it must be noted that the data must always be compared with results from well-known TM complexes to avoid misinterpretations.

In efficient TM organic complexes low lifetimes are usually observed (in order of about a few to hundreds of μs) in contrast with those typical of a merely organic emitter (up to some ms). These TM complexes are remarkable useful for application in OLED active layers.

Under electrical excitation, the process has a markedly different excitation mechanism, as previously noted in chapter 4 for the RE organic complexes. Some aspects are similar in particular same need to have a bipolar nature on the active layer of the OLED to allow the transport of electrons and holes (not necessarily with the same physical properties). When such bi-polarity is not minimally balanced, the emitting complex must the embedded as a dopant into a conductive host matrix.

In an electroluminescent device, the carriers are injected from the electrodes (electrons from cathode and holes from the anode) and recombines in the active layer comprising the emitting molecules. This exciton recombination provides excitation to the emitting molecule, with further emission of light (electroluminescence) as described for the photoluminescence. Figure 5.12 shows a schematic electroluminescence process for the $Ir(ppy)_2ACAC$.

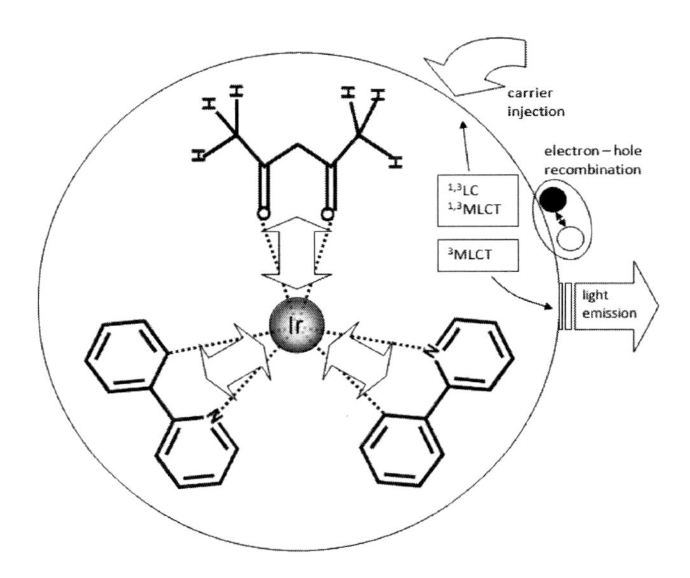

Figure 5.12 A typical electroluminescence process for the $Ir(ppy)_2ACAC$. Under an applied voltage electrons (and/or hole) are injected into the electronic energy levels of the molecule, followed by a recombination and light emission.

As expected, the electron–hole pairs promoting the electroluminescence are not specifically interacting with any of the molecule part, although they can be "confined" in some organic fragments.

In the system described above, the TM ion can be viewed as the modulated on of energy level as the contribution of its *dd** orbitals will changes all the electronic levels diagram, in conjugation with the organic molecular part.

The choice of the organic ligands, besides fulfilling the requirements on the location of the energy levels and on the magnitude of the ligand field, is also commonly made to achieve the color tuning effect. For instance, in the case used for the previous discussion, the $Ir(ppy)_2ACAC$, we can easily compare the well-known results with the "parent" homoleptic[d] complex, $Ir(ppy)_3$. The substitution of ppy group by an ancillary ligand, gives the heteroleptic[e] complex with a general chemical formula of $Ir(ppy)_2LX$, bringing differences on two emission features. The first is a more precise location of emission band peak, which becomes less dependent on the surrounding media of the emitter; the second, and perhaps the most interesting, is the change of the emission maximum with the type of ancillary ligand used [41]. This is associated with the field strength of the ancillary ligand (and more stable to the TM complex emitter environment condition). For instance, emission of the heteroleptic $Ir(ppy)_2pic$ (where pic stands the *picolinate*) is slightly blue shifted when compared to the emission of the heteroleptic $Ir(ppy)_2ACAC$ [42], due to the higher field strength of the ACAC ligand compared to the pic one. The high field strength of an anionic ligand tends to lower the bandgap, reducing the emission wavelength. Compared to the homoleptic complex, $Ir(ppy)_3$, the ACAC complex features no apparent differences pointing in a first and crude approximation, to the absence of bandgap changes and overall suggesting the same field strength for both ligands.

A final note on the TM complex luminescence efficiency will now be given. Due to the triplet harvesting concept and the formation of whole molecular energy levels with contributions from the organic levels and the TM ion, the appearance of a triplet emitter with MLCT character affords a relatively high phosphorescence efficiency (quantum yield), as high as the stronger the MLCT character of the triplet emitter. In practice,

---

[d] A homoleptic complex is an organic complex where the metal ion is bonding in a coordination sphere composed by only one kind of anionic group. In the case given, is the $(ppy)^{-1}$.

[e] A heteroleptic complex is then a complex where the metal ion in bonding in a coordination sphere composed by more than one kind of anionic group.

when designing a TM complex molecular structure, the known pathways for increasing the efficiency should be taken into account, specifically the choice of the organic ligands versus the TM ion (and the respective ligand field strength) and the related final derived energy levels for the spin–orbit coupling effects (necessary) and the also related high energy located levels interference (to avoid); and the final molecular geometry.

As regarded, this is not an easy work. Nevertheless, phosphorescence efficiencies up to 100% (complete triplet harvesting) have been achieved [43] allowing the fabrication of the most bright OLED known (luminance over 10^5 cd/m^2) with very low driving voltages (typically in the range of 3–6 V) [44].

5.3.2 Considerations About the Chemical Structure of the Transition Metal Organic Complexes

Besides the physical questions addressed to the organic ligands–TM ion match necessities, the practical interests, for specific applications include (i) color tuning, (ii) high efficiency (for even high brighter OLEDs), and (iii) chemical stability; and naturally the "interception" of all these areas (physical limitations/ practical interests).

The discrimination of the distinct organic ligands type is not straightforward. For example, a β-diketonate can be used as an ancillary ligand as we have seen in the previous $Ir(ppy)_2ACAC$ example (the ACAC) but a completely different organic cyclometalate ligand (without the typical basic structure of the β-diketonate) can be used in such role, for instance, *phosphane* in some ruthenium complexes. The origin of such difference may lie in the distinct field strength needs for a feasibility of the triplet emitter efficiency. Considering these points, a brief description about the most common organic ligands for TM complexes will be given.

As referred in the section 5.2.2, the basic mechanisms of TM complexes luminescence can be explained considering the organic ligands orbitals $\pi\pi^*$ and the TM ion orbitals dd^*. From such orbitals, four levels are established: $HOMO_\pi$ and $LUMO_{\pi^*}$ and $HOMO_d$ and $LUMO_{d^*}$. These orbitals, responsible for the luminescence properties of the molecule (e.g., determining the ground energy level and the lowest triplet emitting state) are usually called as *frontier orbitals.* The organic ligands must thus have proper frontier orbitals and simultaneously be easily to handle. Moreover, they must attend to the several physical and practical constrains described before.

The cyclometalate ligands of greater interest for TM complexes belong to the four indicated types, namely the C^C, C^N, C^P and N^N. In practice, these cover the most important pre-requisites for successful use in a TM complex, particularly those related with the internal molecule efficiency phosphorescence maximization, color tuning facilities and is relatively good coordination abilities.

Typically we can resume the properties of cyclometalate ligands in two main points:

- The usual ligand strong field with TM ions leads to an increase of the bandgap from the *dd** orbitals to a further reduced probability of having populated high energy *d** states. As discussed before, such population is responsible for the typical phosphorescence quenching by promoting of the non-radiative decay;
- The existence of LC transitions in a near energy location will help the color tuning process.

The choice of the appropriate cyclometalate ligand also depends on the TM ion and/or the path to efficiency maximization and the desired color emission.

The typical differences between the four kinds of cyclometalates ligands are the following (for more detailed information see [45]):

- C^C: typically exhibits a large ligand field strength with further destabilization of the *d** energy states of the MC system, which in general becomes inaccessible to the emitting triplet. This reduces the quenching and can also be useful for designing blue emitters;

- C^N: are interesting for the $\pi\pi^*$ bandgap modulation as the conjugation length can easily be increased or decreased (decrease or increase the bandgap, accordingly). Some anionic cyclometalates ligands can gives strong metal–carbon interactions and a very well stabilized ligand field strength;
- C^P: contain a PPh_2 group to increase MLCT phosphorescence energy. In parallel, they can lead to a relatively strong ligand field strength with some further possibility of destabilization of the d^* energy states of the MC system;
- N^N: have a bandgap from the $\pi\pi^*$ orbitals (LC system) higher than the C^N type. They are also easier to coordinate, which is good for the TM ions with valence 2^+, usually less reactive than the TM ions with valence 3^+.

In Figs. 5.13(A, B, C, and D) structures of the most common cyclometalates are shown.

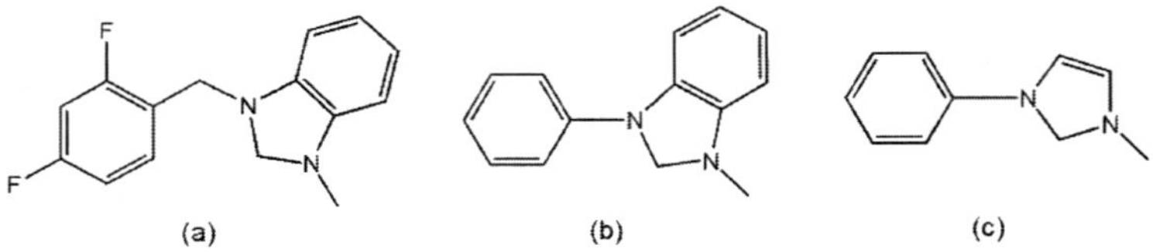

F
F
N
N
(a)
N
N
(b)
N
N
(c)

Figure 5.13 (A) Typical cyclometalates used in transition metals organic complexes, belonging to C^C group. (a) dfb-mb, (b) pmb and (c) pmi.

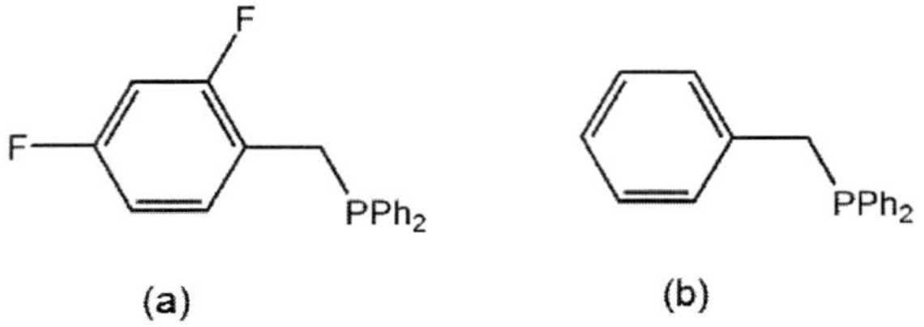

F
F
PPh_2
(a)
PPh_2
(b)

Figure 5.13 (B) Typical cyclometalates used in transition metals organic complexes, belonging to C^P group. (a) dfb–PPh_2 and (b) bz–PPh_2.

Figure 5.13 (C) Typical cyclometalates used in transition metals organic complexes, belonging to N^N group. (a) fppz, (b) fptz and (c) fbpz.

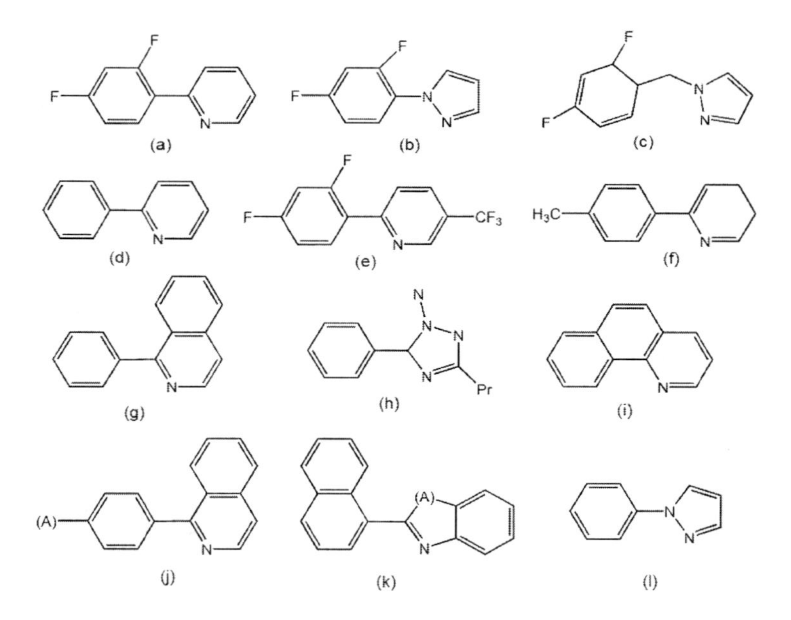

Figure 5.13 (D) Typical cyclometalates used in transition metals organic complexes, belonging to C^N group. (a) dfppy, (b) dfppz, (c) dftb–pz, (d) ppy, (e) dftpy, (f) tpy, (g) nazo, (h) mpptz, (i) bzq, (j) pic (A=H), pic-F (A=F), (k) bon (A=O), α-bsn (A=S) and (l) ppz.

The molecular modification for further emission improvement and/or color tuning can be made by substituting a cyclometalate ligand by an ancillary one. The most common application of organic ligands with this classification is found for iridium complexes; for other TMs, it is accepted that such a specific distinction may not exist and the TM ions organic coordination sphere is composed of ligands belonging to the two groups with particular incidence on the ancillary ones. The coordination sphere is sometimes completed by monodentate ligands (not bridged) with the purpose of increasing the bandgap of the MC transition. In Fig. 5.14 the usual organic ligands acting as ancillary are shown.

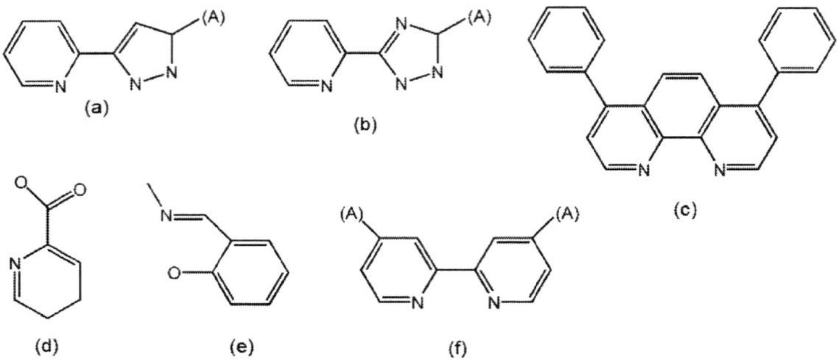

Figure 5.14 Usual organic ligands that can acts as ancillary or be used in the main coordination sphere of some TM complexes. (a) fppz (A=CF_3), bppz (A=*t*Bu), (b) fptz (A=CF_3), bptz (A=*t*Bu), (c) ph_2phen, (d) pic, (e) sal and (f) dpphen (A=Ph), tbbpy (A=*t*Bu). Adding to this list, the ACAC and DBM (Figure 4.9) and bipy and phen (Figure 4.10).

Specific cases of others ligands less known will be described later.

The modulation of the luminescence behavior TM complex, i.e., the overall energy levels location and the respective interactions (spin–orbit coupling and the interference of the MC states) are strongly dependent on the organic ligands employed. Nevertheless there is not a simple way of knowing, *a priori*, which ligand to use. Currently, the most realistic theories presents face to collected experimental data, point to a few basic rules, followed by major group of chemists in the preparation of new TM complexes. They can be summarized as follows:

- TM ions with weak ligand field must be coordinated with organic ligands with strong ligand field. This is not just a matter of chemistry; it is also useful to help increasing the MC transitions bandgap;

- Electrically charged ligands will provide a stronger ligand field. In our case, as the TM has a positive valence, the best organic ligands must be anionic. This question is easy to understand (basically these are electrical charges interactions) and further helps move the energy of the corresponding d^* levels to higher values. Also, by electrostatic attraction, these ligands are expected to establish a stronger more stabilized bond with the TM ion (typically towards it). As an example, ppy (Figure 5.13 (B)) is expected to have a higher field strength than bipy (Figure 4.10);
- If the organic ligands do not provide a ligand field strongest enough to raise the energy levels of the d^* orbitals (increase the MC bandgap), ancillary ligands can be specifically used for that. Such ancillary ligands must have a strong ligand field (as naturally expected);
- In TM ions possessing low ligand field strength, we expect that the bandgap of the MC transition to be relatively low. If even organic ligands with strong ligand field cannot yield a high-efficiency phosphorescence at high energy spectral emission (i.e., towards blue), one way to modify the energy levels of the whole TM complex is to use organic ligands which can acts as acceptors (please note that this does not mean a cationic ligand, in general, they are electrically neutral like carbon monoxide);
- C^C cyclometalates tend bind stronger than the other kind of cyclometalates; but once again such characteristic may be conditioned by the TM ion and the eventual presence of an ancillary ligand.
- The limits for the organic ligand field strength are, at least, two: (1) the influence of their LC bandgap (that interferes with the whole TM complex molecular energy levels) and the desired color emission; and (2) the instability of the $LUMO_{d^*}$ orbitals and its influence on the phosphorescence efficiency (decreasing of the MLCT character of the triplet emitter). Others questions, as the molecular geometry and chemical stability must also be taken into account.

There are specific tables to be used in this process correlating all aforementioned topics. As our knowledge only some standard models will be available. For the unknown or unusual ligands, only the experimental data (eventually supported by theoretical models computation) can be really useful.

5.3.3 The Photophysics of Transition Metal Organic Complexes

5.3.3.1 The general concept

The interest of the TM complexes photophysics is undeniable. This study is absolutely necessary due to the extremely complex energy absorption/emission processes of TM complexes, in particular, the identification of the molecular energy, involved in the transitions, as well the lifetime measurements from time resolved spectroscopy to get insight on the transition probability magnitude (even if only in a qualitative way). All the collect data from all the photophysics experiments will be critical to evaluate the phosphorescence efficiency and to gather ideas for further OLED development.

Currently the photophysics of the TM complexes becomes clearly more important than the corresponding studies in the RE complexes, as the mechanisms for high efficient complex emitter are remarkable more intriguing in the TM complexes as their more sensibility to "small" factors like molecular geometry, organic ligand field strength and reactivity, as well the molecular levels location and the respective quantum physical properties.

The TM complexes, due to the fact that their photophysics process are dependent on energy levels formed by the "mixture" between organic and metal ion orbitals, exhibits in both absorption and emission phenomena large bands, typically unstructured bands revealing the organic character involved. The spectral shapes are clearly featureless, although some structure can be observed corresponding to the absorption/emission from the LC and MLCT energy states. But even such "featured" bands are always wide although clearly distinguishable in the whole spectral band. In certain conditions, typically at low temperature, some vibronic progression can be observed but for the assessment of the interest for OLEDs (working at room temperature) these spectral features aren't important (and often not observable).

The emission spectrum (both in the photoluminescence and the electroluminescence mechanisms) will be governed by the lowest excited triplet state (the triplet emitter) and is usually ascribed to a LC/MLCT mixed character. However, regarding practical applications, the two most important features are the energy of the emission (i.e., the emission color) and the efficiency. For the former point the OLED structure will be important.

5.3.3.2 Absorption and emission spectra

This section will show representative photo-physical data concerning the most interesting TM complexes for application as active layers in visible spectral range OLEDs. Special attention will be given to the color tuning possibilities and the related molecular structure. Special (and unusual) TM complexes will also be mentioned, in the section dedicated to the OLEDs itself.

One of the most interesting TM ion to be used in OLEDs (and much because of its use in LECs) is ruthenium in the valence 2^+, the Ru(II) (only matched by the interest in iridium). The typical molecular geometry is octahedral. The usual emissions of Ru(II) complexes changes from a yellow–orange to a deep red. Nevertheless the most common emission (with strong efficiency) is the reddish–orange.

In Figure 5.15 the absorption spectra of two Ru(II) complexes are shown. These complexes are $[Ru(bipy)_3]^{2+}$ and $[Ru(ph_2phen)_3]^{2+}$ using PF_6 as counterion to make the complexes electrically neutral and contributing to the rearrange of their levels and to an increased phosphorescence efficiency (as discussed before). Thus the final Ru(II) complexes will be $[Ru(bipy)_3](PF_6)_2$ and $[Ru(ph_2phen)_3](PF_6)_2$.

The absorption bands in the near UV–visible spectral region are not much different but we can observe that for $[Ru(ph_2phen)_3](PF_6)_2$ a broader band is observed. As this broadening also occurs towards the low spectral energy region (i.e., towards red) we can conclude from this low energy absorption that the emitting levels will have a higher MLCT character than in $[Ru(bipy)_3](PF_6)_2$. In fact, it is widely accepted that the Ru(II) complexes have an emitting triplet energy level essentially of MLCT character. The ph_2phen ligand seems to stabilize such nature more. The LC absorption bands are not visible and must fall below 350 nm.

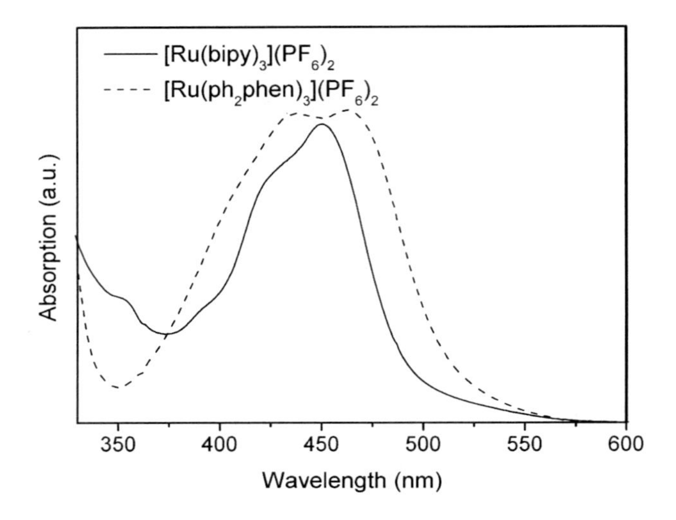

Figure 5.15 Absorption spectra of the $[Ru(bipy)_3](PF_6)_2$ and the $[Ru(ph_2phen)_3](PF_6)_2$ complexes at room temperature. Is only observed the absorption ascribed to the MLCT energy states. Reprinted with permission from Santos, G., et al. (2008) Development and characterization of lightemitting diodes (LEDs) based on ruthenium complex single layer for transparent displays, *Phys. Stat. Sol. (A)*, **205**, 2057–2060.

The photoluminescence spectra are shown in Fig. 5.16. The emission is also a broad band (clearly featureless) attributed to a triplet emitter with a very strong MLCT character. Some authors, inclusively do not distinguish any level character and simply describe the Ru(II) triplet emitter especially for complexes employing the bipy as organic ligand, as the actual case, as being the ^{3}MLCT level [47].

The photoluminescence spectra (excited at 450 nm, i.e., in the maximum of the absorption bands shown in Figure 5.15) reveal that the luminescence of $[Ru(ph_2phen)_3](PF_6)_2$ is slightly shifted to lower energy spectral region in comparison with $[Ru(bipy)_3](PF_6)_2$. As discussed before this represents a stronger ligand field, tending to lower the bandgap and to shift the emission towards the red spectral region. Although in a small amount, but not changing, in an observable way, the emission CIE color coordinates when employed in an OLED, this clearly shows how different ligands can change the emitted color. In any case, such variation of the energy location of the triplet emitter is expected in both Ru(II) complexes because a ph_2phen (see structure in Figure 5.15) has a more extended conjugation length than the bipy (refer to Figure 4.10), though both are electrical neutral ligands. From this last observation, we can conclude that the ligand field of these ligands should be weaker (in agreement with the typical Ru(II) relatively low MC bandgap) than that of the electrically charged ones (i.e., anionic); and the more high conjugation length of the *ph_2phen* compared to the bipy predicts a less molecular bandgap for its Ru(II) complex.

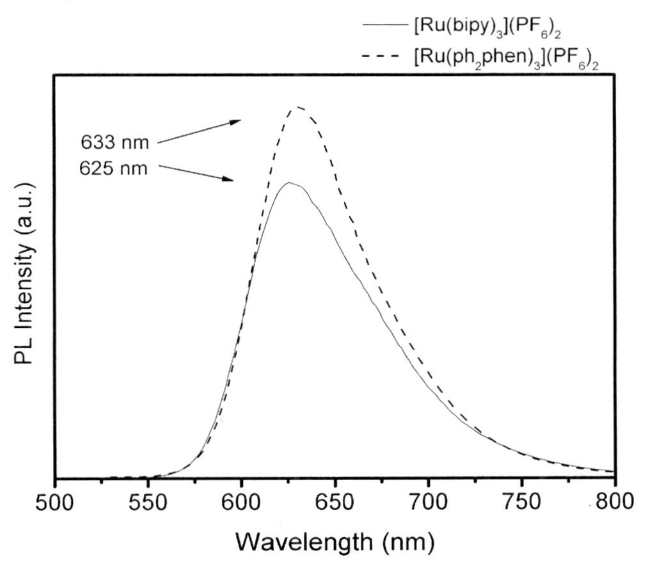

Figure 5.16 Photoluminescence spectra of the $[Ru(bipy)_3](PF_6)_2$ and the $[Ru(ph_2phen)_3](PF_6)_2$ complexes at room temperature. Excitation wavelength of 450 nm. The PL intensity should not be directly compared.

Another interesting TM element for OLEDs applications is rhenium. It usually occurs in the valence 1^+, i.e., Re(I) and forms complexes with octahedral geometry.

One typical Re(I) complex is the *fac*-[ClRe(CO)$_3$(bipy)] (a *facial* isomer) with ligands which are electrically neutral but have the strongest ligand field used in TM complexes (the *CO* group) together the well know *bipy*. Figure 5.17 shows the absorption/photoluminescence spectra.

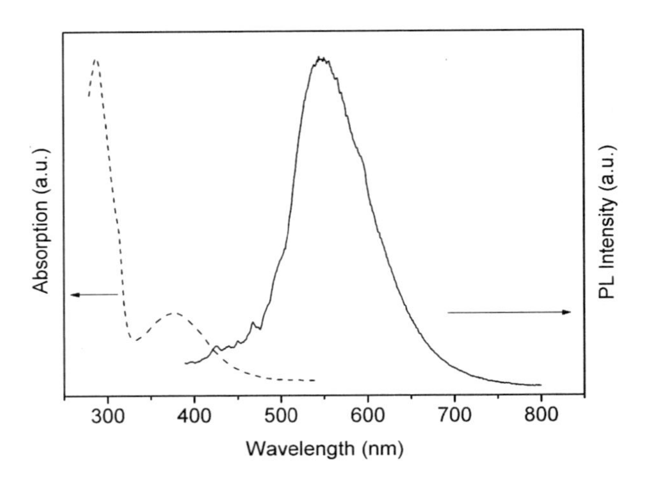

Figure 5.17 Absorption (dashed line) and photoluminescence (full line) spectra of a fac-[ClRe(CO)$_3$(bipy)] complex. The photoluminescence spectrum was obtained under excitation at 370 nm at room temperature. Reprinted from *Journal of Non-Crystalline Solids*, 354(19–25), G. Santos, F.J. Fonseca, A.M. Andrade, A.O.T. Patrocínio, S.K. Mizoguchi, N.Y. Murakami Iha, M. Peres, W. Simões, T. Monteiro, L. Pereira, Opto-electrical properties of single layer flexible electroluminescence device with ruthenium complex, 2571–2574, Copyright (2008), with permission from Elsevier.

The figure shows the typical MLCT states absorption (in the 330–450 nm) with absorption increase when the wavelength decreases. This high energy absorption component can be attributed (as in the previous example of Ru(II)) to LC absorption and perhaps, for the small shoulder at near 325 nm, to the absorption of the ^{1}MLCT energy state. In particular, the strong absorption at near 290 nm is widely accepted to be result from excitation of the LC energy states. The photoluminescence spectra show the typical broad emission band, peaked at near 570 nm (yellow) and commonly attributed to a triplet emitter with a strong MLCT character. As in the case of the Ru(II) complexes, several authors only attributes such emission to the ^{3}MLCT energy state.

Although the Fig. 5.17 shows the usual yellow color associated with Re(I) complexes it's also common to observe a red emission due to the use of organic ligands with higher field strength than bipy. These cases will be explored in the TM-based OLEDs section.

Without any doubt, the iridium-based TM complexes are the most extensively studied due to the easy color tuning by changing the organic ligands, allowing emissions from the blue to red, i.e., all the main spectral visible region. The iridium is typically coordinated with a valence of 3^+ i.e. Ir(III) in an octahedral molecular geometry. Nevertheless, complexes with valences 2^+ and even 1^+ are not unusual.

Perhaps the most known Ir(III) complexes are homoleptic ones, as $Ir(ppy)_3$ and the well studied FIrpic. The later exhibits an interesting blue emission and is often used in OLEDs active layers. An interesting result was recently obtained by [48] using an heteroleptic Ir(III) organic complex to show the influence of the ancillary ligands on the color tuning process. The Ir(III) complexes synthesized were $[Ir(ppy)_2(bipy)](PF_6]$, $Ir(ppy)_2(ACAC)$ and $[Ir(ppy)_2(trpy)](PF_6)$, allowing the color emission to change from red, on the first one, to blue (the last one) returning to the red when the ancillary ACAC ligand was used. The absorption and photoluminescence spectra are shown in Fig. 5.18.

The absorption spectra did not shows many differences between the complexes on with other TM organic complexes, indicating that a similar behavior will be found in the whole molecules LC/MLCT absorption mechanisms. As usual, the absorption low spectral energies must correspond to the MLCT absorption region (less energy and less intensity while the high energy spectral region corresponds to the LC levels absorption.

From the photoluminescence spectra, and considering the previous discussion, it is evident that the trpy complex must have the triplet emitter level of higher LC character (shifted towards the blue spectral region) while the complex with bipy must have the triplet emitter level of higher MLCT character (shifted towards the red spectral region). Moreover, the ligand field strength seemed to be ordered as trpy $<$ ACAC $<$ bipy, as previously discussed.

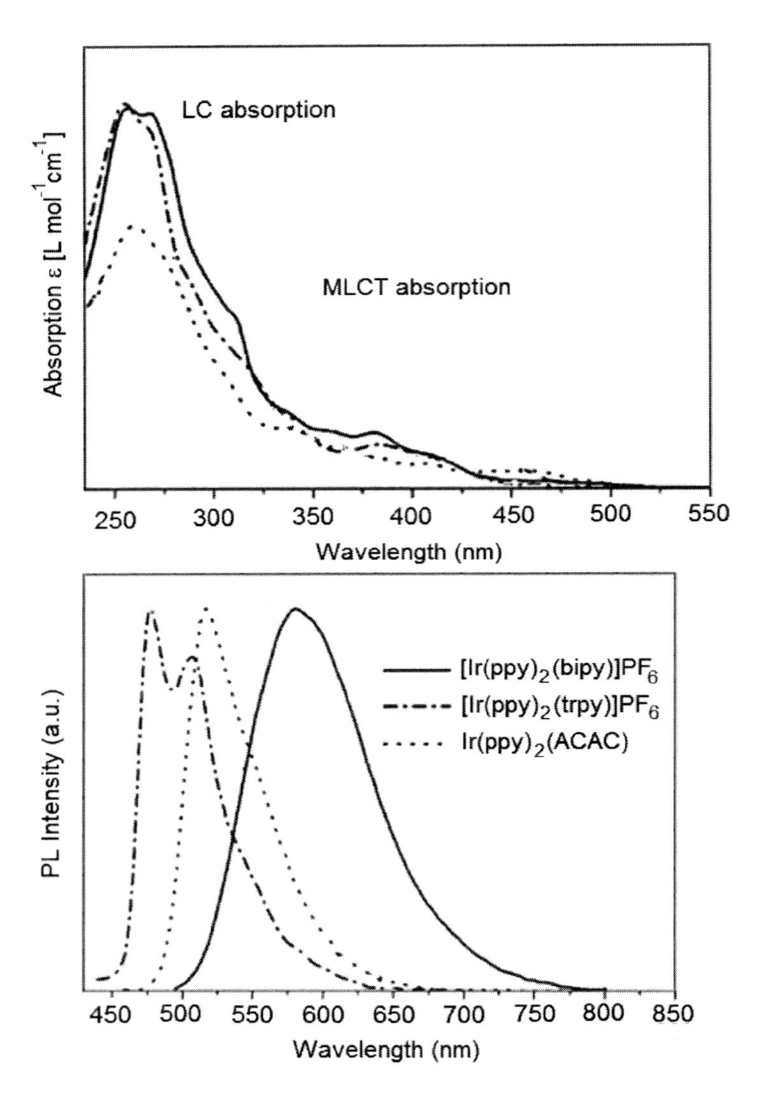

Figure 5.18 Absorption and photoluminescence spectra of three heteroleptic Ir(III) complexes in CH_2Cl_2. Although without significant differences (besides the intensity of the absorption in the LC region), the photoluminescence spectra shows clearly the effect of the different ancillary ligands in the color tuning process. Room temperature. Reprinted with permission from Beyer, B. (2009) Phenyl-1H-[1,2,3]triazoles as new cyclometalating ligands for iridium(III) complexes, *Organometallics*, **28**, 5478–5488. Copyright 2009 American Chemical Society.

Two other TM elements that have been gained a superior interest in recent years are the osmium and platinum. After some disperse work in past, the novel TM organic complexes based on these ions exhibits relatively high phosphorescence emission efficiencies, although not as high as the common TM ions like Re(I), Ru(II) and specifically the Ir(III), that are enough to bring a new interest on them.

Concerning the osmium, the most common valence is 2^+, i.e. Os(II) with a few interesting Os(III) complexes, and as usual with octahedral molecular geometry. The most common emission from Os(II) complexes is the red one. In Fig. 5.19 the absorption and the photoluminescence spectra of an $Os(bptz)_2(PPh_2Me)_2$ are shown [49].

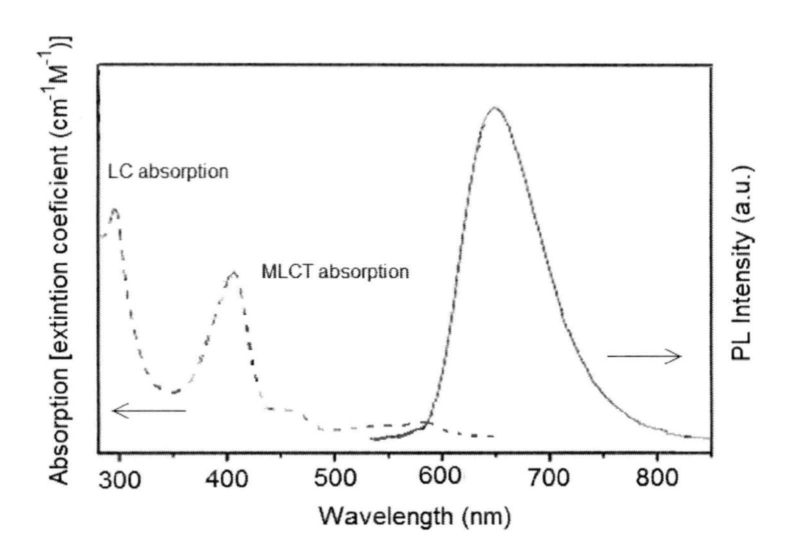

Figure 5.19 Absorption and photoluminescence spectra of an Os(II) complex, the $Os(bptz)_2(PPh_2Me)_2$. Room temperature. Reprinted with permission from Tung, Y.-L., et al. (2004) Highly efficient red phosphorescent osmium(II) complexes for OLED applications, *Organometallics*, **23**, 3745–3748. Copyright 2004 American Chemical Society.

The absorption spectrum shows the typical LC (UV spectral region) and MLCT (near UV-visible spectral region) absorption bands. The authors attribute the UV absorption band to the ^{1}LC absorption transition (as referred before it is allowed by spin multiplicity) and the absorption at near 400 nm is attributed to the ^{1}MLCT absorption transition (also spin allowed). The absorption bands extended to the visible spectral region (at near 450 nm and 550 nm) are attributed to the ^{3}MLCT absorption transitions. As a matter of fact, there aren't many overall differences in the absorption behavior of all TM organic complexes of interest. This question is not a surprise if we take into account the discussion concerning the whole TM complex molecular energy levels formation.

The photoluminescence spectrum corresponds to a red emission centered at near 650 nm. From the authors such emission is due to a triplet emitter with a strong MLCT character (once again is simply designed as emitting MLCT level). Such high MLCT character of the lowest triplet emitting level is expected from an emission in the red spectral region, and is also in agreement with the previous discussed emissions from others TM complexes emitting is this color.

Some Os(II) complexes (and in some cases the OS(III)) complexes, too) can emit in the yellow or green with relative ease. Usually such high energy emissions are achieved through the LC emitting level and depends much on the spin–orbit coupling effect (introducing some MLCT character) to achieve a useful efficiency.

Also less studied, platinum is becoming an element of interest for TM complexes, by its relative ease in obtaining a blue–green emission as well the yellow, yellow–orange and red colors. Such apparently wide spectral color emissions have been attracting much attention of the OLEDs researchers.

Besides the well known PtOEP (a typical red emitter), some homoleptic complexes with a wide color tunic capabilities were recently studied. Contrarily to the previous TM complexes, the platinum, typically coordinated in a valence 2^+, i.e. Pt(II)) doesn't present an octahedral molecular geometry but belongs to the D_{4h} point group, i.e. forms planar (quadratic) complexes. Recalling the previous discussion on this molecular geometry, the Pt(II) organic complexes are expected to have a low spin–orbit coupling and therefore the triplet emitting level will have less MLCT character. Nevertheless, the results are very promising. And considering that even the octahedral complexes often present distorted geometries, the eventual negative influence of a planar geometry on the emission may not be of great relevance.

In Fig. 5.20, the absorption and photoluminescence of two interesting homoleptic Pt(II) complexes, the $Pt(bzpz)_2$ and $Pt(fppz)_2$ are shown [50].

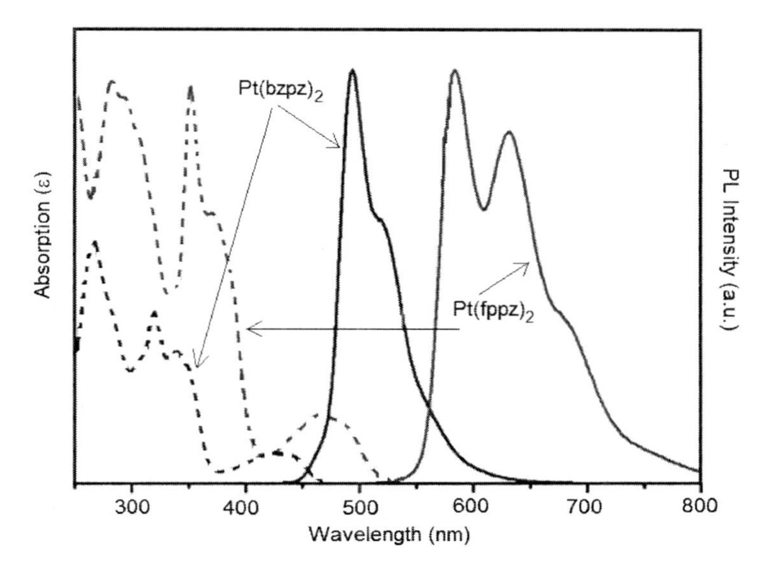

Figure 5.20 Absorption and photoluminescence spectra of two homoleptic Pt(II) complexes in THF. The spectra show clearly the effect of the different cyclometalate ligands in the color tuning process. Room temperature. Reprinted with permission from Chang, S.-Y. (2006) Platinum(II) complexes with pyridyl azolate-based chelates: synthesis, structural characterization, and tuning of photo- and electrophosphorescence, *Inorg. Chem.*, **45**, 137–146. Copyright 2006 American Chemical Society.

In this figure, and according to the author's interpretation, the absorption bands that are observed at low energy, around 416 nm for $Pt(bzpz)_2$ and 460 nm for $Pt(fppz)_2$ are ascribed to the $^1MLCT/^3MLCT$ mixing states, incorporating also an extension of the ^3LC energy states. At high energy, the expected ^1LC absorptions of both Pt(II) complexes are observed. Photoluminescence data was measured in THF solution, which may cause small differences in the band shape and—more importantly—a slight peak shift due to the chromatic effect of the environment rigidity. In spite of that, the effects of the ligands in color emission were obvious, as the $Pt(bzpz)_2$ emitted in the green spectral region, while the $Pt(fppz)_2$ emitted in the red. The authors consider explain this behavior, as the combined result of several factors, namely the vibronic feature observed and the higher emission lifetime of the red emitter compared to the green one (and with other complexes emitting in intermediate regions, not shown in the last figure), pointing to an emitting state with a significant ^{3}LC character. As a matter of fact the more structured, the emission band, the stronger the LC character of the emitting level.

Another relevant photo-physical feature of TM emitting complexes for OLEDs fabrication is the efficiency, related with the emission lifetime (and, in turn, to the LC/MLCT character of the emitting energy level). Under electrical excitation the correspondent electroluminescence is typically weaker than the photoluminescence (in particular due to the constrains in device operation) and strongly dependent on the OLED structure. So, maximizing the complex emission efficiency is indispensable.

Among the many different ways to increase the luminescence efficiency and helps on the color tuning process, we should highlight an interesting powerful tool called molecular engineering. It has shown a fast-growing use due to the increasing precision of theoretical calculations and models implementation. Molecular engineering is a tool to modulate the ligands HOMO and LUMO levels individually, by the use of specific substituents or fragments of each individual organic ligand. In the practice, changing the LUMO level appears to be easier due to the nature of the organic ligands—neutral or anionic. This substitution is performed on a segment of the organic ligand possessing the energy level we desire to change. Let us consider an anionic organic ligand divided into two segments, one neutral and another anionic, i.e. where the negative charge is located. It is not difficult to understand that the $LUMO_{\pi^*}$ level must be located at the neutral segment whereas the $HOMO_{\pi}$ level will be located at the negative segment. Making substitutions in the neutral segment will inevitably produce changes in the $LUMO_{\pi^*}$ level without any modification in the $HOMO_{\pi}$ level from the untouched segment. The final result is a change in the whole organic ligand bandgap, with further influence on the LC and MLCT energy levels. Modifications in the ^{3}LC and ^{3}MLCT levels in particular will change the emission energy, i.e. the color. As expected, not all organic ligands can be handled in such way. And, in complexes with a relatively high molecular complexity, discovering the organic segments where the $LUMO_{\pi^*}$ and $HOMO_{\pi}$ levels are located can be hard work, which will have to relay in the theoretical calculations.

5.4 The Light Electrochemical Cell Concept

The Light Electrochemical Cell (LEC) is a different concept for organic light emitter that has gained a strong interest in the past twenty years due to the successfully use of some TM organic complexes in the active layer. These devices are spectacularly simple to fabricate (just one layer sandwiched between two electrodes) and offer extreme brightness output at low applied voltages. Besides a few reported results with iridium complexes, the preferable choice for LECs is ruthenium complexes. Here, a simple description on these devices will be presented. Specific applications will be described in the next section.

The original LEC concept was introduced in 1995 [51]. The idea was to have an active layer formed by a blend of a semiconducting polymer (or conjugated polymer), an ion conducting polymer and a salt. Two distinct theories were developed to explain the LEC mechanism. The first considers that, under high enough applied voltage, the salt ions distributed in the active layer help the oxidation/reduction of the semiconducting polymer at the electrodes interfaces; this redox process creates a p-doped polymer at the anode and an n-doped polymer at the cathode leading to the appearance of ohmic contacts. The second model is based on the formation of high electrical fields near the electrodes due to the redistribution of ionic charges. The contacts then become ohmic due to the electrical field and, contrarily to the first model, no polymer doping near the electrodes occurs. The redox model (the first) is usually more accepted for the description of the LEC mechanism involved in TM complexes.

Besides the low operating voltages and high brightness, the LECs are naturally independently on the electrodes work-function as the mechanism presupposes the formation of ohmic contacts. Interestingly, light emission can be observed under both forward and reverse bias, although such designation appears to be incorrect for a device with both ohmic contacts. Moreover, the LEC driving voltage is often independent on the active layer thickness, but with some constrains. The first one is the device stability under the applied voltages; if it is higher than the imposed limits for electrochemical stability (depending on the active layer composition) device degradation will occur. Another point is the time delay between the voltage application and the stabilized light emission. Such delay (than in some cases can be of dozens of seconds) is directly attributed to the necessity of the ions redistribution in order to support the redox phenomena at the electrodes.

Application of the LEC principle to TM complexes is not difficult but not all complexes are suitable for these devices. LECs require electrically charged TM complexes, typically cationic by use of TM cation and a neutral organic ligand. This electrically charged complex is surrounded by mobile counterions to balance its charge and allowing the redox mechanism, both in solution and in solid state. Naturally, the TM complex will easily be oxidized at the anode (loose an electron) and reduced at the cathode (gain an electron), acting, in a way like a polymer LEC. Moreover, the mobile counterions are already present in the active layer, so additional compounds for electrolyte functions are no longer needed. In addition, the triplet emitter nature of the TM complexes can be viewed as an enormous advantage due to the high luminescence efficiency combined with a very bright emission.

The process can be summarized as the following reactions:

$$\left[\mathrm{TM(II)L_3}\right]^{2+} \rightarrow \left[\mathrm{TM(III)L_2}\right]^{3+} + \mathrm{e} \tag{5.2}$$

$$\left[\mathrm{TM(II)L_3}\right]^{2+} + \mathrm{e} \rightarrow \left[\mathrm{TM(II)L^{-}L_2}\right]^{+} \tag{5.3}$$

$$[\mathrm{TM(III)L_3}]^{3+} + [\mathrm{TM(II)L^{-}L_2}]^{+} \rightarrow {}^{*}[\mathrm{TM(III)L^{-}L_2}]^{2+} + [\mathrm{TM(II)L_3}]^{2+} \tag{5.4}$$

$$^{*}[\mathrm{TM(III)L^{-}L_2}]^{2+} + [\mathrm{TM(II)L_3}]^{2+} \rightarrow 2[\mathrm{TM(II)L_3}]^{2+} + h\nu \tag{5.5}$$

In these equations, TM, L, e, and $h\nu$ represent the TM ion, the organic ligand, the electron, and the light emission, respectively. The explanation of the mechanism can be summarized as follows:

1. At the anode, the TM complex is oxidized (Eq. 5.2) increasing its valence from 2^+ to 3^+ by loss of an electron;
2. At the opposite cathode, the TM complex suffers a reduction process (Eq. 5.3), where an injected electron brings a negative charge the organic ligands (L^-);
3. Both species hop trough the active layer and react with each other to generate light emission. The first step is the transference of one electron from the reduced species to the oxidized one, which becomes in an excited energy state (Eq. 5.4). The other species returns to its initial condition;
4. The final step is the recombination in the excited species with energy emission ($h\nu$) and its return to the initial condition (Eq. 5.5);
5. The cycle continues until the applied voltage is turned off.

Figure 5.21 shows a simple diagram of the LEC working mechanism.

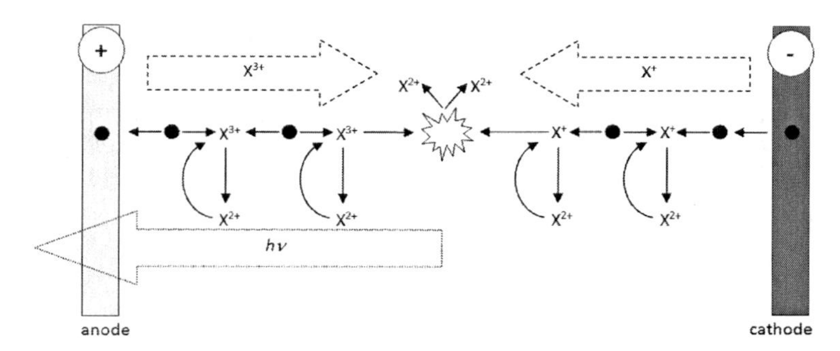

Figure 5.21 Schematic diagram of the Light Electrochemical Cell process considering an entity X with a 2^+ electrical charge.

In the particularly case of TM organic complexes and considering these belong to the definition of "small molecules" given in chapter 2 (even the larger molecules with extended organic ligands) the fabrication methods can include both evaporation technique and a wet process. The best results, to our knowledge, are those obtained by wet process that avoids a possible fragmentation of ligands in the more extended molecules under the thermal evaporation processing. In the case of wet process choice, it's necessary to include a polymer that allows the correct thin film formation. The best usual choice is PMMA (*Poly-methyl-methacrylate*), a dielectric polymer, i.e. it does not interfere in the successive electrical phenomena, and is known to be excellent host for the formation of a perfect film.

Due to the simplicity and high performance, the TM-based LECs (especially those based on Ru(II)) are one of the more interesting fields of TM complexes research. We will focus is point in the next section.

5.5 Transition Metal-Based OLED: The Devices

The TM complexes exhibit a very strong luminescence when the molecular internal efficiency is high enough to achievement of the triplet harvesting. If such property can be use in OLEDs fabrication, and with a correctly projected device structure with sufficient electrical carrier injection and proper confinement in the active layer, a very bright emission is expected As usual in OLED fabrication, the precedent physical study of the optical and electrical properties of the emitting complex is needed. Data will be used in planning the optimal device structure, starting with the selection of the best emitting complex. In this section we will addresses the TM complex-based OLEDs (and LECs) as well the device structure analysis.

5.5.1 The Basic Structure

As expected, the basic structure of OLED employing TM complexes is not much different from those previously discussed, particularly the general structure presented in chapter 2. It also will not be much different from that of RE complexes. Nevertheless, new details are expected, as the electro-optical nature of the TM complexes used in the active layers is quite different. The absolute exception will be the LECs.

In this basic structure of TM complex-based OLEDs we still play with the electrodes work-function, the energy barriers between them and the organic semiconductors, the different electrical carrier mobility and the obvious need to promote electrical carrier confinement. All these points lead to a typical OLED structure employing, three organic layers, the ETL, HTL and EL. Further improvement can be achieved by incorporating hole and electron blocking layers, to guarantee a better efficiency. We still also work based on the previously assumption, that a strong luminescent TM complex will not necessarily give a very bright OLED though, for an inefficient complex the probability of having an efficient OLED is practically zero.

The choice of materials for the ETL, HTL, electron and hole blocking layers will be indexed to their energy levels and those of the active layer, starting the study by the TM complex. The most common materials for sandwich in the active layer are the typical ones as indicated for RE complexes as shown in Figs. 4.19, 4.20, and 4.21. The brief discussion therein presented (section 4.4.1) still applies now, with a few further comments. Considerations on the choice and use of the ETL, HTL, electron and hole-blocking layers, can also be reported to section 4.4.1. In particular, we must stress that, in a real-case scenario, it is impossible to obtain an efficient single layer OLED based on a TM complex (exception for LECs). Also for completely new complexes, not knowing the HOMO/LUMO levels (in the sense of the bandgap physical concept) is an additional problem in the definition of the device structure. Moreover, for the TM complexes, the local molecular conformation in a solid state film after being deposited (regardless of the deposition method) can lead to significant molecular distortions with much more significant influences on the exciton recombination, those traditionally observed for the RE complex-based OLEDs. This implies the refinement of the deposition techniques that nowadays could not be a special problem but needs to be considered.

5.5.2 General Cases of Transition Metal-Based OLEDs: Ruthenium, Rhenium, Iridium, Osmium, and Platinum

Some TM complexes have attracted the attention of several research groups due to their superior performance when used in OLEDs active layers and also their simple synthesis and/or ease in color tuning handling. In this group, ruthenium, rhenium, and, above all, the iridium, are the most studied. Osmium and platinum has begun to appear in the most recent years. And we must not forget the efficient employment in LECs, especially ruthenium. Accordingly, the published results have grown exponentially in the last years requiring a strong filter for literature compilation. For that, and following the already scheme used for RE complexes, we will focus on the main typical results, consisting in a "basic" experimental data, to allow the further understanding of the similar or derived structures.

5.5.2.1 General considerations about using transition metal organic complexes in OLEDs: energy transfer from a host matrix

The principle of the energy transfer from a host matrix to the desired final emitting complex is practical universally consigned. Even in LECs, the use of a host polymer to make a blend with the emitting complex is widely used. Since the fabrication pathway of TM complex-based OLEDs is so disseminated, a few general considerations must be given.

The use of an OLED active layer constituted only by a TM complex is very sparse. As a matter of fact, the technological applications require more and more efficient emitting materials for efficient and bright OLEDs. Under the running conditions of an OLED with its associated constrains (high electrical field disruption, degradation under high electrical power, etc.) one of the best ways to improve the light emission by partial overcoming these drawbacks, is to dope a host matrix (constituted by an organic semiconductor both polymeric or small molecule type) with the emitting TM complex. An active layer only constituted by the complex does not guarantee a good efficiency and light emitting efficacy, except for a few particular cases, but in these the electrical carrier mechanisms and potential barriers can be a problem.

The physical nature of these blended structures can address several issues. The first, and most important, is to promote the energy transfer from the host to the emitting complex, and the transference must be large enough to afford efficient emission, simultaneously avoiding the host emission. The second, is to helps balance the electrical carriers (both holes and electrons) in the active layer increasing the recombination and (in most cases) lowering the driving voltage. Finally, when sandwiched with proper HTL and ETL, it can contribute to the electrical carrier confinement in the active layer, with recognized benefits.

The choice of this TM complex active layer blend can also be supported by a simple finding. Observing the general absorption spectra for practically all TM complexes, is easy to see that the typical MLCT absorption bands region falls in the near UV to visible spectral region (typically blue, blue–green and less commonly green). Recalling the considerations for the energy transfer mechanism in RE complex-based OLEDs, an efficient energy transfer first requires good overlap between the emission band of the host material and the absorption band of the final emitting material. In the case of TM complexes, the general rule is the opposite of which happens with RE complexes (that essentially have absorption almost in the UV and exceptionally in the UV-near visible spectral region), the TM complexes can be easily excited with very common and popular organic semiconductors like PVK and CBP which emit in the blue spectral region, exactly over the TM complexes MLCT absorption bands (compare Figs. 4.13 and 4.15 to 4.18 with Figs. 5.15 and 5.17 to 5.20). Naturally, we are not regarding the LC absorption as it falls clearly into the UV region. Nevertheless, if a strong population the MLCT energy states are obtained, we expected an increase in the phosphorescence emission (excluding the emitting triplet states with a strong LC character but such states have low transition probability as discussed before). So, this situation clearly opens many possibilities to explore the energy transfer process in these complexes, and, may we stress, with good practical results.

The efficiency is also strongly dependent on the TM complex doping concentration, so many efforts in physics research of TM complexes-based OLEDs are presently focused on determining the efficient doping concentration for the OLED emission.

As discussed in chapter 4, the energy transfer from the host material to the emitting one can be made by Förster and Dexter (or both) mechanisms. In some situations the electrical carriers can also be trapped into the final emitting complex after injection/ transport by means of the host material.

Also, the basic mechanism associated with the Förster energy transfer process is the dipole–dipole interaction, which, in organic materials, involves the energy transfer via singlet excited energy states. In the Dexter mechanism, related with tunneling effects (directly related with the overlapping of the wavefunctions of both energy states involved) both singlet–singlet and triplet–triplet pathways can contributes for the energy transfer.

On the other hand, the electrical carrier trapping mechanism can be viewed as a sequential hole (first) and electron (after) trapping in the emitting complex. For this mechanism to be efficient the HOMO level of the emitting complex must be located above (low energy) the host HOMO level and, naturally, the LUMO level (of the emitting complex) must be located bellow (high energy) the host LUMO level. In fact, when a hole is trapped in the HOMO level of the emitting material, the molecule enters an excited stated with 1^+ electrical charge, becoming a cationic excited entity, acting as an excellent trap for electrons from the host material. As expected, this electrical carrier trapping requires a good overlap between both host and emitting complex orbitals.

In general, both Förster and energy trapping on the emitting complex are the most common mechanisms of the energy transfer form a host material to a TM complex. The study of the OLED electroluminescence with the concentration of the emitting complex into the host matrix is an excellent tool to get insight on the energy transfer mechanism as we expect, for electrical carrier trapping and relatively low concentration of the emitting complex, to observe the host emission when the applied voltage increases as a consequence of the triplet emitting level of the TM complex becomes saturated.

5.5.2.2 Ruthenium-based OLEDs

Considered by many authors as one of the most interesting TM element for organic devices applications (OLEDs, LECs and dye-sensitized photovoltaics) application of ruthenium in light emitting devices ascends to the 1980 decade [52]. The versatility of Ru(II) complexes in both OLEDs and LECs employments allied with the simple devices structures (even in OLEDs) with strong bright and high efficiency, makes Ru(II)-based light emitting devices serious candidates for solid state lighting.

The most wide spread application of the Ru(II) in light emitters is on LECs or as dye for doping a device active layer. Since the best known experimental data are about the LECs, we starts this discussion about Ru(II)-based devices with that kind of light emitter.

The simplest Ru(II) LEC device is formed by a single layer sandwiched by two electrodes. As discussed before, the work-function of such electrodes is not important as the redox mechanism near the electrodes will provide an electrical ohmic behavior.

One well known Ru(II) complex that is very suitable for electroluminescence is the $[Ru(bipy)_3](PF_6)_2$, usually blended with the dielectric polymer PMMA to improve the thin film layer formation by wet process (and at laboratory typically by spin-coating although for many educational purposes a spin-casting technique is also viable). Under the experimental conditions used [46] the total layer thickness was around 100 nm. The general LEC scheme is shown in Figure 5.22.

The electroluminescence and photoluminescence spectra are shown in Fig. 5.23. For electroluminescence, the CIE color coordinates are included.

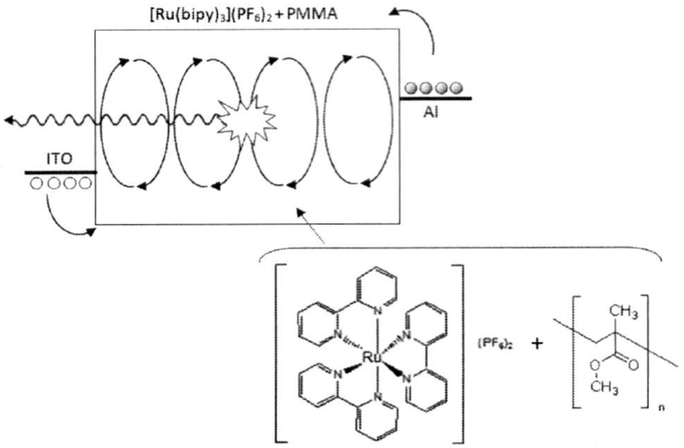

Figure 5.22 Scheme of a Light Electrochemical Cell with an active layer formed by $[Ru(bipy)_3](PF_6)_2$ (with PMMA for film formation). Bellow the molecular structures of the Ru(II) complex and PMMA. Although the differences between the work-functions of ITO and Al (indicated) there is no influence on the LEC behavior.

Several comments must be made. Firstly, the electroluminescence spectrum reproduces the photoluminescence spectrum with exception of a very small shift towards the high energy spectral region (less than 5 nm) without special significance. As discussed, this broad band emission is essentially ascribed to an emitting triplet state with a strong MLCT character. This can be confirmed by the high radiometric wall-plug efficiency obtained. Secondly, a low driving voltage is obtained (about 3.5 V, typically 5 V) and a large operating region without degradation, indicating a good electrochemical stabilization. This data is of major interest for potential applications in solid state lighting. Thirdly, a curious result about color purity has obtained. On the CIE color coordinate chart, we can see that the (x, y) locations (0.61, 0.39) indicate a very pure color. Although we are in the presence of a large emission band (and therefore not "monochromatic") this result brings an additional interest for this device. Finally, and on the other hand, the response or delay time (the time between the external voltage is applied and the light emission stabilizes) is about 1 minute (and between 30 and 40 seconds for the LEC starts to emit). This is a typical time, though high values can be obtained, and for specific applications it can be a real problem.

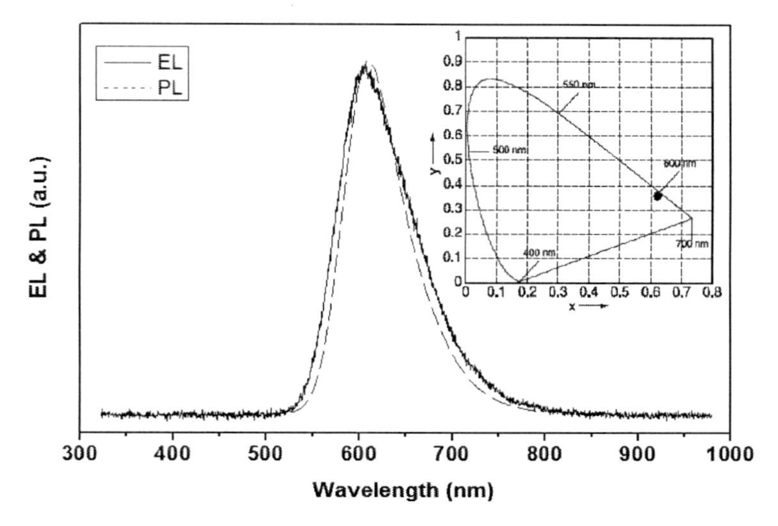

Figure 5.23 Electroluminescence spectrum of the ITO/[Ru(bipy)$_3$](PF$_6$)$_2$/Al LEC under a driving voltage of 5 V. The photoluminescence of the Ru(II) complex is shown for comparison. Inset: the CIE color coordinates. Reprinted from *Journal of Non-Crystalline Solids*, 354(19–25), G. Santos, F.J. Fonseca, A.M. Andrade, A.O.T. Patrocínio, S.K. Mizoguchi, N.Y. Murakami Iha, M. Peres, W. Simões, T. Monteiro, L. Pereira, Opto-electrical properties of single layer flexible electroluminescence device with ruthenium complex, 2571–2574, Copyright (2008), with permission from Elsevier.

In Fig. 5.24 the optical power vs. electrical power (again obtained only for the electroluminescence maximum wavelength near 625 nm) is shown. The radiometric wall-plug efficiency can be extracted from this plot as described in chapter 3.

The wall-plug efficiency for the maximum emission wavelength is of 3×10^{-2}%, a value one order of magnitude higher than the usual efficiencies on RE complexes-based OLEDs. Integrating the entire electroluminescence band a value near 0.17% is achieved. It must be noted that this wall-plug efficiency is measured considering only the light emitted from the normal LEC surface (and as previously discussed all the others contribution are not computed) considering such surface a Lambertian shape emitter. With an integrating sphere, the radiometric efficiency would clearly be much higher and the total emitting optical power could easily achieve some dozens of mW. This simple extrapolation data evidences the superior interest of Ru(II) complex-based light-emitting device.

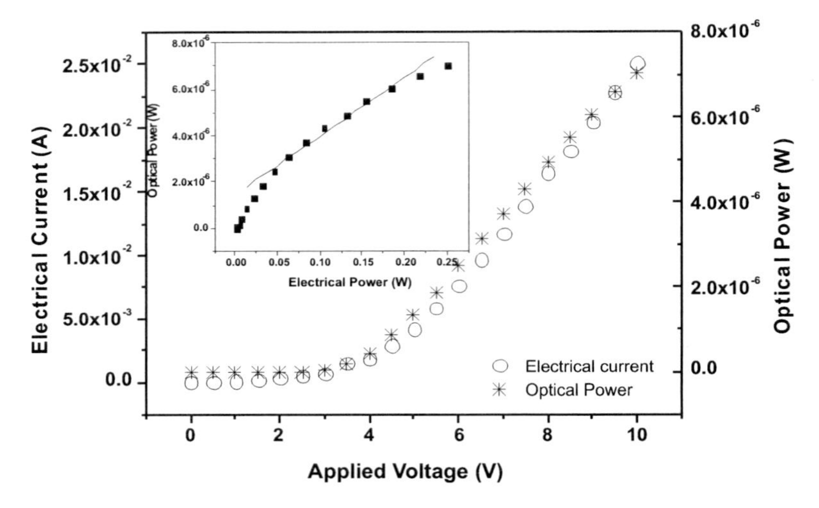

Figure 5.24 Electrical current and optical power (@ 625 nm) for the ITO/$[Ru(bipy)_3](PF_6)_2$/Al LEC with the applied voltage. Inset: the optical power vs. electrical power. The straight line is the linear fit in the dynamic LEC region allowing the calculation of the radiometric wall-plug efficiency. Reprinted with permission from Santos, G., et al. (2008) Development and characterization of lightemitting diodes (LEDs) based on ruthenium complex single layer for transparent displays, *Phys. Stat. Sol.* (*A*), **205**, 2057–2060.

The main advantage of Ru(II) complexes for LEC applications is their anionic nature, with the corresponding surrounding counterions. In the previous complex, the charged molecules, i.e. the $[Ru(bipy)_3]^{2+}$ are surrounded by the $(PF_6)^-$ anions. Much of the necessary time for the emitting light stabilization (the aforementioned delay time) depends on the redistributions of such ions (to promote the redox mechanism at the electrodes) and different anions can be used as we will see later.

Others Ru(II) complexes have also been extensively studied. An interesting example is the $[Ru(ph_2phen)_3](PF_6)_2$. Using the same device structure, a low wall-plug efficiency is obtained [46]. In Fig. 5.25 the molecular structure and the electroluminescence spectrum (maximum wavelength near 630 nm) is shown.

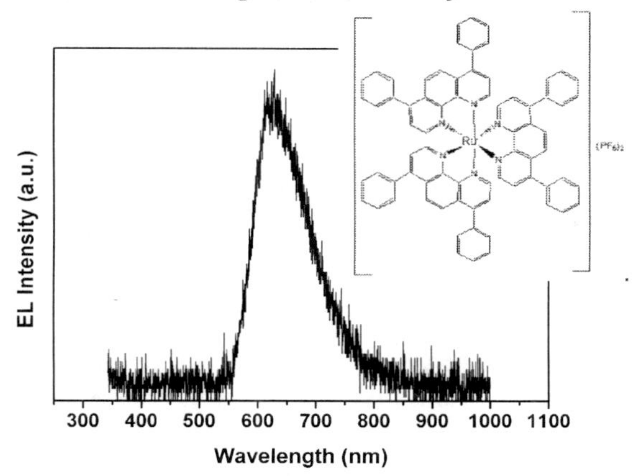

Figure 5.25 Electroluminescence spectrum of the $ITO/[Ru(ph_2phen)_3](PF_6)_2/Al$ LEC under a driving voltage of 3 V. Inset: the Ru(II) molecular structure.

Again the electroluminescence spectrum reproduces the photoluminescence one. The CIE color coordinates is (0.60, 0.38) indicating also a very good color purity. The lowest electroluminescence intensity (not only in the spectrum shown in last figure due to the low applied) but also observed in general and reverberated into the radiometric wall-plug efficiency (near one half than the obtained for $[Ru(bipy)_3](PF_6)_2$) indicates that Ru(II) complex with ph_2phen organic ligand has a low overall performance. Nevertheless, the electrochemical stability remains excellent. Concerning the electroluminescence performance, we expect, as discussed for the photo-physical properties, the ph_2phen to have slightly higher phosphorescence efficiencies than bipy, due to the stronger MLCT character of the emitting triplet. Nevertheless, the expected behavior was not observed under electrical field excitation (contrarily to our experiments on photoluminescence, which appeared to follow the prevision). The explanation can be, in a first attempt, addressed to the different molecular conformation in the thin film due to the deposition technique, or other parameters with influence on the internal phosphorescence efficiency. As discussed before, this is somewhat common. Moreover the $[Ru(ph_2phen)_3](PF_6)_2$ molecule is clearly larger than the $[Ru(bipy)_3](PF_6)_2$ allowing more pronounced effects of the thin film molecular conformation differences.

The extraordinarily simple device structure opens new possibilities for applications. One of them is naturally the flexible light emitter. Interesting results were obtained with the $[Ru(bipy)_3](PF_6)_2$ as active layers [53]. In spite of low luminous efficiency (3×10^{-2} lm/W) and luminance (near 1 cd/m^2 but under electrical current values within the electrochemical stability range) the idea was successful launched. Moreover, the measurements were done in the normal light emission direction without an integrating sphere, so the data must be carefully compared to the other bibliographic results.

One of disadvantages pointed to the Ru(II) LECs is the slow response time (delay time) and therefore much attention has been paid to overcome this constrain without losing efficiency or adding too much complexity to the device structure (that is in last analyses an important technical question on that kind of light emitters). As this large delay time results of the need for the spatial redistribution of the counterion density for the electrochemical reactions, introducing an ionic conductivity level, the differences approaches to reduce this time have been focused directly or indirectly on the counterions.

An interesting work which played directly with the counterions nature, used as base the typical $[Ru(bipy)_3]^{+2}$ complex [54, 55], and studied small and large counterions. The first group comprised BF_4^- and ClO_4^- while the other group included the PF_6^- and AsF_6^-. Not surprising, the small counterions provided short delay times, the most interesting results derived from a comparison between two devices fabricated with the same complex and counterion but under different atmospheres, that is under air and inside a gloves box. The delay time was clearly shorter for the device fabricated and operating on normal atmospheric conditions, which the authors have interpreted as the result of some incorporation of water, increasing the counterion mobility. Still, the size of the counterions was the most important factor, the best result being obtained with BF_4^-.

A different approach to reduce the delay time is to increase the electrical charge injection with an organic semiconductor layer between the Ru(II) film and the electrode. One attempt used a thin layer of BBOT (*2,5-bis(5-tert-butyl-2-benz-oxazolyl) thiophene*), a known electron transport polymer, between the active Ru(II) layer and the aluminum cathode [56]. The authors observed a reduction of the delay time (typically from three to six times smaller) when compared to one the measured for the single layer Ru(II) complex. It must be noted that this work, although employing the same $[Ru(bipy)_3](PF_6)_2$ complex, had an active layer solely composed by the Ru(II) complex, without any polymer host. The explanation for the much lower delay time observed was ascribed to the better speed of electron injection onto the Ru(II) active layer.

A similar approach was used with a $[Ru(dphphen)_3](ClO_4)_2$ (*ruthenium(II)-tris(4,7-diphenyl-1,10-phenanthroline*) using Alq3 as electron injection/transport layer and Ag as cathode [57]. The authors found a short delay time (only 2 seconds until reach the maximum luminance near 1300 cd/m^2) but the driving voltage increased from near 2.5 V to 7 V. The explanation is the same as given before, i.e. and increase of the electron injection speed at the Ru(II) complex/ETL interface provided by the electrical field inside the Alq3 layer, improving the electrical charge injection balance in the Ru(II) emissive layer. This electrical field (that corresponds to a drop voltage across the Alq3 layer) also accounts for the more high driving voltage. It must be noted that the success in decreasing the delay time is attributed to the accelerated electron injection instead to the injected electron density itself.

An important question concerning the stability of Ru(II)-based LECs is the degradation of the Al cathode, a serious problem for technological applications. The origin is usually attributed to a diffusion of aluminum into the active layer when the device is not working, i.e. when it is in the off state. The aforementioned addition of PMMA to the Ru(II) complex, can decrease the degradation rate. In the analyses of the electrical current bellow the diving voltage (in our case between 3 V and 5 V) we can observe an unipolar charge injection due to the presence of the mixed-valent R(II)/Ru(I) states caused by the aluminum cathode. This manifests as a relatively high electrical current at this applied voltage range. A way to investigate unipolar injection is to perform a fast voltage seep, avoiding the redistribution of the counterions (that are slow). In this case the electron hoping between the Ru(I) and Ru(II) species is controlled by a uniform electrical field across the active layer. This mechanism can be given by the following simplified expression [21]:

$$J = J_0[\exp(\beta V) - \exp(-\beta V)] \quad (5.6)$$

where J_0 is the exchange current density flowing between the mixed-valent sites and β a parameter dependent on the physical properties of the material and temperature. Figure 5.26 shows the model fit to our results with the ITO/[Ru(bipy)$_3$](PF$_6$)$_2$:PMMA/Al LEC.

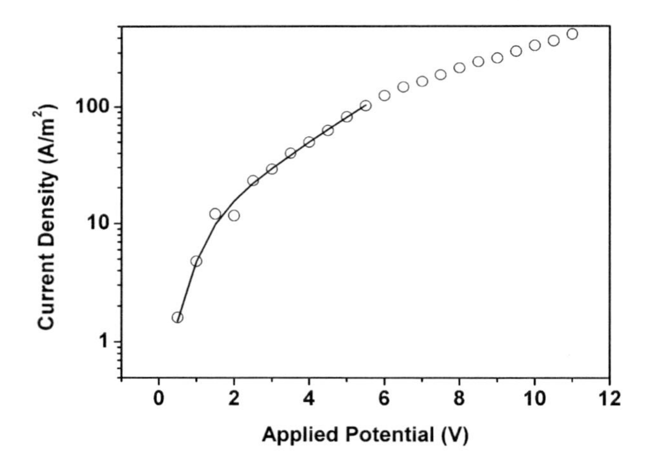

Figure 5.26 Electrical current density vs. applied voltage for the ITO/[Ru(bipy)$_3$](PF$_6$)$_2$:PMMA/Al LEC. Full line: fit to the model derived from unipolar injection. Reprinted from *Journal of Non-Crystalline Solids*, 354(19–25), G. Santos, F.J. Fonseca, A.M. Andrade, A.O.T. Patrocínio, S.K. Mizoguchi, N.Y. Murakami Iha, M. Peres, W. Simões, T. Monteiro, L. Pereira, Opto-electrical properties of single layer flexible electroluminescence device with ruthenium complex, 2571–2574, Copyright (2008), with permission from Elsevier.

From that fit we obtain a value for J_0 near 10 mA/cm^2, which is in agreement with the device high current density. This result points to the hypothesis that the mixed-valent Ru(I)/Ru(II) states caused by reaction with the cathode yield an unipolar charge injection at applied potentials below the driving voltage. This result was also previously observed [58] and may explain the LEC performance degradation. One solution proposed is to have a cathode made of a more electrochemical stable metal, as Ag, for instance.

Of particular interest is the idea of building a light emitter which comprises both LEC and OLED and the emitting color will be sensible to the applied voltage. The light emitter structure is relatively simple: ITO/[Ru(bipy)$_2$(dimbpy)](PF$_6$)$_2$/Alq3/NPB/Ag [59]. The device mechanism shows a clearly time dependent response in forward bias. Initially the spectrum is almost green (from the Alq3 emission) and the characteristic orange–red emission will appear, together with the green one, some time after increasing the applied voltage, corresponding to the expected delay time discussed before or to the stabilization of LEC emission. As a matter of fact, under reverse bias, only a weak orange–red emission is observable, obviously corresponding to the Ru(II) emission that is expected to be almost independent on the bias polarity (it's a LEC device). Although interesting, the color tuning by changing the applied voltage is not very practical because under forward bias the color change depends on the time, which, of course, cannot be controlled in a continuous operation.

A different approach was used comprising a light emitting device with a single layer of a new Ru(II) complex, the dinuclear complex [Ru(ph)$_4$Ru](PF$_6$)$_4$, mixed with a PPV (*polyphenylenevinylene*) derivate polymer [60]. Under forward bias (4 V) emission features the known red color form some Ru(II) complexes, while under reverse bias (−4 V) a green emission is observed. The authors have established a model to explain the reverse bias behavior which includes the hypothesis of the reduced Ru(II) complex to promote the electron transfer to PPV. This gives an interesting "picture": in forward bias the PPV emits in the green region; the Ru(II) complex reduced gives a "near blue" emission and oxidized Ru(II) complex gives the orange–red emission.

Several others Ru(II) complexes have been synthesized, allowing color tuning from the orange–red to pure red. In spite of the many well succeeded complexes prepared, with strong phosphorescence from the yellow (band peak at 596 nm) to the far red (band peak at 718 nm) [61] further applications in OLEDs are little disseminated.

Attempts to fabricate OLEDs based on Ru(II) complexes were done but almost the efficient devices required energy transfer from a host material. Besides the use of a single active layer with a dielectric polymer for film formation, other Ru(II) emitting devices have been exhaustively studied. The most relevant, reports on a Ru(II) in a semiconducting polymer blend, a two or three layer device; the additional layer was used to improve one type of electrical carrier injection, for a successful reduction of the device delay time. In all the examples bellow, the delay time is practically inexistent; but the device structure is necessarily more complex and the driving voltages are quite higher, though this is not necessarily bad for some applications.

A complete work was done with a [Ru(4,7-Ph_2-phen)$_3$](ClO_4)$_2$ (*ruthenium(II)4,7-diphenyl-1,10-phenanthroline(4,7-Ph_2-phen)*) blended with a host of one of three blue emitting polymers (of wide band-gap, so that the emission band overlaps the MLCT absorption bands of the Ru(II) complex), namely the PVK (*polyvinylcarbzole*), the PDHF (*polydihexylfluorene*) and LPPP (a modified ladder-like *polyphenylene*) [62]. According the authors, the main idea is to promote the Förster energy transfer from host polymeric matrix to the Ru(II) complex, the guest. The device structure employed was ITO/host + Ru(4,7-Ph_2-phen)/LiF/Al with variable relative percentages of polymeric host and Ru(II) guest (0.2%, 1%, and 5% of Ru(II) in mass). The best result was achieved with PVK as host polymer, on which further improvement was made by adding extra layers to help the electrical carrier injection and confinement. The tested devices had a structure of ITO/PVK:Ru(4,7-Ph_2-phen)$_3$(ClO_4)$_2$ (5%)/PBD/Alq3/LiF/Al. The device presented a characteristic broad band in the red spectral region, with a maximum at 612 nm, and CIE color coordinates of (0.62, 0.37). The driving voltage was near 10 V with a luminance changing from 140 cd/m^2 to 500 cd/m^2 (applied voltages of 21 V and 26 V, respectively). Although this is not a great result, it had the merit of show experimental evidence for one important pathway.

The use of PVK as the polymeric host for Ru(II) complexes was also successfully employed in the complex [Ru(bipy)$_2$DIM](ClO_4)$_2$ (*ruthenium(II)bis(2,2'-bipyridine)(4,7-dimethyl-1,10-phenanthroline)*) in a device structures of ITO/PVK: [Ru(bipy)$_2$DIM](ClO_4)$_2$/Al and ITO/PVK: [Ru(bipy)$_2$DIM](ClO_4)$_2$/BCP/Alq3/Al [63]. Results were curious, as the first and simplest structure with 6% a Ru(II) mass yields an emission peak at 600 nm, with a driving voltage of 16 V, whist with 36% the emission peak is located near 620 nm, with a driving voltage near 3 V. For this last Ru(II) mass percentage, a luminance near 40 cd/m^2 (applied voltage of 7 V) was achieved, though with low external efficiency (10^{-2}%). Further improvement for the same Ru(II) percentage, could be obtained using a more complex structure, in which the electron injection and transport was increased by a BCP/Alq3 bi-layer (remember that BCP is a hole blocking layer and Alq3 a ETL). In this structure (and although the driving voltage was increases up to 20 V due to the presence of more organic layers—i.e. increasing the resistance) a luminance of near 95 cd/m^2 with an external efficiency of 0.14% was obtained. The authors stated that, firstly the Ru(II) complex was reduced to Ru(I) most likely on bipy organic ligand, leading to an electron location in the π^* energy level; further recombination with an hole at $d\pi$ energy level (admixture between LC and MLCT levels) from the subsequent oxidized Ru(III) entity would have produced the electroluminescence. The different on the emission wavelength obtained for the different mass percentage of the Ru(II) complex in PVK were attributed to different molecular environments (a possibility the we have already discussed).

The simplest device structure employing Ru(II) complexes able to take advantage on the energy transfer process is the PVK + Ru(II) blend, and this was, obviously one the first tested with positive results. Immediate improvements were done by the incorporation of an ETL material, typically PBD into the blend for a better balance of the electrical carrier due to the HTL of PVK. One of the firstly interesting tests, with this blend involved three different Ru(II) complexes in a PVK:PBD host matrix [64]: [Ru(ph$_2$bipy)$_3$](Cl)$_2$

(ruthenium(II) (4,4′-diphenyl-2,2′-bipyridine)$_3$), [Ru(diyp$_2$bipy)$_3$](Cl)$_2$ *(ruthenium(II) (2,2′-bipyridyl-4,4′-dimethyl)$_3$)* and [Ru(phen)$_3$](Cl)$_2$ *(ruthenium(II) (1,10-phananthroline)$_3$)*. The device structure was therefore, ITO/PVK:PBD:Ru(II)/Mg/Ag. The concentration of the host matrix was 100:40, respectively, whereas the concentration of Ru(II) complexes changed from 0.5% to 10% (in mass). Although no information is available about the emission efficiency and optical power or luminance, the results evidenced that Ru(II) complexes with more planar ligands (ph$_2$bipy and particularly the phen) exhibited better electrical carrier transport for high Ru(II) complexes percentages in the host matrix. Also, such complexes also exhibited a broader and more intense electroluminescence band shifted towards red. This easy approach also gives a very low delays time (or not measurable by eye response) and can be employed for different Ru(II) complexes.

Much more work has been done concerning the Ru(II) complexes for light emitting devices (see [65–67]). From in the technological point of view, the single layer device appears to be of particular interest. Much effort needs to be done to overcome their principal problem: the delay time.

5.5.2.3 Rhenium-based OLEDs

Rhenium is a particularly interesting TM element for OLEDs active layers due to its characteristic and very bright yellow emission, considered by many researches one the most beautiful along with the "sky-blue" of some iridium-based OLEDs.

Rhenium belongs to the first row of Periodic Table TM elements and its coordination complexes usually comprise Re(I). Its relatively low ligand field strength suggested the employment of organic ligands with a strong ligand field, as CO. Having typically a yellow emission, some results in orange–red and near read emissions were obtained, although these cannot compete with Ru(II) complexes, so interest of this element remains in the yellow emission.

The relative difficulty of processing the Re(I) complexes by thermal evaporation methods, as well as the energy absorption bands, typically extended from the UV to near UV–blue spectral region (cf. Figure 5.17) make them good candidates for wet OLED process fabrication, simultaneously incorporating a polymeric host matrix (alone or with another organic semiconductors) to provide the energy "pump." Nevertheless, there are successful reports of OLEDs with an active layer formed only by the Re(I) complex. In our work, a *fac*-[ClRe(CO)$_3$(bipy)] (*rhenium(I) (2,2-bipyridine) Cl(CO)$_3$*) was used to fabricate an OLED completely by thermal evaporation with Re(I) complex as active layer. The complex has neutral organic ligands (the bidentate *bipy* and the monodentate CO) but its charge is balanced by the mono anionic the Cl^- counterion; as expected, it employs strong ligand field organic entities like CO.

One of the more problematic issues lies the determination of the HOMO and LUMO levels of such complex. Considering the proximity of the similar complexes based on ClRe(CO)$_3$ part, we can use, for a crude estimation, a value between 2.9 eV and 3.3 eV for the LUMO level and near 6.4 eV for the HOMO level. Naturally, that these estimated values are not very precise, but they are still very useful in a first approach.

The OLED structure is shown in Fig. 5.27 and corresponds to a simple by-layer structure with 20 nm tick TPD acting as HTL and *fac*-[ClRe(CO)$_3$(bipy)] (50 nm tick).

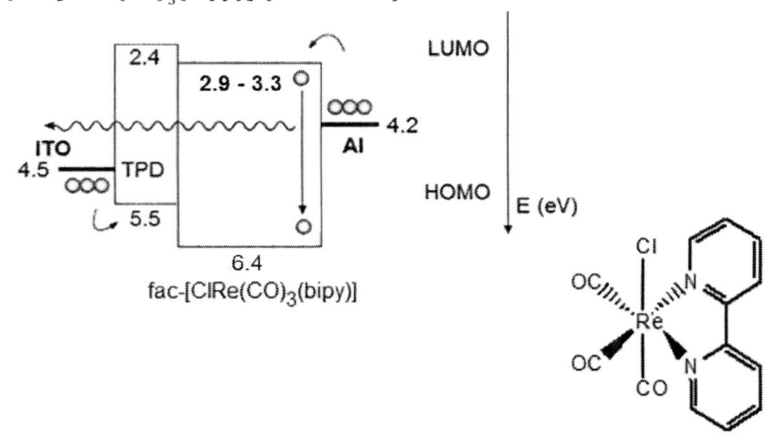

Figure 5.27 Energy levels diagram of a ITO/TPD/*fac*-[ClRe(CO)$_3$(bipy)]/Al OLED. The molecular structure of the Re(I) emitter is shown. The represented thicknesses of the organic layers are not in the same scale in the diagram.

The approximation of the Re(I) complex LUMO level to the known LUMO levels of the most commonly used ETL like Alq3 and butyl-PBD, or simply PBD (with a difference of about only ± 0.2 eV) predict, face to Al cathode, a lower electron injection barrier. Moreover, the Ru(II) and Re(I) TM complexes are known to exhibit a bipolar electrical nature and hence, they could facilitate both the charge injection and transport processes. The absence of any barrier for hole injection at the TPD / Re(I) complex was important to allow a useful electrical carrier injection. Besides such considerations the LUMO level of TPD allowed the material to acts as electron blocking layer which was necessary due to the absence of ETL.

The electroluminescence spectrum (with CIE color coordinates) is shown in Fig. 5.28.

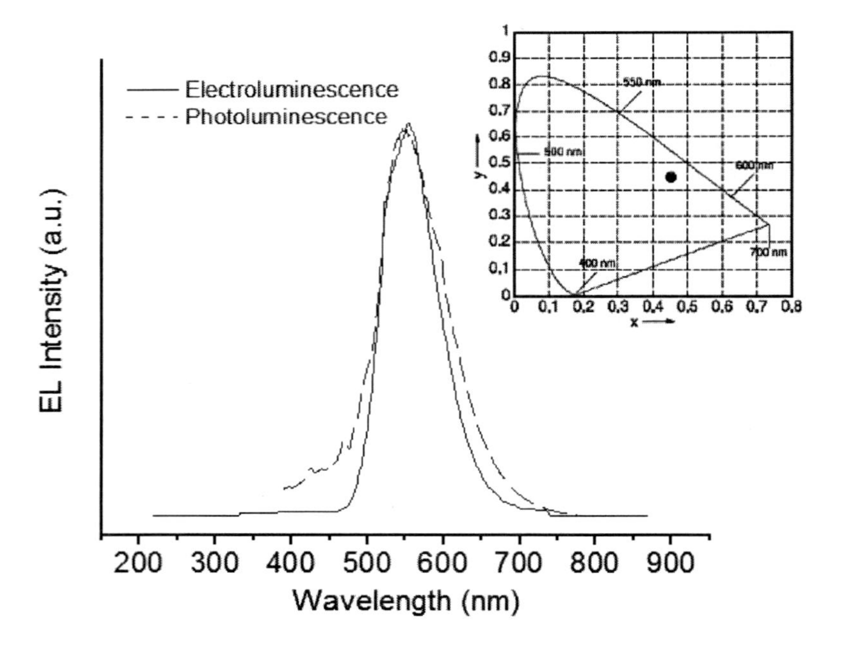

Figure 5.28 Electroluminescence spectrum (full line) of the ITO/TPD/*fac*-[ClRe$(CO)_3$(bipy)]/Al OLED under an applied voltage of 14 V. For comparison the photoluminescence of the Re(I) complex thin film is shown (broken line). The spectra are no at same scale. Inset: The CIE color coordinates of the electroluminescence.

For comparison, the photoluminescence spectrum is also presented. Both luminescence bands appear to be similar, particularly in the peak position a general featureless shape, but the photoluminescence band is more broadened. This difference is not usual (as the previously reported results have shown) but it may be correlated with the experimental setup (a more precise one was used for electroluminescence) and/or the use of a thin film for photoluminescence, contrarily to the usual powder. As a consequence, we can expected the CIE color coordinates to change a little, for both luminescence data, although remaining within the yellow range color. In fact, for electroluminescence, the CIE color coordinates are (0.45, 0.44) and for photoluminescence are (0.43, 0.43). So, there are not significant changes. The emission band is peaked at near 560 nm.

Note that the electroluminescence spectrum has no significant changes when the applied voltage changes, (apart the intensity of course) with minimum variations on the CIE color coordinates. From this coordinates we can also conclude that this device exhibits good color purity but not as high as the previous Ru(II)-based devices) indicating that Re(I) complexes are naturally interesting yellow emitters.

In Fig. 5.29, the electrical current and optical power at electroluminescence maximum vs. applied voltage are shows. The driving voltage is about of 11 V, a bit over the ideal value, but this may be due to device configuration, using only one HTL and Al for the cathode. A possible improvement is to use another HTL instead of TPD, as m-MTDATA (HOMO level at 5.1 eV and LUMO level at 1.9 eV). Improvements to the hole injection into Re(I) complex layer are limited by the much low LUMO level of m-MTDATA due to the high barrier formed between both layers (in last instance we can erroneous suppose that such barrier will be a very good electron blocking but a disruption is the most probable). In theory a good choice would be to have the m-MTDATA between the TPD and the ITO electrode.

Further improvements can be made, in particular the introduction of a more hole-blocking ETL but perhaps the best choice is to transfer energy from a conductive host matrix, as our results have shown (and will be discussed later).

The radiometric wall-plug efficiency obtained from the power plot show in the last figure is near 5×10^{-3}% (obtained only for the emission wavelength of the maximum electroluminescence band), one order of magnitude lower that the one measured for Ru(II) complexes. This value increases up to 0.09% when the emission band is integrated, and the maximum optical power (within the OLED dynamic operating region) falls to mW. All electroluminescence measurements are collected under the normal direction from the OLED emitting surface. This further predicts a significant increase of these values (both efficiency and optical power) if an integrating sphere were used.

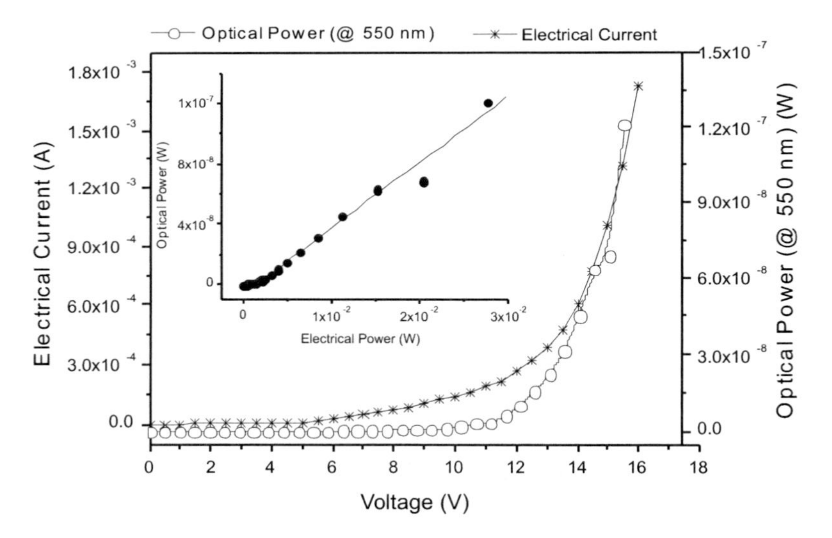

Figure 5.29 Electrical current and optical power (@ 550 nm) for the ITO/TPD/*fac*-[ClRe(CO)$_3$(bipy)]/Al OLED with the applied voltage. Inset: the optical power vs. electrical power. The straight line is the linear fit in the dynamic OLED region allowing the calculation of the radiometric wall-plug efficiency.

Regardless of these questions, the overall discussed approach proved correct and allows, to test the TM complex used as OLED active layer.

In practice, there is not much work done using only Re(I) in the OLEDs active layer. As found for Ru(II) complexes, and perhaps even more widely used, Re(I) complexes are employed in OLEDs as doping guests in a host matrix, typically a polymeric (further OLED fabrication by a wet process) or organic small molecule can also be used and in that case, the typical procedure for OLED fabrication is the thermal co-evaporation of the mixture. This pathways probably involves most Re(I)-based OLEDs. The benefits are several as referred before.

CBP is the most employed organic semiconductor host in Re(I) OLEDs prepared by thermal evaporation. One of the first successful OLED structures using this host material comprise two Re(I) complexes from the same family the $ClRe(CO)_3$dmfbipy (*Rhenium(I) (4,4′-dimetyl format-2,2′-bypyridine) Cl(CO)$_3$*) and $ClRe(CO)_3$dbufbipy (*Rhenium(I) (4,4′-dibutyl format-2,2′-bypyridine) Cl(CO)$_3$*) in an OLED structure of ITO/NPB/CBP:Re(I)/BCP/LiF/Al. NPB is used as HTL whereas the BCP is the hole blocking layer [68]. The authors found an optimal percentage in mass of $ClRe(CO)_3$dbufbipy near 2% (582 cd/m^2 and 1.3 cd/A) and another near 6% (739 cd/m^2 and 2.5 cd/A) for $ClRe(CO)_3$dmfbipy, before the decrease of the efficiency perhaps due to the triplet–triplet annihilation. The electroluminescence is typically red with a broad band centered at near 625 nm for both Re(I) complexes.

Using a similar OLED structure, with different Re(I) complexes, $ClRe(CO)_3$phen and $ClRe(CO)_3$dmphen the authors achieved a much better performance that the previous Re(I) complexes [69]. The results show for $ClRe(CO)_3$phen a luminance near 2800 cd/m^2 (efficiency of 6.67 cd/A) and for $ClRe(CO)_3$dmphen a luminance near 3700 cd/m^2 (efficiency of 7.15 cd/A). The optimal Re(I) complex mass percentage in the CBP host was found around 6% for $ClRe(CO)_3$phen and near 10% for $ClRe(CO)_3$dmphen. Concerning the energy transfer mechanism, the authors suggests an efficient Förster process supported by the OLED electrical behavior, which had no clear differences on the electrical current vs. applied voltage with different concentrations of the Re(I) complex, excluding the electrical carrier trapping into those complexes, and by the good overlap between the CBP emission spectral band and the ^{1}MLCT absorption band of the Re(I) complexes. The overlap promotes the host-to-guest singlet ^{1}MLCT energy transfer with further intercrossing system to the ^{3}MLCT state which the authors claims to be the triplet emitter. Both devices emits in the yellow spectral region with broad emission bands peaked at near 550 nm.

More complex OLEDs structures employing Re(I) complexes into a CBP host can be found in the literature. For instance, a six layer device of ITO/m-MTDATA/NPB/CBP:Re(I)/Bphen/Alq3/LiF/Al, employing a BrRe(CO)$_3$TPIP (*rhenium(I) (3-ethyl-2-4'-triphenylamino)imidazo [4,5-f]1,10-phenanthroline)(CO)$_3$Br*) a yellowish green emission (broad band peaked at 555 nm) is obtained with high luminance and good efficiency (6500 cd/m^2 and 16.7 cd/A) [70]. Unfortunately the OLED structure is too complex.

An interesting result, was a green electroluminescence obtained from Re(CO)$_3$(bipy)(btpz) (*rhenium(I) (2,2'-bipyridine) (3,5-bis (trifluoromethyl) pyrazolate)(CO)$_3$*) in a device structure of ITO/NPB/CBP/TPBI: Re(CO)$_3$(bipy)(btpz)/TPBI/Mg:Ag, where the TPBI denotes *2,2',2"-(1,3,5-benzenetriyl)tris[1-phenyl-1H-benzimidazole]* and the percentage of the Re(I) complex was 2% [71]. The emission band of TBPI overlaps the MLCT absorption bands of the Re(I) complex so the OLED electroluminescence spectrum is a broad band centered at near 530 nm with a luminance of 2300 cd/m^2 and an efficiency of 0.75 lm/W.

Besides the traditional CBP host for Re(I) complexes, polymeric hosts as the PVK may also be used allowing a wet deposition of the active layer (although not so widely as the CBP). In our recent work, we studied the electroluminescence behavior of OLEDs with an active layer formed by PVK: *fac*-[ClRe(CO)$_3$(bipy)] and different mass percentage of the Re(I) complex. The device structure was ITO/PEDOT:PSS/PVK+ *fac*-[ClRe(CO)$_3$(bipy)]/butyl-PBD/Al and the Re(I) complex was percentages under study were 3%, 5%, 10%, 20%, and 40%. Figure 5.30 shows the device structure, with the energy levels diagram.

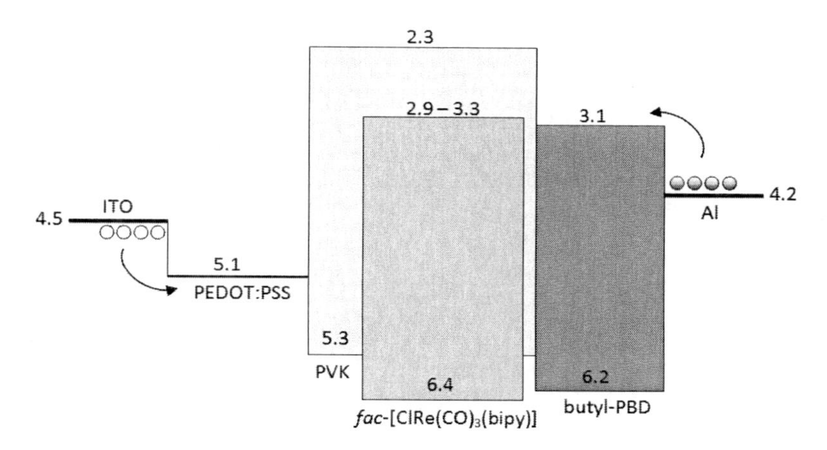

Figure 5.30 Energy levels diagram of a ITO/PEDOT:PSS (80 nm)/PVK:*fac*-[ClRe$(CO)_3$(bipy)] (120 nm)/butyl-PBD (20 nm)/Al OLED. The represented thicknesses of the organic layers are not in the same scale in the diagram.

The choice of the HTL and ETL is not difficult to understand. Moreover, the butyl-PBD, although small hole blocking layer, helps in the electrical carrier confinement inside the active layer.

The electroluminescence behavior strongly depended on the Re(I) complex concentration. First, the luminance measured under the normal at the OLED surface and not integrated, exhibited a maximum for a Re(I) complex concentration of 20%, a value clearly higher than the usual percentage employed with CBP as the host and secondly, the electrical current vs. applied voltage also depended strongly on the Re(I) complex mass percentage as the electrical current density increased with the complex mass. In the luminance measured conditions, the best power efficiency (for Re(I) complex concentration of 20%) was 0.22 lm/W. Figures 5.31 and 5.32 show the luminance and the electrical current vs. applied voltage.

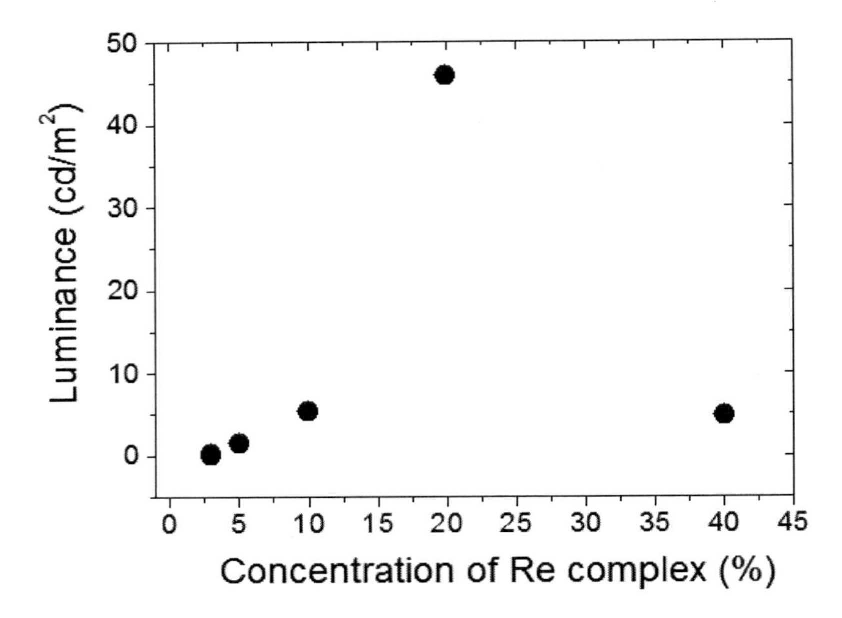

Figure 5.31 Luminance as a function of Re(I) complex concentration. Applied voltage of 30 V.

It must be noted that when high mass percentage of the Re(I) complex, were used, the formed film had poor quality, bringing an additional contribution to the decrease of luminance and efficiency. Still it is evident that a strong increase in the luminance was obtained when the Re(I) complex concentration changed from 10% to 20%.

Concerning the electrical current vs. applied voltage (Fig. 5.32) a strong increase in the current density (for a fixed applied voltage and under OLED dynamic region) was also observed when the Re(I) complex concentration changed from 10% to 20%. At higher mass percentage tested (40%), the current density increased a bit less. Again, the physical conditions of the film formation may account for such result. These results on high concentration of the guest in the host matrix pose an interesting question about the apparent absence of the quenching effects. Perhaps it is not as important as the poor film formation.

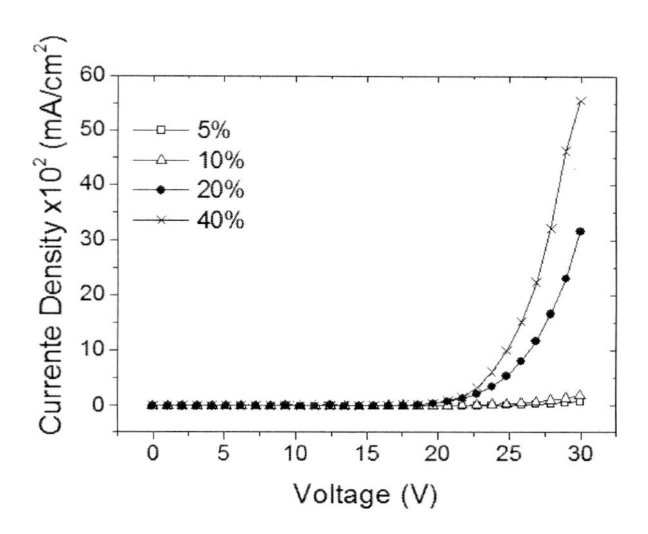

Figure 5.32 Electrical current density vs. applied voltage as a function of Re(I) complex concentration.

The electroluminescence spectra are shown in Fig. 5.33.

A few important points of the OLED electroluminescence results must be focused. First, the driving voltage is relatively high, and almost independent from the Re(I) complex concentration. Typical values of 21–25 V were measured, the lower ones observed for higher guest concentration. Such values aren't entirely surprising because of the use of butyl-PBD, as previous results on different kind of OLEDs have shown. Also, the active layer had no electron transport material in the blend. Secondly, all electroluminescence data showed a small blue band of the PVK, showing that the energy transfer occurs was not complete. Finally, the CIE color coordinates changed little with the Re(I) concentration keeping the known yellow color with coordinates near (0.45, 0.45).

The overlap between the PVK emission band and the Re(I) complex MLCT absorption band observed in photoluminescence may be an indication of an efficient energy transfer via Förster or Dexter mechanisms, but the strong changes in the electrical current vs. applied voltage and the fact that the expected LUMO level of the *fac*-[ClRe(CO)$_3$(bipy)] lie bellow the LUMO level of the PVK (i.e. the $LUMO_{Re}$ > $LUMO_{PVK}$) may also suggest an energy transfer by sequential electrical carrier trapping by the Re(I) complex.

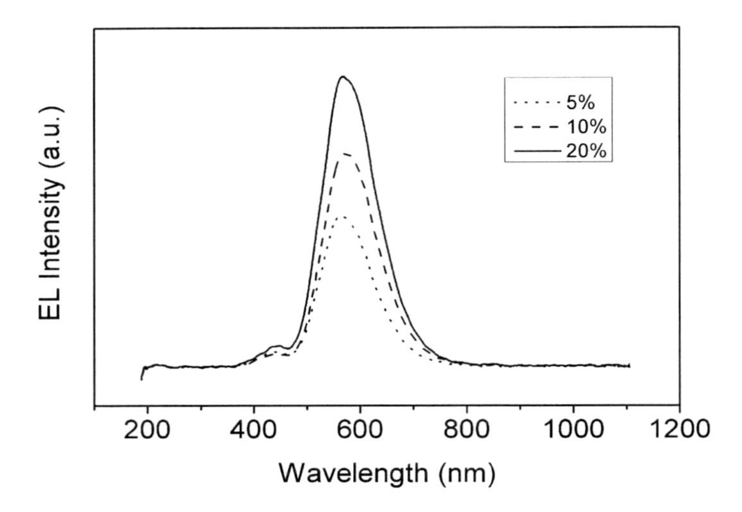

Figure 5.33 Electroluminescence spectra dependence on the Re(I) complex concentration. Applied voltage of 30 V.

Moreover, the electroluminescence spectra under different applied voltages show that the PVK small emission varied with the applied voltage, being almost absent at low applied voltages, this suggesting that under high applied voltage the Re(I) triplet emitter becomes saturated as expected from the trapping of electrical carriers in the Re(I) complex. Still, we cannot exclude the Dexter energy transfer due to the clearly high Re(I) complex concentration needed to achieve the best efficiency, compatible with an additional triplet–triplet energy transfer besides the singlet–singlet process that is the only pathway for the Förster mechanism. The emission is, as in all Re(I) complexes-based OLEDs, ascribed to a triplet emitting with a strong MLCT character. Exploring this structure, by changes in the ETL and/or incorporating and ETL into the emissive blend, should improve the performance.

Finally, in a particularly interesting work, a Re(I) complex was used in an attempt to fabricate a WOLED, $ClRe(CO)_3dmbpy$ (*Rhenium(I) (4,4′-dimetyl-2,2′-bypyridine) Cl(CO)*$_3$) was used, in a device structure of ITO/NPB/CBP: $ClRe(CO)_3$dmbpy/BCP/LiF/AL [72]. The OLED emission combined yellow contribution from the Re(I) complex with the blue from the NPB. The final color could be tuned by the applied voltage or by the thickness of the CBP:$ClRe(CO)_3$dmbpy layer. The Re(I) concentration (in mass) is 2% and the optimal driving voltage for white emission was of about 15 V with a luminance and efficiency of 1400 cd/m^2 and 2.9 cd/A. The best CIE color coordinates were (0.30, 0.37) (remember that white emission is (0.33, 0.33)).

5.5.2.4 Iridium-based OLEDs

Iridium organic complexes are undoubtedly the most widely studied materials for use as active layers of TM complexes-based OLEDs. To the typical features of TM complex OLEDs, the iridium-based ones add high efficiency, high brightness at low driving voltages, and colors covering all the visible spectrum from blue to red [73, 74]. Such high versatility is very difficult, if not impossible, to obtain with non-iridium TM complexes.

Perhaps the most interesting OLED based on Ir(III) complex is the blue emitter. Two well-known Ir(III) complexes emitting in the blue spectral region are FIr6 (*bis(4′,6′-difluorophenylpyridinato) iridium(III) tetra(1-pyrazolyl)borate*) and FIrpic (*bis(4′,6′-difluorophenylpyridinato)-iridium(III) picolinate*) featuring sky–blue and greenish–blue colors, respectively [75]. Several other iridium complexes has also been synthesized and tested for blue electroluminescence. Nevertheless, the search for a "bluer" emitter, by playing with the organic ligands, never ends. A recent work has successfully use one isomer of an Ir(III) complex to emit in a "real blue" color [76]. The complex is Ir(dfppy)$(fppz)_2$ (*2-(2,4-difluorophenyl) pyridine iridium(III) bis 5-(2-pyridyl)-3-trifluoromethylpyrazole*) where the isomeric difference is in the functional group of both fppz ligands that is located trans to the *dfppy* ligand. The OLED structure employed was ITO/NPD (30 nm)/TCTA (30 nm)/CzSi (3 nm)/CzSi: Ir(dfppy)$(fppz)_2$ (10%) (25 nm)/UGH2: Ir(dfppy)$(fppz)_2$ (10%)(3 nm)/UGH2 (2 nm)/TAZ (50 nm)/LiF (0.5 nm)/Al, where TCTA is (*4,4′,4′-tris(N-carbazolyl) triphenylamine*), CzSi is (*9-(4-tertbutylphenyl)-3,6-bis(triphenylsilyl)-9H-carbazole*), UGH2 is (*p-bis (triphenylsilyl) benzene*) and TAZ is (*3-(biphenyl-4-yl)-4-phenyl-5-(4-tert-butylphenyl)-1,2,4-triazole*). Both CzSi and UGH2 are wide bandgap organic semiconductors, which can be employed as host materials. TCTA is a HTL and TAZ an ETL. The device structure also employs CzSi and UGH2 layers as electron blocking and hole blocking layers, respectively. The most interesting point of this device is the use of two emissive layers with different host materials (the CzSi and UGH2) claimed by the authors to improve the electrical carrier balance (between holes and electrons injection/transport) and to move the exciton formation zone away from the HTL and ETL interfaces (including the blocking layers). The electroluminescence spectrum is shown in Fig. 5.34.

The OLED figures of merit are really interesting: luminance of 4000 cd/m^2 (at 16 V) with a power efficiency of 8.5 lm/W.

Considering this double emitting layer structure concept, further applications can be made even using known blue Ir(III) emitters. Such approach was successfully applied to the FIr6 complex [77]. The authors employed an OLED structure of ITO/PEDOT:PSS/NPB/Ad-Cz:FIr6/UGH2:FIr6/TPBI/Al. The best concentration of FIr6 complex was established at 3% (in mass). Ad-Cz (*2,2-bis(4-carbazolyl-9-ylphenyl) adamantane*) is a wide bandgap organic semiconductor providing along with UGH2, a good host matrix for Ir(III) blue emitting complexes. The device working mechanism is quite simple: a percentage of the holes accumulated in the Ad-Cz/UGH2 interface (HOMO levels of 5.8 eV and 7.2 eV, respectively) are injected into the FIr6 complex in the UGH2:FIr6 layer. In parallel, the recombination of holes and electrons transported into the UGH2 material gives rise to excited triplet energy states in the FIr6 complex with further light emission. Since the HOMO level of FIr6 is 6.1 eV, a direct hole injection into this complex in UGH2 can be made. Concerning the Ad-Cz:FIr6 layer, excited states of the Ad-Cz generated in the Ad-Cz/UGH2 interface will provide the FIr6 emission by energy transfer. Inside the UGH2 layer, a direct FIr6 emission is expected. This structure gives clearly two distinct ways to obtain the Ir(III) emission , giving a distributed emitting region. The two emitting layers system helps maximize the efficiency, as the host emission, as well as some quenching processes can be minimized without loss of the total emission. For the aforementioned device, a maximum efficiency of 14.3 lm/W was obtained and, in fact, no electroluminescence of the host materials was found

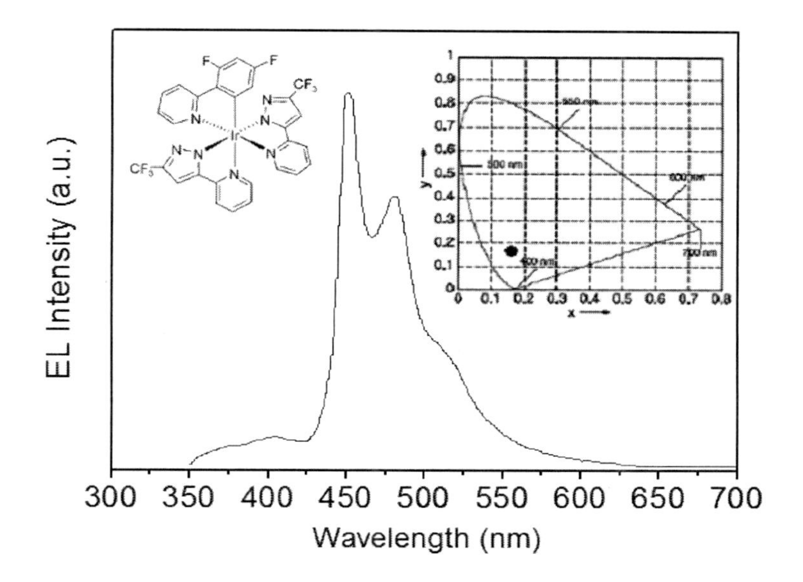

Figure 5.34 Electroluminescence spectrum of the ITO/NPD/TCTA/CzSi/CzSi: Ir(dfppy)(fppz)$_2$/UGH2: Ir(dfppy)(fppz)$_2$/UGH2 (2 nm)/TAZ/LiF/Al OLED. Inset: the CIE color coordinates and the Ir(III) complex molecular structure. Reprinted with permission from Yang, C.-H., et al. (2007) Blue-emitting heteroleptic iridium(III) complexes suitable for high-efficiency phosphorescent OLEDs, *Angew. Chem.*, **46**, 2418–2421.

Before discussing the typically green color of Ir(III)-based OLEDs, an important result must be presented, concerning a greenish-blue emission, on the *fac*-[Ir(F$_2$-mppy)$_3$] complex (*iridium(III) tris(5-(2,4-difluophenyl)-10,10-dimethyl-4-azatricycloundeca-2,4,6-triene*) [78]. The OLED structure was ITO/NPB/CBP: *fac*-[Ir(F$_2$-mppy)$_3$]/BCP/Alq3 /Mg:Ag, with NPB and Alq3 being the traditional HTL and ETL, respectively, and BCP as hole blocking layer. The host material was CBP and the Ir(III) complex (guest) was used at 10%. This conventional structure yields an electroluminescence in the blue–green spectral region (CIE color coordinates of (0.18, 0.36)) with an interesting luminance of 9000 cd/m^2, for an applied voltage of 12.5 V, and an efficiency of 9.58 lm/W. This result clearly shows the strong influence of the BCP hole blocking layer in "trapping" the holes in the host material to afford an excellent electrical carrier confinement.

Green Ir(III) complexes are extensively applied for OLEDs with high brightness and efficiency. Contrarily to those of "pure" blue emitters, the device structures are much simpler (see the aforementioned blue–green Ir(III)-based OLED), which, in parallel with the efficiency of the green Ir(III) complexes available, contributes to their extensive study.

A green Ir(III) complex-based OLED was reported employing a layer of Ir(ppy)$_3$ in a CBP host and comprising an HTL (TPD) and pOXD, a copolymer which includes an oxadiazole unit and has electron transporting and hole blocking properties [79]. The OLED structure was ITO/PEDOT:PSS/pOXD:TPD:CBP:Ir(ppy)$_3$/Ca(acac)$_2$/Al:Li:Al. The mass percentage concentrations of the different materials in the solution for spin coating deposition were 0.3% for CBP, 0.06% for Ir(ppy)$_3$, 0.044% for TPD, and 0.34% for pOXD. The driving voltage for a luminance of 100 cd/m^2 was of about 4.7 V, whereas with an applied voltage of 12.2. V a luminance of 10000 cd/m^2 was obtained. A maximum luminance of 33000 cd/m^2 with an efficiency of 55 cd/A was achieved. The electroluminescence spectrum showed a broad band, (full width at half maximum near 70 nm), centered at 510 nm.

Using a much more simpler device structure, the green Ir(III) emitter, Ir(DPPF)$_3$ (*tris[9,9-dihexyl-2-(phenyl-4′-(pyridine-2″-yl)) fluorene] iridium(III)*) was used to fabricate a high-brightness OLED [80]. The structure employed was the ITO/PEDOT:PSS/PVK:PBD:Ir(DPPF)$_3$/Ca:Ag, with a in mass percentages of PBD and Ir(DPPF)$_3$ in the host (PVK) of 40% and 1%, respectively. As PVK is a hole transport material, the electron transport material, PBD, was added to the emissive layer to form a blended host of PVK:PBD, able to easily transport both electrons and holes. This work was an interesting study on the optimal guest concentration to profit from energy transfer mechanisms. This optimal value for Ir(DPPF)$_3$, was of 0.1%; in addition, at this mass percentage no electroluminescence contributions of the PKV and PBD were observed, which suggest that electrical carrier trapping in the guest was the mechanism involved in the energy transfer from the host. This result is academically interesting as the PVK photoluminescence emission band overlaps all of the Ir(III) complex MLCT absorption bands, suggesting a Förster mechanism. Of interest was also the maximum luminance, near 3500 cd/m^2 for an applied voltage of 30 V, although with relatively low efficiency (3.5 lm/W) when compared to others Ir(III) complexes in a PVK host. Figure 5.35 shows the electroluminescence spectrum and the electrical current and brightness vs. applied voltage.

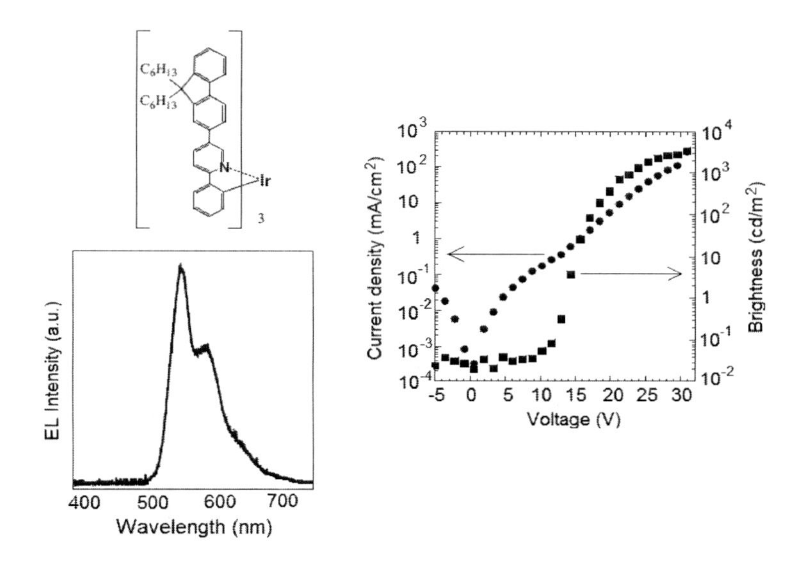

Figure 5.35 Electroluminescence spectrum of the ITO/PEDOT:PSS/PVK:PBD:Ir(DPPF)$_3$/Ca:Ag OLED for an applied voltage of 9 V and the current density and brightness vs. applied voltage. The structure of the Ir(III) emitter is also shown. Reprinted with permission from Gong, X., et al. (2003) Electrophosphorescence from a polymer guest–host system with an iridium complex as guest: Förster energy transfer and charge trapping, *Adv. Func. Mater.*, **13**, 439–444.

Addressing the yellow, an efficient OLED was made using the (SBFP)$_2$Ir(acac) (*iridium(III)bis(2-(9,9′-spirobi[fluorene]-7-yl)pyridine-N,C$^{2'}$)* acetylacetonate) [81]. The device structure is particularly interesting because it includes another Ir(III) complex, the Ir(ppz)$_3$, not as an emitter but as an electron blocking layer, thus bring composed of ITO / m-MTDATA (60 nm) / Ir(ppz)$_3$ (10 nm) / mCP:(SBFP)$_2$Ir(acac) (30 nm) / TPBI (40 nm) / LiF (0.5 nm) / Al with 20% $(SBFP)_2Ir(acac)$ in mass in a host matrix of a mCP (*N,N-dicarbazolyl-3,5-benzene*). The idea of using the $Ir(ppz)_3$ as electron blocking layer is easily understood by looking its HOMO and LUMO levels, at 5.0 eV and 1.6 eV, respectively, while the HOMO and LUMO levels of the mCP are 5.9 eV and 2.4 eV, respectively. This way a relatively high barrier for electrons was expected at the $Ir(ppz)_3$/mCP interface. Moreover, $Ir(ppz)_3$ has a triplet energy state of about 3.1 eV lying 0.9 eV above that of $(SBFP)_2Ir(acac)$ which is enough to keep the excitons created in the emissive layer from diffusing to the $Ir(ppz)_3$ layer. The device with this structure shows very good results, as can be noted from the electroluminescence data present in Fig. 5.36.

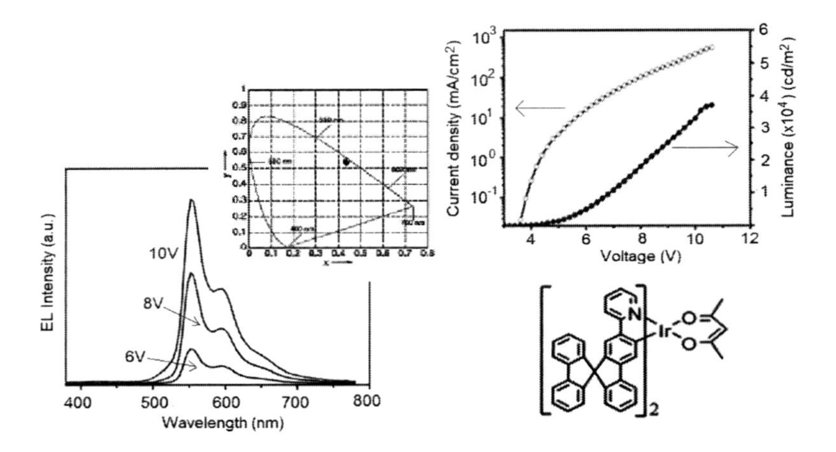

Figure 5.36 Electroluminescence spectrum of the ITO /m-MTDATA/ $Ir(ppz)_3$/mCP:$(SBFP)_2Ir(acac)$/TPBI/LiF/Al OLED for different applied voltages (bottom left) and the current density and brightness vs. the applied voltage (top right). The structure of the Ir(III) emitter is also shown. Inset: the CIE color coordinates. Reprinted with permission from Fei, T., et al. (2009) Highly efficient pure yellow electrophosphorescent device by utilizing an electron blocking material, *Semicond. Sci.Technol.*, **24**, 105019/1–5.

The most interesting feature of this yellow Ir(III) complex-based OLED is its brightness raising about 37000 cd/m^2 for an applied voltage of near 10 V; it also had a good efficiency, 50.6 cd/A for a brightness of 1000 cd/m^2. The driving voltage was only 3.2 V and the CIE color coordinates were (0.46, 0.53), evidencing a very high purity for the yellow color.

In the orange–red spectral region one particular kind of Ir(III) complexes was successfully developed. Several others attempts, were also only succeeding to synthesize the photoluminescence emitter, and without further known work of OLEDs fabrication [82, 83]. The working OLEDs were prepared with a set of $(C^{\wedge}N)_2Ir(acac)$ complexes, where C^N was cyclometalate ligands belonging to the *2,4-diphenylquinoline* family, and all exhibited strong orange–red electroluminescence [84]. The typical device structure employed was ITO / NPB (40nm) / CBP:$Ir_{complex}$ (30 nm) / BCP (10 nm) / Alq3 (30 nm) / LiF (1 nm) / Al, where the $Ir_{complex}$ was one of the four distinct Ir(III) complexes. In general the best concentration (in mass) of the Ir(III) complex was 3%. Figure 5.37 shows the relevant optical and electrical information as well the Ir(III) molecular structures.

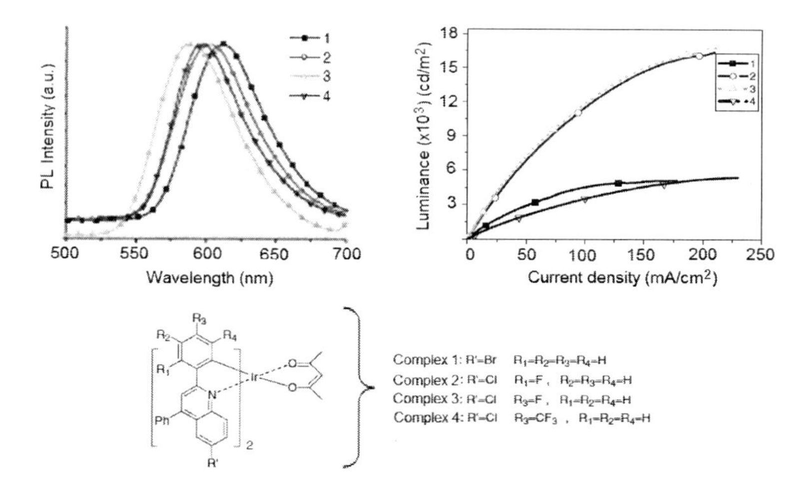

Figure 5.37 Photoluminescence spectra of four (C^N)Ir(acac) orange–red emitters and the luminance vs. electrical current density of the ITO/NPB/CBP:$Ir_{complex}$/BCP/Alq3/LiF/Al OLEDs for each kind of complex. According to the authors, the electroluminescent spectra were similar to herein shown photoluminescence ones. The general molecular structure of the Ir(III) complexes is shown. Reprinted from *Journal of Organometallic Chemistry*, 691(20), X. Zhang, J. Gao, C. Yang, L. Zhu, Z. Li, K. Zhang, J. Qin, H. You, D. Ma, Highly efficient iridium(III) complexes with diphenylquinoline ligands for organic light-emitting diodes: Synthesis and effect of fluorinated substitutes on electrochemistry, photophysics and electroluminescence, 4312– 4319, Copyright (2006), with permission from Elsevier.

The results show that all four OLEDs featured broad-band electroluminescence spectra, peaked between 593 nm and 616 nm, leading to an orange–red emission. An important structure-activity relation pointed by the authors of this study is that the introduction of a fluorinated substituent on the cyclometalate ligands reduces the triplet–triplet annihilation, as evidenced by the current efficiency plots. The maximum luminance obtained was of about 5100 cd/m^2, 16400 cd/m^2, 17000 cd/m^2 and 5400 cd/m^2 for complexes denoted by 1, 2, 3, and 4, respectively (see molecular structure in last figure). In the same order, the power efficiency was of about 1.39 lm/W, 2.25 lm/W, 2.23 lm/W and 0.88 lm/W, for a current density of 100 mA/cm^2.

In the yellow–orange spectral region a work concerning the two Ir(III) complexes, a homoleptic one and other one which was heteroleptic replacing a cyclometalate ligand by an ancillary one, was carried to show the influence of this substitution in the emission spectral shift from yellow to orange [85]. The complexes were the homoleptic Ir(DPA-Flpy)$_3$ (*tris (9,9-diethyl-7-pyridinylfluoren-2-yl)diphenylamine iridium(III)*) and the heteroleptic Ir(DPA-Flpy)$_2$(acac) (*bi (9,9-diethyl-7-pyridinylfluoren-2-yl)diphenylamine acetylacetonate iridium(III)*)). The first one presented a yellow emission while the second one had an orange emission, both with high luminance.

The device structure was ITO/NPB/CBP:Ir(III)$_{complex}$ /TPBI/LiF/Al with a typical active layer with CBP as host and the Ir(III) complex as guest. The ideal guest concentration was found around 5% for both complexes. The OLED figures of merit were quite interesting: a luminance of about 8300 cd/m^2 for both devices and for an applied voltage of 12 V, and the efficiency reached nearly 21 lm/W on the homoleptic complex at an applied voltage of 4.5 V, and almost 12 lm/W for the heteroleptic complex at an applied voltage of 5 V. This work is an interesting example of simple color tuning via organic ligands substitution. In fact, substituting one DPA-Flpy cyclometalate ligand for ACAC, the emission shifted towards the red, indicating an energy lowering of the triplet emitting state, in agreement with the strong ligand field of ACAC. The DPA-Flpy ligand is an electron donating ligand known to increase the HOMO level of the emitting complex while simultaneously rendering it with a hole transporting character. So, the observed changes upon substitution were quite expectable.

Concerning the red spectral region, there are also a few Ir(III) complexes with confirmed potential of applications. Some of them are quite simple to synthesize and can be easily incorporated into OLEDs active layers. The "receipt" is always the same, and consists of the ubiquitous host matrix for the guest Ir(III) complex. Interesting results were obtained a set of four Ir(III) complexes bearing ACAC as the ancillary ligand (as expected, to help the emission spectrum shift towards red). The complexes were $(piq)_2Ir(acac)$ (*bi 1-phenylisoquinoline acetylacetonate iridium(III)*), $(1\text{-}niq)_2Ir(acac)$ (*bi 1-(1′-naphthyl)isoquinoline acetylacetonate iridium(III)*), $(2\text{-}niq)_2Ir(acac)$ (*bi 1-(2′-naphthyl)isoquinoline acetylacetonate iridium(III)*) and the $(m\text{-}piq)_2Ir(acac)$ (*bi 1-phenyl-5-methylisoquinoline acetylacetonate iridium(III)*) [86]. The general OLED structure is well known: ITO/NPB/CBP:Ir(III)complex/ BCP/Alq3/Al all with 6% (mass of the Ir(III) complex guest). Figure 5.38 shows the electroluminescence spectra and the molecular structures of the complexes.

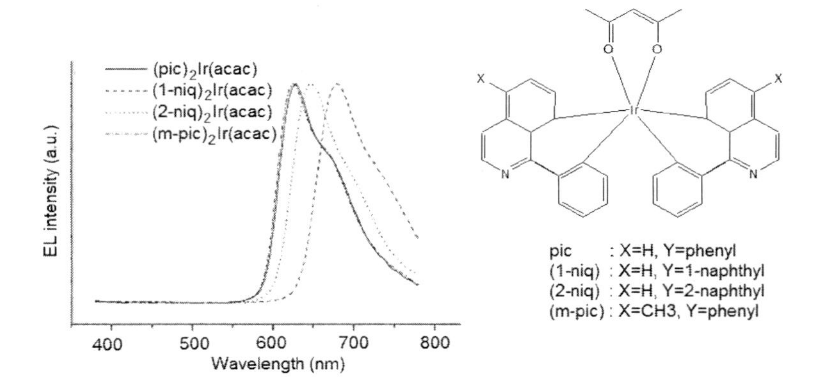

Figure 5.38 Electroluminescence spectrum of the $ITO/NPB/CBP:Ir_{complex}$/ BCP/Alq3/Al OLED for an electrical current density of 50 mA/cm^2. The structures of the Ir(III) complexes are also shown. C.-H. Yang, C.-C. Tai, I.-W. Sun (2004) Synthesis of a high-efficiency red phosphorescent emitter for organic light-emitting diodes, *J. Mater. Chem.*, **14**, 947–950. Reproduced by permission of The Royal Society of Chemistry.

The CIE color coordinates of the different OLEDs active layers were (0.68, 0.32) for $(piq)_2Ir(acac)$, (0.70, 0.27) for $(1\text{-}niq)_2Ir(acac)$, (0.70, 0.30) for $(2\text{-}niq)_2Ir(acac)$ and (0.68, 0.32) for $(m\text{-}piq)_2Ir(acac)$, all in the deep red region. Typically the (1-nic) and (2-niq) cyclometalates lead to a significant red shift (although with not much influence on the CIE color coordinates) with low OLED efficiency. The maximum brightness was near 15600 cd/m^2, 600 cd/m^2, 7900 cd/m^2 and 17200 cd/m^2 for the $(piq)_2Ir(acac)$, $(1\text{-}niq)_2Ir(acac)$, $(2\text{-}niq)_2Ir(acac)$ and $(m\text{-}piq)_2Ir(acac)$ active layers-based complexes, respectively, with respective efficiencies of 7.8 cd/A (8.7 V), 0.24 cd/A (11.9 V), 3.8 cd/A (10.7 V) and 8.9 cd/A (9.6 V). The influence of the cyclometalate ligands were clearly evidenced in these results: the *phenyl* group gives a much better OLED performance that the *naphthyl* one.

The use of combinations of cyclometalates and β-diketonates as ancillary ligands for color tuning with Ir(III) complexes is very common. In an interesting work, several of such Ir(III) complexes were synthesized and tested in OLEDs active layers [41]. The results for a general OLED structure of ITO/NPB/CBP:$Ir(III)_{complex}$/BCP/Alq3/Mg:Ag/Ag were emissions in the green, yellow and red spectral regions. Another interesting work with ACAC as ancillary ligand successfully obtained green and yellow Ir(III)-based OLEDs employing PVK as host instead CBP, in a device of structure of ITO/PVK:$Ir(III)_{complex}$/F-TBB/Alq/LiF/Al [87].

An interesting application of the blue–green and red Ir(III) complexes emitters was made with $(CF_3ppy)_2Ir(pic)$ (*bis[2-(3,5-bistrifluoromethy phenyl)-pyridinato-N,C2']iridium(III)picolinate*) (blue–green) and $(btp)_2Ir(acac)$ (*bis[2-(2'-benzo[4,5-a]thienyl) pyridinato-N,C3']irridium(III)acetylacetonate*) (red), in an attempt to obtain the desired WOLED [88] by double emission mechanism, with an OLED structure of ITO/PEDOT:PSS/NPB/CDBP:$(CF_3ppy)_2Ir(pic)$/BAlq/CDBP : $(btp)_2Ir(acac)$/BAlq/LiF/Al. The guest mass was 3% for $(CF_3ppy)_2Ir(pic)$ and 1% for the $(btp)_2Ir(acac)$ and the host was CDBP (*4,4'-Bis(9-carbazolyl)-2,2'-dimethyl-biphenyl*), a derivate of the known CBP. As expected, the device emission comprises two main broad bands, strongly dependent on the emissive layer thickness and on the guest concentration, centered at 472 nm and 620 nm. The CIE color coordinates were (0.35, 0.36) and the maximum luminance, for the "white" emission, was near 3600 cd/m^2. The CIE color coordinates didn't change significantly upon this luminance indicating that the BAlq layer between the two emissive layers effectively blocked hole migration from the blue–green emissive layer to the red one, preventing the diffusion of excitons generated in the first emissive layer into the second one. This technique is very useful in multi emissive layers OLEDs as the different emissions can be confined in their respective layers and unbiased by the others emissive layers. An interesting observation on this WOLED, is that the white emission was not achieved from the typical sum of RGB emission but rather by attempting to locate it in the straight line which connecting the blue - green with the red region in the color coordinates chart. This approach has become widely studied in the last years.

Others solutions for WOLEDs-based on Ir(III) complexes have been developed. One simple concept used green and red emitting Ir(III) complexes in the same layer and added another with a blue emitting Ir(III) complex [89]. The device structure was ITO/PEDOT:PSS/PVK:PBD:TPD:Ir(mppy)$_3$:Ir(pic)$_2$/PVK:FIrpic/Al, where the Ir(mppy)$_3$, Ir(pic)$_2$ and FIrpic are the green, red and blue Ir(III) emitters. The resulting CIE color coordinates were (0.32, 0.42) with a maximum luminance of about 3100 cd/m^2 for an applied voltage of 13 V.

The LEC concept has been studied for Ir(III) complexes. As reported for Ru(II) complexes, the basic mechanism is the redox process occurring at the device cathode and anode. The active layer of the device must thus comprise an ionic complex with mobile counterions. One work used [Ir(F-mppy)$_2$(dtb-bipy)](PF$_6$) (*iridium(III) bis 2-4′-fluorophenyld-5-methylpyridine 4,4′-di-tert-butyl-2 ,2′ bipyridine hexafluorophosphate*) in a single layer device of ITO/[Ir(F-mppy)$_2$(dtb-bipy)](PF$_6$)/Au [90]. The emission is green with a broad band centered at near 550 nm. The delay time is incredibly high—a minimum of near 40 minutes—but apparently typical for Ir(III) non-fluorinates-based LECs [91]. In order to reduce this delay time the authors included an ionic liquid into the device active layer. Since the delay time is governed by the ionic conductivity in the active media, it should decrease upon addition of extra mobile counterions. In fact, the inclusion of BMIM(PF_6) (*1-butyl-3-methylimidazolium hexafluorophosphate*) deceases the delay time to near 25 minutes, which is still a high value, but proved, nonetheless, the practicability of the solution.

Another LECs based on Ir(III) was also tested, with the complexes Ir($ppy)_2$sb](PF_6) (*iridium(III) 2-phenylpyridine 4,5-diaza-9,9′-spirobifluorene hexafluorophosphate*) and [Ir($dFppy)_2$sb](PF_6) (*iridium(III) 2-(2,4-difluorophenyl)pyridine 4,5-diaza-9,9′-spirobifluorene hexafluorophosphate*) employed in a LEC structure of ITO/Ir($III)_{complex}$/Ag [92]. The [Ir($ppy)_2$sb](PF_6)-based LEC showed an orange emission (broad band centered at 580 nm) while the LEC based on [Ir($dFppy)_2$sb](PF_6) showed a green emission (a broad band centered at 535 nm). The driving voltages were quite low, typically about 2.6–2.8 V, but the delay time remains very high even after adding the ionic liquid BMIM(PF_6): about 170 minutes for the LEC based on the first complex and about 90 minutes for the second. Again, the principle of LEC could be applied to these Ir(III) complexes, but much work still to be done until acceptable delay times can be achieved. A more recent work using [Ir($ppy)_2$(Hpbpy)](PF_6) [93] features a delay time of days (!).

5.5.2.5 Osmium- and platinum-based OLEDs

Osmium and platinum TM complexes less studied than the previously TM, as they have been over shadowed by iridium and ruthenium. The full color spectra obtained by the Ru(II), Re(I) and Ir(III) complexes, their technological interest and well known device structures are major distractions to the opening of new focuses in TM complexes-based OLEDs. Nevertheless, the osmium and platinum complexes could also be useful for these devices.

Concerning the osmium element, a red emitter device will be the most targets. An interesting work tested fourteen Os(II) complexes in an OLED structure of ITO/BTPD-PFCB/AL/Ca where *AL* denotes the active layer composed by PVK:PBD:Os(II)$_{\text{complex}}$ or PVN:PBD:Os(II)$_{\text{complex}}$ [94]. The BTPD-PFCB (*tetraphenyldiamine-containing perfluorocyclobutane*) is a hole transporting polymer and PVN (*poly(2-vinylnaphthalene*) is a possible substituent of PVK. The Os(II) complex concentration was 3% (in mass). Two kind of basic Os(II) complexes were used, with a general formula Os(N^N)$_2$LX(Y) where N^N is the cyclometalate ligand, LX the ancillary ligand and Y a counterion. The best result was obtained with (Os(O-Me$_2$phen)(ph$_2$P$_2$)](Ts)$_2$ *([osmium(II) (4,7-bis(p-methoxyphenyl)-1,10-phenanthroline)$_2$ cis-1,2-bis(diphenylphosphino)ethylene](Ts)$_2$*) with a red emission (broad band centered at 611 nm) with a luminance about 1430 cd/m^2 and a driving voltage (for a luminance of 1 cd/m^2) of 7.6 V. Although the maximum brightness was obtained with PVK:PBD host, the maximum external efficiency was measured when the PVN:PBD host was employed. As a matter of fact, the emission spectra of PVN:PBD has a much better overlap with the MLCT absorption bands of the Os(II) complexes than that of PVK:PBD. Thus, and considering that the host to guest Förster energy transfer mechanism depends on this overlap, this mechanism is to expected be predominant in PVN:PBD hosts. On the contrary, for PVK:PBD hosts, an electrical carrier trapping in the Os(II) complexes was proposed by the authors as the main energy transfer process. Interestingly, and in spite of its low external efficiency, this mechanism provided OLEDs with substantial higher brightness. The emission is assumed to be originated at a ^{3}MLCT energy state.

Different Os(II) complexes were also tested for an OLED with red emission. The complexes were Os(fppz)$_2$(PPh$_2$Me)$_2$ (1), Os(fppz)$_2$(PPhMe$_2$)$_2$ (2) and Os(bptz)$_2$(PPh2Me)$_2$ (3), where fppz and bptz are *pyridyl pirazolate* and *pyridyl triazolate* derivatives, respectively, and PPh$_2$Me and PPhMe$_2$ are *dipheylmethyl* and *phenyldimethyl phosphines* (see on the right side of Fig. 5.39 for further details on the ligands) [49]. The Os(II) complexes were tested as OLED active layers in a very device simple structure of ITO/PVK:Os(II)$_{\text{complex}}$/PBD/LiF/Al, all of 10% (in mass). In this

OLED structure, PVK simultaneously functions as host and HTL. In Fig. 5.39 the electroluminescence spectra and the molecular structure of the Os(II) complexes are shown.

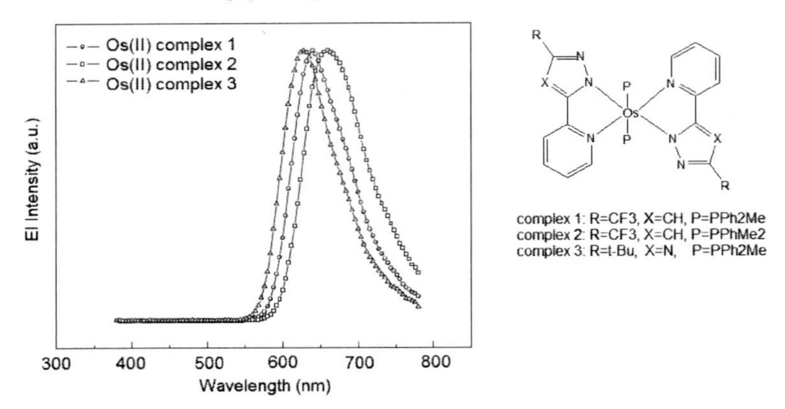

Figure 5.39 Electroluminescence spectra of the ITO/PVK:Os(II)complex/PBD/LiF/Al OLED. The general molecular structure of the Os(II) complexes is also shown. Reprinted with permission from Tung, Y.-L. ,et al. (2004) Highly efficient red phosphorescent osmium(II) complexes for OLED applications, *Organometallics*, **23**, 3745–3748. Copyright 2004 American Chemical Society.

The electroluminescence spectra all showed a deep red emission (clearly featureless, typical of a triplet emitter of high MLCT character) with a red shift depending on the Os(II) molecular structure. This shift can be associated with differences on the electron donating character of the PPh_2Me and $PPMe_2$ ligands. Moreover, these *phenylphosphine* ligands contributed to raise the energy of the $HOMO_d$ level of the Os(II) complexes. The maxima wavelength emission were around 626 nm, 640 nm and 658 nm, for Os(II) complexes indicated as 1, 2 and 3 in the figure, respectively. The corresponding measured luminance were of about 3300 cd/m^2, 2200 cd/m^2 and 800 cd/m^2 with driving voltages of 5.5 V, 8.5 V and 8.5 V (for an emission of 1 cd/m^2), respectively.

Os(II) complexes were also used in attempting applications for white OLEDs. An interesting approach was made with an orange emitting Os(II) complex, quite similar to complex 3 from previously example, but without extra *trifluoromethyl* group on the bptz ligand, to give, the $Os(bpftz)_2(PPh_2Me)_2$, where bpftz is, thus *CF_3-bptz.* The device structure was ITO/PEDOT:PSS/ PVK:PBD:DPAVBi: $Os(bpftz)_2(PPh_2Me)_2$/TPBI/Mg:Ag where the DPAVBi *(4,4′-bis[2-{4-(N,N-diphenylamino)phenyl}vinyl]biphenyl)* is a blue emitting dye [95]. The concept was to achieve an emission comprising two independent emissions opposite points of the CIE color coordinates chart, or, in simple terms, complementary colors (blue and orange) which can be connected by a straight line passing in the white region. The results were very promising as the CIE color coordinates were (0.33, 0.32) and the OLED featured a luminance of over 11300 cd/m^2 (for an applied voltage of 16 V) with an efficiency of 13.2 cd/A.

Platinum-based OLEDs have also stimulated research in the past years due to very promising results with high-brightness devices. The most efficient Pt(II) complexes exhibit a triplet emitter state which may be of LC or MLCT character. And as, for almost all TM complex-based OLEDs, the Pt(II) complex is embedded as a guest into a host matrix to form the device's active layer. The predominant emitting colors are yellowish green, yellowish orange and red.

An important work on the application of Pt(II) complexes in OLEDs was made using a set of cyclometalated organic ligands of the C^N and N^N types which featured a molecular planar geometry [96]. The complexes had the generic formula Pt(a-bypy)$(C{\equiv}C)_nR$) (*platinum(II) 6 - aryl - 2,2 ′ - bipyridine)$(C{\equiv}C)_nR$*), n = 1–4 and R = (*aryl, alkyl* or *trimethylsilyl*). Of the considerable set of Pt(II) complexes synthesized, six of them appeared to be interesting for light emitting devices. The authors also found that a discerning replacement of the cyclometalated ligands with ligands of the *σ-alkynyl* family could tune the color from yellowish green to red. In fact, this *σ-alkynyl* ligands acts in specific ways to condition the Pt(II) complex luminescence, in particular by altering the energy state of the complex triplet emitter, typically with a strong MLCT character, due to the π orbital conjugation. Further modification of the C^C and C^N cyclometalate ligands will help on the color tuning by changing (as previous discussed) the location of the related π^* orbitals energy states. The OLED structure employed was ITO/NPB/CBP:Pt(II)$_{complex}$/BCP/Alq3/Mg:Ag. The Pt(II)*complex* concentration was tested between 2% and 6% (in mass), though, as expected, the higher doping concentrations lowered the device performance and efficiencies by quenching effects. In Fig. 5.40 the electroluminescence spectra and the Pt(II) complexes molecular structures are shown.

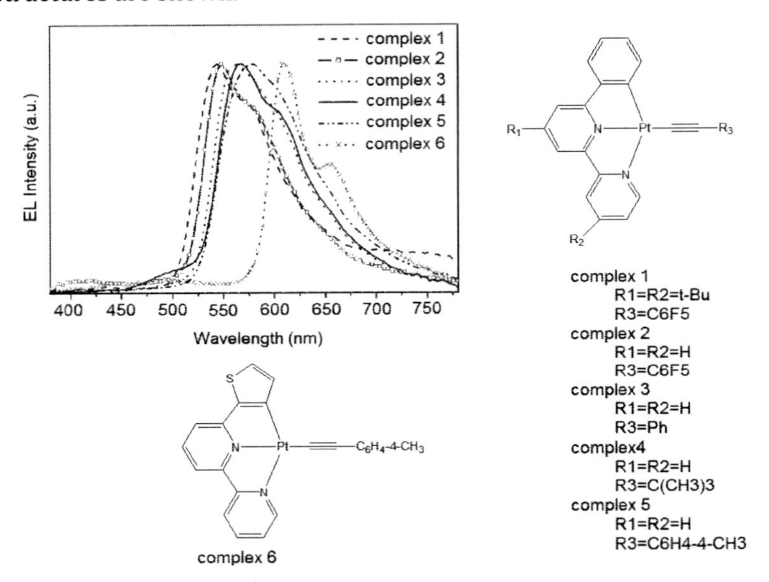

Figure 5.40 Electroluminescence spectra of the ITO/NPB/CBP:Pt(II) complex/Alq3/Mg:Ag OLED. The general molecular structure of the Pt(II) complexes is also shown. Reprinted with permission from Lu, W., et al. (2004) Light-emitting tridentate cyclometalated platinum(II) complexes containing σs-alkynyl auxiliaries: tuning of photo- and electrophosphorescence, *J. Am. Chem. Soc.*, **126**, 4958–4971. Copyright 2004 American Chemical Society.

The obtained results are very promising. An OLED prepared with 2% (mass) of the complex 1, exhibited a maximum luminance of about 10000 cd/m^2 and the maximum efficiency was near 4.2 cd/A (for an electrical current density of 10 mA/cm^2). The general driving voltage was between 3–4 V. It must be noted that a very low emission from the CBP can be observed in the 400–430 nm spectral region (refer to Fig. 5.40). The CIE color coordinates were (0.42, 0.53), (0.44, 0.51), (0.48, 0.48), (0.48, 0.47), (0.51, 0.47), (0.55, 0.31) for the complexes named 1, 2, 3, 4, 5, and 6, as identified in the figure. They represent emission bands with maximum wavelengths in 545 nm (green–yellow), 548 nm (green–yellow), 564 m (yellow), 567 nm (yellow), 580 nm (yellow) and 612 nm/656 nm (orange–i.e. ranging from green–yellow to orange–red spectral region. It is not difficult to conclude that these OLEDs emissions had relatively poor color purity.

Another general work concerning Pt(II) complexes-based OLEDs was done with a family of $Pt(N^\wedge N)_2$ complexes comprising several different N^N cyclometalate ligands, namely bppz, bzpz, bmpz , bqpz, fppz, hppz, bptz and hptz [50] . This work is particularly interesting because it allows a direct comparison of the organic ligands in a homoleptic molecular structure. Of these cyclometalate ligands the authors used the $Pt(bppz)_2$ (*platimum(II) bis 2,6- bis(pyrazol-1-yl)pyrazine*) and $Pt(fppz)_2$ (*platimum(II) bis 3-trifluoromethyl-5(2-pyridyl)pyrazole*) to fabricate OLEDs with a structure given by ITO / NPB (40 nm) / $CBP{:}Pt(II)_{complex}$ (30 nm) / BCP (10 nm) / Alq3 (30 nm) /LiF (1 nm) / Al with guest concentrations (in mass) of 6–7%, 20%, and 50%. A device with 100% of Pt(II) complex in the active layer was also tested. The results, along with the ligands comparison, allow to study the influence of the host matrix on electroluminescence.

In the OLEDs with the $Pt(bppz)_2$ complex, the electrical current and luminance data vs. applied voltage showed a current density increase with the guest concentration increase. This behavior suggests that the host to guest energy transfer has followed the electrical carrier trapping mechanism [80], by the simple assumption that by increasing the guest concentration more trap sites will be available. Additionally, the electroluminescence spectra shifted towards low energy (red) as the guest concentration increased (and the emission band has broadened). The best result was observed for a guest concentration of 6% with a luminance of about 27100 cd/m^2 for an applied voltage of 15 V, and an efficiency of 8.6 cd/A for an electrical current density of 20 mA/cm^2. The OLED typically emits in the green–yellow spectral range, with the emission maxima ranging from 502 nm (6% of guest concentration) to 556 (100% of the $Pt(bppz)_2$ complex in active layer). Considering the $Pt(fppz)_2$ emitter, in the same device structure, emissions were typically in the yellow–orange spectral region, also dependent on the guest concentration, with maxima peaking from 552 nm (7% of guest concentration) to 616 nm (100% of $Pt(fppz)_2$ complex in active layer). The best result was obtained for a guest concentration of 20%, with a luminance of 41000 cd/m^2 (applied voltage of 15 V) with and efficiency of 19.7 cd/A (for an electrical current density of 20 mA/cm^2). In this example, the main character of the triplet emitter wasn't the typical MLCT/LC but was attributed to the MMLCT or metal–metal to ligand charge transfer, common in a binuclear emission, in which the metal ions interact (in this case Pt–Pt) due to their close proximity. Although this formal mechanism is not much different from MLCT, it involves a π–π stacking and the emission is consequently red-shifted, i.e. the triplet energy level is lower. This effect was more pronounced for $Pt(fppz)_2$ than for $Pt(bppz)_2$ and in both cases these was strong influence of the planar molecular geometry. Moreover the fppz ligand will increase such interaction.

Some others interesting results of OLEDs employing Pt(II) complexes has been reported in the literature and clearly show a gradual interest on such TM complexes for light emitter devices [97, 98].

5.6 Final Considerations about Transition Metal-Based OLEDs

In this chapter the fundamental concepts about the use of TM complexes in OLEDs active layers were focused. Not surprisingly their use is simultaneously attractive but difficult to optimize from the physical point of view. Fortunately, and except a few particular cases, the main pathways are well established, helping to design the OLED with account of the photo-physical properties of the light emitting complex. The constant development of novel organic ligands is an important additional tool for the improvement of TM-based OLEDs. Now, some remarks can be focused.

The recognized superior interest demonstrated of TM complexes-based OLEDs relies on fulfillment of three important requirements: high brightness, high efficiency and low driving voltage. Nevertheless, as demonstrated in this chapter, combining all three features in a single OLED is not an easy task. Still, the current state of the art, with several excellent results reported, opens new perspectives for future developments.

TM complex-based OLEDs typically present a large emission bands, making it difficult to obtain pure-color displays for use in a color matrix (with some exceptions). This may, their extreme bright emission is now also being directed towards more practical applications such solid state lighting. Still, even for displays applications, it must be noted that some of the herein presented OLEDs exhibits an industrial quite reasonable color purity, showing that the fine-tuning of the structure of the emitting complex may, in theory, produce the required TM complexes for displays.

A side from the specific examples presented on, there are a few topics representing the main pathways for the successful integration of TM complexes onto light emitting devices, which can now be highlighted. These are, in a brief description, the following:

- MT complex-based OLEDs often have an active layer, based on the universal principle of the host → guest energy transfer. The host matrix must be a hole transport material or a blend formed by a hole and an electron transport material. This structure is designed to achieve an efficient energy transfer to the guest TM complex by a predominant Förster, Dexter or electrical carrier trapping mechanism. This "energy pumping" methodology has proved extremely efficient by a vast number of positive results from different authors;
- Further improvement can be obtained using layers to block the electrical carriers, in particular the electron blocking layers. In that point, no special differences is observed for the others kind of active OLED layers;
- The particular nature of energy levels of a TM complex molecule requires a special perspective concerning their application. In fact, emission occurs at *molecular energy levels* ascribed to the whole molecule and not to the individual contributions of the organic ligand or metal ion. Both entities, with their respective orbitals-defined energy levels will

define the structure of the molecular energy levels and the luminescence will be governed by the whole molecule;

- The main emissions are originated at the MLCT and LC energy levels defined in last point, leading to broad and typically featureless, emission bands, particularly those originated at MLCT levels; simple vibronic structures can be observed in some LC originated emissions. The very extended absorption bands allows to identify the LC absorptions (typically in the UV spectral region) and the MLCT absorptions (typically in the near UV-visible spectral region); this last and considering that for almost TM complexes of interest have their absorption bands in more or less the same spectral region, allows the use of practically the same host materials for all TM complexes;
- The main advantage of the TM complexes is the well-known triplet harvesting which allows a theoretical internal efficiency of 100%, which is actually almost observed in a few complexes. This property is an exceptional advantage for the preparation of high efficient and brightness OLEDs;
- The chemical modifications of existing organic ligands or the synthesis of new ones could introduce significant changes in the TM complex molecular levels and therefore, easily help tune the emitted color. The case of iridium is emblematic: all the main visible colors can be achieved; even for other common TM elements, a relatively wide range of colors is available. Overall, it is not difficult to find a TM complex of a particular color for a high-brightness OLED;
- Finally, some ionic TM complexes (mostly ruthenium-based or, less often and with fewer practical application, iridium-based), can be used in the LECs, simple high-brightness emitters with an enormous application potential in technological areas. The high delay time associated with that kind of light emitter is the main problem to overcome;

As a final note, the role of phosphorescence TM complexes in the path of OLEDs development is well recognized, in parallel with RE complexes. Moreover, the high brightness and efficiencies, allied with the simple OLED structures (for which the LEC concept is absolutely indispensable) are very attractive for flexible light emitters.

Perhaps one of the major drawbacks of TM complex-based OLEDs is the typical saturation of the light emitting sites in the active layer due to the specific nature of the emitters (triplet–triplet annihilation, long lifetimes, when used as guest the high concentration quenching, etc.), which can be addressed by the synthesis of particular ligands, not always easy as demonstrated by the present examples. This shows the extreme importance of the chemical synthesis optimization, including the necessary adjustments for the color tuning. And naturally much work needs to be done to optimize the "game" of the molecular energy levels, that in all internal process, governs all the complex luminescence behavior and by consequence, the OLED employing such complexes.

These considerations can be regarded as the main "rules" which govern the performance of a TM complex-based OLED. As previously described, the questions involving the points focused must be taken into consideration when the device is designed. And naturally we must be aware that not every good light-emitting TM complex can efficiently open a new path to "high-brightness OLEDs" as the discussion has indicated.

References

1. Yersin, H. (2004) *Top. Curr. Chem.* **241**, *Transition Metal and Rare Earth Compounds*, "Triplet Emitters for OLED Applications. Mechanisms of Exciton Trapping and Control of Emission Properties," Springer, Germany, pp. 1–26.

2. Schüring, J., Schulz, H. D., Fischer, W. R., Böttcher, J. and Duijnisveld, W. H. (eds.) (1999) *Redox: Fundamentals, Processes and Applications*, Springer-Verlag, Germany.

3. Wu, A. (1998) Light emitting electrochemical devices from sequentially adsorbed multilayers of a polymeric ruthenium (II) complex and various polyanions, *Thin Solid Films*, **327–329**, 663–667.

4. Bernhard, F. (2002) Electroluminescence in ruthenium(II) complexes, *J. Am. Chem. Soc.*, **124**, 13624–13628.

5. Tarr, D. A. and Miessler, G. L. (1991) *Inorganic Chemistry* 2nd edn., Prentice Hall, USA.

6. Chan, W. K. (1999) Light-emitting multifunctional rhenium(I) and ruthenium(II) 2,2'-bipyridyl complexes with bipolar character, *Appl. Phys. Lett.*, **75**, 3920–3922.
7. Wu, A. (1999) Solid-state light-emitting devices based on the tris-chelated ruthenium(II) complex: 3. high efficiency devices via a layer-by-layer molecular-level blending approach, *J. Am. Chem. Soc.*, **121**, 4883–4891.
8. Kalyuzhny, G. (2003) stability of thin-film solid-state electroluminescent devices based on tris(2,2'-bipyridine) ruthenium(II) complexes, *J. Am. Chem. Soc.*, **125**, 6272–6283.
9. Lundin, N. (2006) Synthesis and characterization of a multicomponent rhenium(I) complex for application as an OLED dopant, *Angew. Chem.*, **118**, 2644–2646.
10. Duan, Y. (2005) High-efficiency red electrofluorescence devices employing a rhenium complex as phosphorescent sensitizer, *Opt. Quantum Electron.*, **37**, 1121–1127.
11. Liu, C. B. (2007) A multicomponent rhenium-based triplet emitter for organic electroluminescence, *Chem. Phys. Lett.*, **435**, 54–58.
12. Tokito, S. (2003) High-efficiency white phosphorescent organic light-emitting devices with greenish-blue and red-emitting layers, *Appl. Phys. Lett.*, **83**, 2459–2461.
13. Adachi, C. (2000) High-efficiency organic electrophosphorescent devices with tris(2-phenylpyridine)iridium doped into electron-transporting materials, *Appl. Phys. Lett.*, **77**, 904–906.
14. Fujii. K. (2005) Bright and ultimately pure red electrophosphorescence diode bearing diphenyl-quinoxaline, *IEICE Electron. Express*, **2**, 260–266.
15. Kappaun, S. (2008) Phosphorescent organic light-emitting devices: working principle and iridium based emitter materials, *Int. J. Mol. Sci.*, **9**, 1527–1547.
16. Kim, J. H. (2003) Bright red-emitting electrophosphorescent device using osmium complex as a triplet emitter, *Appl. Phys. Lett.*, **83**, 776–778.
17. Chien, C.-H. (2009) Efficient red electrophosphorescence from a fluorene-based bipolar host material, *Org. Electron.*, **10**, 871–876.

18. Williams, E. L. (2007) Excimer-based white phosphorescent organic light-emitting diodes with nearly 100% internal quantum efficiency, *Adv. Mater.*, **19**, 197–202.

19. DeRosa, F. (2005) Synthesis and luminescence properties of Cr(III) complexes with cyclam-type ligands having pendant chromophores, trans-$[Cr(L)Cl_2]Cl^1$, *Inorg. Chem.*, **44**, 4166–4174.

20. Basu, C. (2007) A novel blue luminescent high-spin iron(III) complex with interlayer O–H...Cl bridging: Synthesis, structure and spectroscopic studies, *Polyhedron*, **26**, 3617–3624.

21. Jernigan, J. C. (1989) Electrical-field-driven electron self-exchange in a mixed-valent osmium(II/III) bipyrldine polymer: solid-state reactions of low exothermicity, *J. Phys. Chem.*, **93**, 4620–4621

22. Zheng, X. Y. (2003) A white OLED based on DPVBi blue light emitting host and DCJTB red dopant, *Displays*, **24**, 121–124.

23. Chen, L. (2009) Carrier transport in zinc phthalocyanine doped with a fluorinated perylene derivative: bulk conductivity versus interfacial injection, *J. Phys. Chem. C*, **113**, 17160–17169.

24. Iwase, T. (2003) Photovoltaic characteristics of copper phthalocyanine-poly(p-phenylene) co-evaporation Films, *Jpn. J. Appl. Phys.*, **42**, 5330–5335.

25. Senthilarasu, S. (2004) Characterization of zinc phthalocyanine (ZnPc) for photovoltaic applications, *App. Phys. A: Mater. Sci. Process.*, **77**, pp. 383–389.

26. Grätzel, M. (2001) Photoelectrochemical cells, *Nature*, **414**, 338–344

27. Jones, C. J. (2002) *d- and f- Block Chemistry*, Wiley-VCH, UK.

28. Rabek, J. F. (1987) *Mechanisms of Photophysical Processes and Photochemical Reactions in Polymers*, John Wiley & Sons, Sweden.

29. Tinkham, M. (1992) *Group Theory and Quantum Mechanics*, 2nd edn., Dover, USA.

30. Ballhausen, C. J. (1979) *Molecular Electronic Structures of Transition Metal Complexes*, McGraw-Hill, USA.

31. Gilbert, A. and Baggot, J (1991) *Essentials of Molecular Photochemistry*, Blackwell Scientific Publications, UK.

32. Shriver D. F, Atkins, P. W. and Langford, C. H. (1991) *Inorganic Chemistry*, Oxford University Press, UK.

33. Yersin, H. and Finkenzeller, W. J. (2008) *Highly Efficient OLEDs with Phosphorescence Materials* (ed. Yersin, H.), Chapter 1 "Triplet Emitters for Organic-Light Emitting Diodes: Basic Properties," Wiley-VCH, Germany, pp. 1–98.

34. Judd, B. R. (1962) Optical absorption intensities of rare-earth ions, *Phys. Rev.*, **127**, 750–761.

35. Cohen - Tannoudji, C., Diu, B. and Laloë, F. (1977) *Quantum Mechanics*, vol. 1, Wiley, USA.

36. Rausch, A. F. (2007) Spin-orbit coupling routes and OLED performance-studies of blue-light emitting Ir(III) and Pt(II) complexes, *Proc. SPIE*–The International Society for Optical Engineering, vol. 6655, 66550F/1.

37. Hay, P. J. (2002) Theoretical studies of the ground and excited electronic states in cyclometalated phenylpyridine Ir(III) complexes using density functional theory, *J. Phys. Chem. A*, **106**, 1634–1641.

38. Colombo, M. G., Andreas, H. and Güdel, H. U. (1994) *Top. Curr. Chem.* **171**, (ed. Yersin, H.) "Competition between ligand centered and charge transfer lowest excited states in bis cyclometalated Rh^{3+} and Ir^{3+} complexes," Springer, Germany, pp. 143–171.

39. Baldo, M. A. (1999) Very high-efficiency green organic light-emitting devices based on electrophosphorescence, *Appl. Phys. Lett.*, **75**, 4–6.

40. Colombo, M. G. (1994) Facial tris cyclometalated rhodium($^{3+}$) and iridium($^{3+}$) complexes: their synthesis, structure, and optical spectroscopic properties, *Inorg. Chem.*, **33**, 545–550.

41. Lamansky, S. (2001) Highly phosphorescent bis-cyclometalated iridium complexes: synthesis, photophysical characterization, and use in organic light emitting diodes, *J. Am. Chem. Soc.*, **123**, 4304–4312.

42. Lamansky, S. (2001) Synthesis and characterization of phosphorescent cyclometalated iridium complexes, *Inorg. Chem.*, **40**, 1704–1711.

43. Lowry, M. S. (2006) Synthetically tailored excited states: phosphorescent, cyclometalated iridium(III) complexes and their applications, *Chem. Eur. J.*, **12**, 7970–7977.

44. Tsuzuki, T. (2003) Color tunable organic light-emitting diodes using pentafluorophenyl-substituted iridium complexes, *Adv. Mater.*, **15**, 1455–1458.

45. Chi. Y. (2010) Transition-metal phosphors with cyclometalating ligands: fundamentals and applications, *Chem. Soc. Rev.*, **39**, 638–655.

46. Santos, G., et al. (2008) Development and characterization of light-emitting diodes (LEDs) based on ruthenium complex single layer for transparent displays, *Phys. Stat. Sol. (a)*, **205**, 2057–2060.

47. Kalyanasundaram, K. (1982) Photophysics, photochemistry and solar energy conversion with tris(bipyridyl)ruthenium(II) and its analogues, *Coord. Chem. Rev.*, **46**, 159–244.

48. Beyer, B., et al. (2009) Phenyl-1H-[1,2,3]triazoles as new cyclometalating ligands for iridium(III) complexes, *Organometallics*, **28**, 5478–5488.

49. Tung, Y.-L., et al. (2004) Highly efficient red phosphorescent osmium(II) complexes for OLED applications, *Organometallics*, **23**, 3745–3748.

50. Chang, S.-Y., et al. (2006) Platinum(II) complexes with pyridyl azolate-based chelates: synthesis, structural characterization, and tuning of photo- and electrophosphorescence, *Inorg. Chem.*, **45**, 137–146.

51. Pei, Q. (1995) Polymer light-emitting electrochemical cells, *Science*, **269**, 1086–1088.

52. Juris, A. (1988) Ru(II) polypyridine complexes: photophysics, photochemistry, and chemiluminescence, *Coord. Chem. Rev.*, **84**, 85–277.

53. Santos, G., et al. (2008) Opto-electrical properties of single layer flexible electroluminescence device with ruthenium complex, *J. Non Cryst. Sol.*, **354**, 2571–2574.

54. Buda, M. (2002) Thin-film solid-state electroluminescent devices based on tris(2,2'-bipyridine)ruthenium(II) complexes, *J. Am. Chem Soc.*, **124**, 6090–6098.

55. Rudmann, H. (2002) Solid-state light-emitting devices based on the tris-chelated ruthenium(II) complex. 4. high-efficiency light-emitting devices based on derivatives of the tris(2,2'-bipyridyl) ruthenium(II) complex, *J. Am. Chem. Soc.*, **124**, 4918–4921.

56. Shiratori, S. (2001) A quick operating organic EL device using Ru complex light-emitting layer with 2,5-bis(4-tert-butyl-2-benz-

oxazolyl) thiophene electron transport layer, *Mater. Sci. Eng. B*, **85**, 149–153.

57. Yang, J. (2003) Light-emitting devices based on ruthenium(II) (4,7-diphenyl-1, 10-phenanthroline)$_3$: Device response rate and efficiency by use of tris-(8-hydroxyquinoline)aluminum, *J. Appl. Phys.*, **94**, 6391–6395.

58. Rudmann, H. (2002) Prevention of the cathode induced electrochemical degradation of †Ru(bpy)$_3$(PF$_6$)$_2$ light emitting devices, *J. Appl. Phys.*, **92**, 1576–1581.

59. Zhen, C. (2005) Incorporation of electroluminescence and electrochemiluminescence in one organic light-emitting device, *Appl. Phys. Lett.*, **87**, 093508 1–3.

60. Welter, S. (2003) Electroluminescent device with reversible switching between red and green emission, *Nature*, **421**, 54–57.

61. Tung, Y.-L. (2005) Organic light-emitting diodes based on charge-neutral Ru(II) phosphorescent emitters, *Adv. Mater.*, **17**, 1059–1064.

62. Xia, H. (2004) Highly efficient red phosphorescent light-emitting diodes based on ruthenium(II) complex-doped semiconductive polymers, *Appl. Phys. Lett.*, **84**, 290–292.

63. Yang, J. (2004) Electroluminescence of ruthenium(II)(4,7-diphenyl-1,10-phenanthroline)$_3$ from charge trapping by doping in carrier-transporting blend films, *Chem. Phys. Lett.*, **385**, 481–485.

64. Chang, S. M. (2006) Red electroluminescence from ruthenium complexes, *Surface & Coatings Technology*, **200**, 3289–3296.

65. Chuai, W. (2004) Novel polypyridyl ruthenium(II) complexes for electroluminescence, *Synth. Met.*, **145**, 259–264.

66. Bolink, H. J. (2005) Observation of electroluminescence at room temperature from a ruthenium(II) bis-terpyridine complex and its use for preparing light-emitting electrochemical cells, *Inorg. Chem.*, **44**, 5966–5968.

67. Bolink, H. J. (2009). Deep-red-emitting electrochemical cells based on heteroleptic bis-chelated ruthenium(II) complexes, *Inorg. Chem.*, **48**, 3907–3909.

68. Li, F. (2003) Red electrophosphorescence devices based on rhenium complexes, *Appl. Phys. Lett.*, **83**, 365–367.

69. Li, F. (2004) Highly efficient electrophosphorescence devices based on rhenium complexes, *Appl. Phys. Lett.*, **84**, 148–150.

70. Liu, C. (2006) Triphenylamine-functionalized rhenium(I) complex as a highly efficient yellow-green emitter in electrophosphorescent devices, *Appl. Phys. Lett.*, **89**, 243511/1-3.

71. Ranjan, S. (2003) Realizing green phosphorescent light-emitting materials from rhenium(I) pyrazolato diimine complexes, *Inorg. Chem.*, **42**, 1248–1255.

72. Li, F. (2003) White-electrophosphorescence devices based on rhenium complexes, *Appl. Phys. Lett.*, **83**, 4716–4718.

73. Tsutsui, T. (1999) High quantum efficiency in organic light-emitting devices with iridium-complex as a triplet emissive center, *Jpn. J. Appl. Phys. Part 2*, **38**, L1502–L1504.

74. Aldachi, C. (2001) High-efficiency red electrophosphorescence devices, *Appl. Phys. Lett.*, **78**, 1622–1624.

75. Holmes, R. J. (2003) Blue organic electrophosphorescence using exothermic host–guest energy transfer, *Appl. Phys. Lett.*, **82**, 2422–2423.

76. Yang, C.-H. et al. (2007) Blue-emitting heteroleptic iridium(III) complexes suitable for high-efficiency phosphorescent OLEDs, *Angew. Chem.*, **46**, 2418–2421.

77. Fukagawa, H. (2008) Highly efficient deep-blue phosphorescence organic light emitting diodes with a double-emitting layer structure, *Appl. Phys. Lett.*, **93**, 133312/1–3.

78. Chew, S. (2006) Photoluminescence and electroluminescence of a new blue-emitting homoleptic iridium complex, *Appl. Phys. Lett.*, **88**, 093510/1–3.

79. Kim, S. W. (2006) Highly efficient green phosphorescent single-layered organic light-emitting devices, *Appl. Phys. Lett.*, **89**, 213511/1–3.

80. Gong, X., et al. (2003) Electrophosphorescence from a polymer guest-host system with an iridium complex as guest: Förster energy transfer and charge trapping, *Adv. Func. Mater.*, **13**, 439–444.

81. Fei, T. (2009) Highly efficient pure yellow electrophosphorescent device by utilizing an electron blocking material, *Semicond. Sci. Technol.*, **24**, 105019/1–5.

82. Hsu, N.-M. (2006) Accelerated discovery of red-phosphorescent emitters through combinatorial organometallic synthesis and screening. *Angew. Chem.*, **118**, 4244–4248.

83. You, Y. (2005) Inter-ligand energy transfer and related emission change in the cyclometalated heteroleptic iridium complex: facile and efficient color tuning over the whole visible range by the ancillary ligand structure, *J. Am. Chem. Soc.*, **127**, 12438–12439.

84. Zhang, X., et al. (2006) Highly efficient iridium(III) complexes with diphenylquinoline ligands for organic light-emitting diodes: Synthesis and effect of fluorinated substitutes on electrochemistry, photophysics and electroluminescence, *J. Organomet. Chem.*, **691**, 4312–4319.

85. Wong, W.-Y. (2006) Amorphous diphenylaminofluorene-functionalized iridium complexes for high-efficiency electrophosphorescent light-emitting diodes, *Adv. Func. Mater.*, **16**, 838–846.

86. Yang, C.-H. (2004) Synthesis of a high-efficiency red phosphorescent emitter for organic light-emitting diodes, *J. Mater. Chem.*, **14**, 947–950.

87. Sun, P. (2006) Synthesis of novel Ir complexes and their application in organic light emitting diodes, *Synth. Met.*, **156**, 525–528.

88. Tokito, S. (2005) High-efficiency blue and white phosphorescent organic light-emitting devices, *Curr. Appl. Phys.*, **5**, 331–336.

89. Tang, K.-C. (2008) Broad band and white phosphorescent polymer light-emitting diodes in multilayer structure, *Synth. Met.*, **158**, 287–291.

90. Slinker, J. D. (2005) Green electroluminescence from an ionic iridium complex, *Appl. Phys. Lett.*, **86**, 173506/1-3.

91. Slinker, J. D. (2004) Efficient yellow electroluminescence from a single layer of a cyclometalated iridium complex, *J. Am. Chem. Soc.*, **126**, 2763–2767.

92. Su, H.-C. (2007) Highly efficient orange and green solid-state light-emitting electrochemical cells based on cationic IrIII complexes with enhanced steric hindrance, *Adv. Func. Mater.*, **17**, 1019–1027.

93. Bolink, H. J. (2008) Long-living light-emitting electrochemical cells—control through supramolecular interactions, *Adv. Mater.*, 20, 3910–3913.

94. Carlson, B. (2002) Divalent osmium complexes: synthesis, characterization, strong red phosphorescence, and electrophosphorescence, *J. Am. Chem. Soc.*, **124**, 14162–14172.

95. Shih, P.-I (2006) Efficient white-light-emitting diodes based on poly(N-vinylcarbazole) doped with blue fluorescent and orange phosphorescent materials, *Appl. Phys. Lett.*, **88**, 251110/1–3.

96. Lu, W., et al. (2004) Light-emitting tridentate cyclometalated platinum(II) complexes containing σs-alkynyl auxiliaries: tuning of photo- and electrophosphorescence, *J. Am. Chem. Soc.*, **126**, 4958–4971.

97. Chan, S. C. (2001) Organic light-emitting materials based on bis (arylacetylide) platinum(II) complexes bearing substituted bipyridine and phenanthroline ligands: photo- and electroluminescence from ^{3}MLCT excited states, *Chem. Eur. J.*, **7**, 4180–4190.

98. Kavitha, J. (2005) In search of high-performance platinum(II) phosphorescent materials for the fabrication of red electroluminescent devices, *Adv. Func. Mater.*, **15**, 223–229.

Chapter 6

Synthesis of Light-Emitting Complexes

Susana Braga[a] **and Luiz Pereira**[b]
[a]*Department of Chemistry and CICECO, University of Aveiro, Aveiro, 3810-193, Portugal*
[b]*Department of Physics and i3N, University of Aveiro, Aveiro, 3810-193, Portugal*
sbraga@ua.pt, luiz@ua.pt

The development of novel rare earth and transition metal complexes for employment in OLEDs, though a remaining challenge to the present day, comprises a series of fundamental routes, quite useful for the preparation of basic precursors and to act as basics of novel synthetic pathways for more elaborate complexes. After the examples and discussions on structure activity relations of these complexes presented in the previous chapters, it is pertinent to introduce some details on their preparative methods. In this chapter, the basics of the chemical syntheses of rare earth and transition metal complexes are briefly described. Routes involving well-known europium and ruthenium complexes are used as examples.

6.1 Lanthanide-Based Emitting Complexes

6.1.1 General Background

Rare earth chemistry was developed in the early decades of the twentieth century, essentially with the isolation, purification, and molecular weight determination of yttrium and the main

Organic Light-Emitting Diodes: The Use of Rare-Earth and Transition Metals
Luiz Pereira

www.panstanford.com

lanthanides. At that time, europium was obtained from minerals such as the Brazilian monazite in the form of a crude oxide. The raw product was subject to a series of purifications and fractionings to eliminate other RE and metal oxides, made reactive with hydrochloric acid, and precipitated to form crystalline salts of $EuCl_3$ [1]. Today, the chloride salt is easily available from commercial suppliers.

6.1.2 General Method for the Preparation of Lanthanide Diketonates

Most complexes used as emitting materials in OLEDs are based on β-diketonates of europium or other lanthanide salts. Their preparation is relatively simple and follows a common procedure established in the 1950s [2]. Posterior modifications were introduced, aiming at reducing the toxicity of the solvents and reactants employed or deeming to meet the solubility requirements of more specific ligands.

Thus, and using europium as a model, the preparation of lanthanide diketonates can be made by treating an aqueous solution of the corresponding salt (in this case, $EuCl_3$) and the desired diketonate (dket) with an alkalizing agent, typically sodium hydroxide; for each mole of salt, three equivalents of base are required, plus the ligand in a slight excess, that is, around 3.3 moles. The resulting product can be recrystallized for further purification. For ligands that are less soluble in water, as for instance *1,1,1-trifluoro-4-phenyl-1,3-butanedione* (HBTA), a 50:50 mixed solution of water and ethanol can be used as the reaction medium [3]; other co-solvents may also be employed, depending on the specific solubilities of the ligands. This procedure leads to the isolation of a europium tris-β-diketonate with two water molecules completing the coordination sphere, i.e. $Eu(dket)_3(H_2O)_2$.

The presence of coordination water molecules in an emitting complex (EC), however, and as mentioned in the previous chapters, is highly undesirable due to the occurrence of non-irradiative relaxation processes. The lanthanide *acqua* complexes can thus be further modified by replacing the waters in the coordination sphere with a non-charged, or neutral, ligand.

Typically, planar aromatic *di-amines* are employed, as these have higher affinity toward the lanthanide ion, easily displacing the bound waters from their sites. The reaction is carried by a simple co-dissolution of equimolar amounts of the lanthanide diketonate and the neutral ligand in an appropriate solvent, usually ethanol, chloroform, or a mixture of both, with stirring for a few hours and solvent removal to isolate the new product.

Monodentate ligands such as aromatic rings with a single amine group can also be used to replace the waters as neutral ligands. In this case, two moles per each mole of $Eu(dket)_3(H_2O)_2$ are obviously required, one for each water position.

6.1.3 Alternative Methodologies and a Comparative Case Study

In recent years, some authors have successfully developed a one-pot method for the preparation of lanthanide ECs. The formed products may present themselves as different polymorphs forms those obtained by the two-step process for the same EC; in fact, polymorphism is common in these complexes and may be strongly influenced by the reaction conditions.

To illustrate this effect, the two different methods for the preparation of a case study, $Eu(NTA)_3(bipy)$, are described below.

(a) *One-pot preparation*. The ligands NTA and bipy in a 3:1 mole ratio were dissolved in the minimum amount of ethanol and slowly added with the stoichiometric amount of NaOH for the neutralization of the acidic proton of the β-diketone. A concentrated ethanolic solution of $Eu(NO_3)_3{\cdot}6H_2O$ (one mole equivalent) was added drop-wise, with stirring, over a period of 30 min. Precipitation was immediate and after 2 h of stirring, the light yellow product was isolated by filtration and dried. Recrystallization from $CHCl_3$/isopropanol yielded crystals with two different shapes, plates, and blocks, corresponding to the unsolvated form and $Eu(NTA)_3(bipy){\cdot}0.5$(isopropanol), respectively [4].

(b) *Two-step preparation*. Preparation of $Eu(NTA)_3(bipy)$ was achieved by first binding *1-(2-naphthoyl)-3,3,3-trifluoroacetonate* (NTA) to europium, and then using this product in a second reaction with *2,2'-bipyridine* (bipy). The first step of the preparation, comprising the diketone protonation and coordination to europium, was adapted from a procedure described by Charles and Perroto for $Eu(dbm)_3$ complexes [5]. In our case, HNTA (1.94 g, 3 × 2.43 mmol) was dissolved in 25 mL ethanol at 50°C and added dropwise with an aqueous solution of NaOH 1 M until a pH = 8 was reached. $EuCl_3{\cdot}6H_2O$ was, in turn, dispersed into 35 mL of water; this suspension was added to the alkalinized solution of the ligand. Upon mixing, a yellow suspension started to form, and vigorous stirring was maintained for another 4 h. The product was isolated by filtration and washed thoroughly with water to remove the remaining paste, allowing the isolation of a yellow powder. In the second step, $Eu(NTA)_3(H_2O)_2$ (0.344 g, 0.35 mmol) was dissolved in 15 mL of $CHCl_3$ and added with bipy (0.055 g, 0.35 mmol), also dissolved in $CHCl_3$ (5 mL); the mixed solution was stirred for 3 h at room temperature. The solvent was then removed and the solid washed with n-hexane and vacuum-dried [6]. Recrystallization from $CHCl_3$/diethyl ether yielding crystals suitable for X-ray diffraction, which showed that the molecule crystallized together with chloroform as $Eu(NTA)_3(bipy){\cdot}0.44CHCl_3$. The distances between the Eu and the nitrogen atoms of bpy (Eu–N = 2.57 Å) and the oxygen atoms of the three NTA groups (Eu–O = 2.315 – 2.437 Å) are similar to those of the other polymorphs. The main difference lies in the geometry of the coordination polyhedron, which is a distorted square antiprism. In the unsolvated polymorph, the polyhedron is a well-defined square antiprism, whereas in the isopropanol solvate the polyhedron is a distorted bicapped trigonal prism.

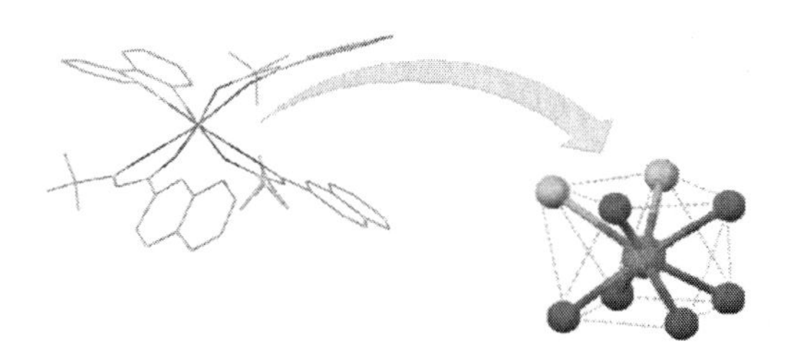

Figure 6.1 Crystal structure for $Eu(NTA)_3(bipy)\cdot 0.44CHCl_3$ (represented on the left) with focus on the coordination polyhedron around Eu^{3+} (right), shaped as a distorted square antiprism; for clarity, only the close-proximity atoms are shown, that is, the oxygens from the diketonates and the nitrogens from bipyridine.

6.1.4 Outlook

The choice of the preparative method for a particular lanthanide emitting complex will rely, ultimately, on the user's own preference. However, the advantages and disadvantages of each alternative must be carefully pondered before making a final choice.

In fact, although the one-pot synthesis method allows a fast preparation of the desired complex, uncertainties about the product purity may arise as the formation of by-products and contamination with unreacted starting materials bringing a higher degree of complexity in purification of the complex. The classical two-step synthesis, in turn, is more time consuming but allows the isolation, purification, and characterization of an intermediate *acqua* complex that will assure a quick and nearly 100% yield reaction with the neutral ligand.

Regarding the formation of polymorphs, the given example has shown that their formation is mostly dependent on the solvent used. Polymorphs are also known to exhibit slightly different luminescent behavior, so it is of vital importance to use the solvent that favors the formation of the most interesting polymorph as a carrier for the deposition of the complex onto an OLED layer. This way, the obtained film will be composed of the polymorph with the best luminescent features.

6.2 Transition Metal Complexes: Ruthenium-Based Emiting Complexes and Their Preparation

OLED research has been driven by the expectations of surpassing LEDs and liquid crystal displays by affording brighter and more flexible displays at a lower cost. The part of these objectives related to bright emission can be achieved using ECs based on Os(II), Ir(III), and Pt(II). Yet, these do not comply with the required low prices, thus being useless for commercialization. Ruthenium complexes have emerged amongst the many possible alternatives to precious metals that have been evaluated, mostly because they combine a lower cost with relatively straightforward syntheses. In fact, their flexibility and ease of use explains why the coordination chemistry of ruthenium, largely developed in the seventies, has continued to grow until the present day. Descriptions on the preparation of a plethora of complexes are available in the literature, including those with inorganic amines, ethylenodiamine complexes, ketonates, and even amino acid coordination complexes. This section focuses on ruthenium(II) complexes of planar aromatic amines, the most widely used as ECs for organic electronics.

6.2.1 Ru(II) Complexes Synthesized From a Precursor

These two-step pathways may be deemed disadvantageous due to the extra preparative work, but also feature some positive points. The dimethylsulphoxide method is ideal for those discouraged by hazardous solvents and features a preparative step that is simple and very effective, yielding a precursor that may be used to prepare a plethora of products (some of which falling far beyond the scope of this book). In turn, the use of microwave radiation as heating source in chemical reactions brings huge reductions in synthesis times, allows the use of non-inert atmospheres and the preparation of very pure products in high yield, a strongly appealing set of advantages.

The two synthetic pathways are described as follows:

Via dimethylsulfoxide. The complex used as synthetic precursor is $RuCl_2(Me_2SO)_2$ (Me_2SO = dimethylsulfoxide or DMSO). It was first prepared by Rempel et al. by the prolonged interaction of ruthenium trichloride ($RuCl_3$) in DMSO under hydrogen at 80°C [7], and its synthesis was later simplified by Evans et al., that merely refluxed $RuCl_3 \cdot 3H_2O$ in DMSO for a few minutes [8]. Reaction of this precursor with two equivalents of 2,2'-bipyridine (bipy) or 1,10-phenanthroline (phen) yields, respectively, $Ru(bipy)_2X_2$ and $Ru(phen)_2X_2$, where X is a leaving group (Cl or DMSO). These complexes can be further modified with another aromatic amine to obtain heteroleptic complexes—metal compounds having more than one type of ligand—or with another equivalent of the same ligand to prepare the homoleptic complexes $[Ru(bipy)_3]^{2+}$ and $[Ru(phen)_3]^{2+}$.

By microwave-assisted synthesis. In this procedure, the $RuCl_3 \cdot 3H_2O$ salt is first converted to $Ru(cod)Cl_2$ (cod = cyclo-octadiene) and this intermediate reacts with bipy in DMF for 45 min under a 600 W microwave irradiation [9]. The downpointistheuseofDMF(dimethylformamide),atoxic(vapors irritanttomucosaandeyes,teratogenic,possiblecarcinogen)and flammable solvent (self-ignition at 445°C).

6.2.2 Preparation of Ru(Ii) Complexes from the $RuCl_3$ Salt

This synthetic pathway permits the direct one-pot preparation of $Ru(bipy)_2Cl_2$ from the commercially available $RuCl_3$, but also requires DMF as the solvent. For each mole of $RuCl_3 \cdot 3H_2O$, 2.02 moles (one percent excess) of bipy are stirred in DMF with heating at reflux for 8 h, in the presence of 0.1% LiCl; reported yields were of 65–70% [10]. The same method can be applied to prepare complexes of the formulae $Ru(NN')_2Cl_2$, where NN' is, for example, *1, 10-phenanthroline, 4,4'-dimethyl-2,2'-bipyridine,* or *3,4,7,8-tetramethyl-1,10-phenanthroline* [11].

6.2.3 Recent Adaptations and Improvements to the One-Pot Method

An alternative method using non-toxic solvents was described for the one-pot preparation of $[Ru(bipy)_3]^{2+}$, involving reduction of Ru(III) from the reactant $RuCl_3{\cdot}3H_2O$ to Ru(II) species, commonly known as "blue ruthenium" [12]. This reaction seems to occur spontaneously under reflux in hydroethanolic solution, forming a "hot blue solution" that is then added with three equivalents of bipy in ethanol. Stirring of this mixture will result, over time, in color shifting from blue to deep red, which indicates completion of the reaction. Precipitation with $NaClO_4$ led to the isolation of $[Ru(bipy)_3][ClO_4]_2$. Replacing bipy with phen in the same procedure allows the preparation of $[Ru(phen)_3][ClO_4]_2$ [13].

References to the ruthenium blue solution are also made in the preparation of other ruthenium complexes. For instance, this method was successfully adapted to the preparation of ruthenium complexes with the tridentate ligands *tert-pyridine* and *biquinolylpiridine* [14].

6.2.4 Other Ru(II) Complexes

Ruthenium can form ECs with a plethora of other ligands, at times by complicated reactions that require inert atmosphere and have numerous intermediate steps.

Interest in these complexes as ECs has lost spark, mostly due to the costly and long-preparative pathways, and perhaps more importantly, because the electroluminescent features are not much different from those of $[Ru(bipy)_3]^{2+}$. For instance, Ru complexation with the monodentate ligand pyridine (py) to yield $[Ru(py)_6]^{2+}$ requires the preparation of $RuCl_2(PPh_3)_3$ (PPh = triphenylphosphine) from $RuCl_3$, followed by reflux in methanol to yield $[RuH(H_2O)_2(methanol)(PPh)_2]^+$. Finally, this intermediate complex reacts over a 24 h period with the pyridine ligand, under nitrogen, to obtain the desired product [15]. However, the electrochemistry of $[Ru(py)_6]^{2+}$ is nearly identical to that of $[Ru(bipy)_3]^{2+}$ ion and its MLCT band (at 345 nm) produces an excited state crudely resembling ruthenium(III) and reduced pyridine. This example illustrates how the search for new Ru(II) ECs can be discouraging when performed by trial-and-error and how the molecular design, based on a careful choice of ligands, is vital for the good luminescent performance of newly synthesized ruthenium ECs.

A successful example of rational molecular design is the neutral complex *trans*$(ifpz)_2Ru(PPh_2Me)_2$ (ifpz = *3-trifluoromethyl-5-(1-isoquinolyl)(pyrazolate* and PPh_2Me = *diphenylmethylphosphine*). The chosen ligands promote an all *trans* coordination geometry which contributes to a bright saturated red emission, both in the solid state (maximum at 632 nm, with Φ= 0.24) and when the complex is incorporated into an OLED. In fact, among other similar phosphine complexes prepared by the same authors, that with *cis* geometry has reduced quantum efficiency and produces an OLED of inferior EL properties [16].

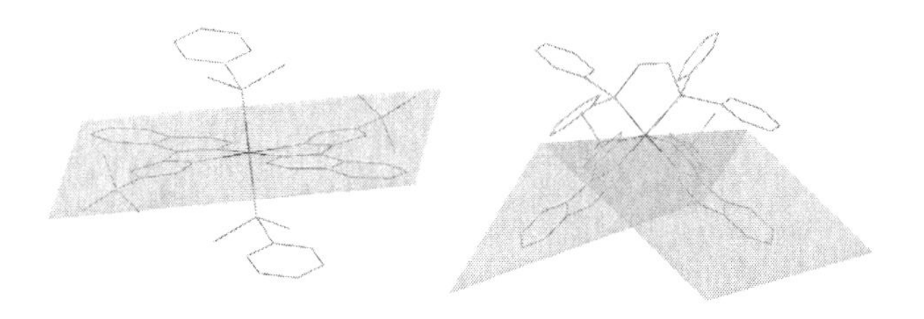

Figure 6.2 Crystal structures of: left, a *trans* complex, $Ru(ibpz)PPhMe_2$, showing the plane formed by the ligands with the phosphine groups in an equatorial position, and right, a similar complex with *cis* coordination, $Ru(ibpz)Ph_2PC{=}CPPh_2$, showing how the ibpz ligand planes are perpendicular to one another, with negative influence on the overall quantum efficiency of the complex; ibpz denotes *3-tert-butyl-5-(1-isoquinolyl) pyrazolate*. Contents of this and the previous figures were drawn from the atomic coordinates of the complexes using the software package mercury [17].

References

1. Meyers, E. L. (1935) Observations on the rare earths. XLIII. I. The atomic weight of europium. II. The specific gravity of europium chloride, *J. Amer. Chem. Soc.*, **57**, 241–243.

2. Moeller, T. (1956) Observations on the rare earths. LXVII. Some observations of non-aqueous solutions of some acetylacetone chelates, *J. Inorg. Nucl. Chem.*, **2**, 64–175.

3. Ismail, M. (1969) Preparation and properties of lanthanide complexes of some β-diketones, *J. In. Nucl. Chem.*, **31**, 1715–1724.

4. Thompson, L. C. (1998) Isomerism in the adduct of tris(4,4,4,-trifluoro-1-(2-naphthyl)-1,3-butanedionato)europium(iii) with dipyridyl, *J. Alloys Compd.*, **275–277**, 765–768.

5. Charles, R. G. (1964) Rare earth dibenzoylmethides preparation, dehydration and thermal stability, *J. Inorg. Nucl. Chem.*, **26**, 373–376.

6. Fernandes, J. A. (2006) β-Cyclodextrin inclusion of europium(III) tris(b-diketonate)-bipyridine, *Polyhedron*, **25**, 1471–1476.

7. James, B. R. (1971) Ruthenium halide dimethylsulphoxide complexes from hydrogenation reactions, *Inorg. Nucl. Chem Lett.*, **7**, 781–784.

8. Evans, L. P. (1973) Dichlorotetrakis(dimethyl sulphoxide) ruthenium(II) and its use as a source material for some new ruthenium(II) complexes, *J. Chem. Soc. Dalton Trans.*, 204–209.

9. Rau, S. (2004) Efficient synthesis of ruthenium complexes of the type $(R\text{-bpy})_2RuCl_2$ and $[(R\text{-bpy})_2Ru(L\text{–}L)]Cl_2$ by microwave-activated reactions (R: H, Me, tert-But) (L–L: substituted bibenzimidazoles, bipyrimidine, and phenanthroline), *Inorg. Chim. Acta*, **357**, 4496–4503.

10. Sullivan, B. P. (1978) Mixed phosphine 2,2'-bipyridine complexes of ruthenium, *Inorg. Chem.*, **17**, 3334–3341.

11. Giordano, P. J. (1978) Excited state proton transfer of ruthenium(II) complexes of 4,7-dihydroxy-1,10-phenanthroline. Increased acidity in the excited state, *J. Am. Chem. Soc.*, **100**, 6960–6965.

12. Mercer, E. E. (1971) The blue ruthenium chloride complexes and their oxidation products, *Inorg. Chem.*, **10**, 2755–2759.

13. Togano, T. (1992) One-pot and selective synthesis of a series of [$RuCl_6$-2nLn] (L= bidentate ligand, n = 0–3) types of complexes with polypyridyl ligands; another example of the synthetic utility of 'ruthenium-blue' solution, *Inorg. Chim. Acta*, **195**, 221–225.

14. Klassen, D. M. (1975) Synthesis and spectroscopic characterization of [2,6-di(2'-quinolyl)pyridine] complexes of ruthenium(II) and osmium(II), *Inorg. Chem.*, **14**, 2733–2736.

15. Templeton, J. L. (1979) Hexakis(pyridine)ruthenium(II) tetrafluoroborate. Molecular structure and spectroscopic properties, *J. Am. Chem. Soc.*, **101**, 4906–4917.

16. Tung, Y. L. (2006) Orange and red organic light-emitting devices employing neutral Ru(II) emitters: rational design and prospects for color tuning, *Adv. Funct. Mater.*, **16**, 1615–1626.

17. Bruno, L. J. (2002) New software for searching the Cambridge Structural Database and visualising crystal structures, *Acta Cryst.*, **B58**, 389–397.

Chapter 7

A Final Touch: New OLEDs — Research Perspectives

Without doubts the rare earth and transition metal organic complexes for OLEDs active layers become more and more attractive far beyond the academic scientific research. Aiming the possible technological applications, the industry looks in a different perspective for such kind of organic light emitters, each of them with specific (and proper) advantages that, in general, are not competitive: the RE complexes are typically focused in the "quasi-monochromatic" emission while TM complexes take the "high brightness" "market. The discussions made in this book are directed for these two pathways, trying to explore the foundations of each one independently. None of them is the better. Both are strongly important in "its niche" of possible applications with specificities that don't allow any overlap.

7.1 Rare Earth and Transition Metal-Based OLEDs —The Desire of New Improovements

Inevitably the improvement of the RE- and TM-based OLEDs will require a strong effort in three directions: novel organic ligands, development of new device structures, and fine tune of the processing methods.

Organic Light-Emitting Diodes: The Use of Rare-Earth and Transition Metals
Luiz Pereira

www.panstanford.com

Regardless of some sparse research, the actual route for OLEDs development must comprise an inter-disciplinary work involving all the aforementioned topics. As a matter of fact, everyday more interconnected fundamental research/technological applications fields tend to adopt that insight. Still, some aspects must be stressed as fundamental and need to be further explored. These aspects can be summarized as follows:

- New luminescent complexes with high internal efficiency and with much high chemical stability. One of the main problems in OLEDs development is precisely the low stability of the emitting complexes that, in several cases, cannot be applied as active OLED layer, besides their strong luminescence and high internal efficiency. This problem arises essentially from the pathway initially employed for OLEDs development and that currently still be largely used: the complexes are firstly synthesized thinking in their superior photoluminescent properties and not in their processing for electroluminescence.
- Simplifying the OLED structure. It is strongly important for technological applications that the device architecture becomes simpler as possible. Naturally, this effort cannot compromise the device lifetime and performance. The optimal architecture is the three layer device and for success, the material synthesis must also addresses the HOMO and LUMO levels, not only for the emitting complexes but also for the HTL and ETL. We must search for better materials for electrical carrier confinement avoiding the efficient but complex architecture employing the electron and/or hole-blocking layers. In such effort, the polymeric materials traditionally used for host matrix in the energy transfer process inside the OLED active layer must also be eventually re-designed, in order to accommodate this new strategy;
- Development of new (or specifically adapted) physical models for OLEDs behavior. Though such field is already

under significant research efforts, much work needs to be done to create effective models that comprise the very complex OLED optical and electrical mechanisms. The optimization of the actual theoretical frameworks must contribute to a better understanding of the OLED physical nature, applicable to a large used device architectures, in order to produce a useful feedback for further developments;

- In spite of novel extraordinary results, almost of them irreproducible, much efforts need to be applied in the fundamental studies concerning important topics, especially for new emitting complexes and the simplest OLED architecture that can be made with them: device degradation, lifetime, structural defects and their influence of device behavior, interfacial effects on both organic material—electrode and organic—organic regions, electrical carrier transport, and electroluminescent mechanisms.
- Improve the device efficient, under the new concept of simpler OLEDs structures, with design and synthesis of novel ETL and HTL materials that simultaneously must help on the electrical carrier balance and device architecture simplification. The goals must also comprise the operating voltage reduction to low as possible and the optimization of the light extraction, this last with strong relationship with the optimization of the thin organic layers deposition.
- For emitting materials that can successful be used in OLEDs, important studies can be made by understanding the internal physical process, aiming the molecular re-design to increase the performance. Typically, such studies comprises the photo-physical process like the energy transfer and annihilation mechanisms, internal molecule mechanisms involving the different spin multiplicity excitation/de-excitation pathways and photo-degradation mechanisms.

- Optimization of the deposition process. Actually many know the influence of the organic thin layers morphology on the device performance, so further improvements can be made on this field. Particularly, everyday more interesting printing methods, only available for wet depositions as already used in technological applications based on polymer-LEDs, must be adapted to the novel emitters. In each more focused application of OLEDs in general and architectural environment lighting, the deposition techniques must produces, in large areas suitable for panel wall and roof illumination, a homogeneous and uniform organic thin film layers. Actually some large-area and fast systems such as R2R (roll-to-roll) and large in-line evaporators are currently under development and test. This problem should now be addressed and clearly is at the top ten priorities for marked applications. One relevant problem associated with this large emitting area that can be physically addressed is the relative unknown effects of large electrical currents and low applied voltages. Here, both experimental and theoretical research is needed.
- Finally, the high desired flexible emitter. If for rigid device the technological application already (in principle!) overcomes the major encapsulating drawbacks, the solutions for flexible devices appear far from the required market needs. Several experimental solutions have been tested but with not much impact. This question is strongly relevant: organic emitters are, perhaps, the unique ones that fulfill all the required properties to be employed in a real high flexible light emitting device; but the fast degradation under normal atmospheric conditions implies stronger efforts to found an encapsulation solution that keeps the device flexibility.

Much of these questions are currently under attention and naturally, novel solutions will appear in the next OLEDs generation.

7.2 The Future of Rare Earth and Transition Metal-Based OLEDs

Rare earth and transition metal-based OLEDs arrived to a relatively mature stage for possible applications. Despite the different methods for brightness measurements, sometimes impossible to directly compare some results, it is accepted that this kind of metallo-organic light emitters will be incorporated in the next OLEDs generation. Several factors, extensively focused in this book, justify such sentence. For now, some new strategies can be suggested, aiming further improvements.

Much of the possible pathways for RE and TM complexes based OLEDs improvements are based essentially on two main areas: (1) general device improvement as focused in topics from last section; and (2) complex specific addressable points. Let us focus on the last one.

Though different mechanisms for light emission are available, both RE and TM complexes intrinsically depend on the organic ligands behavior. This dependence immediately suggests novel ligands.

Concerning the RE complexes the optimization of the ligand triplet T_1 state population must be considered as the main strategy. The energy difference between that level and the resonant RE ion excited level must be relatively low, avoiding the (i) very low population probability if T_1 falls below or (ii) non-radiative pathways if T_1 is much higher. On this molecular design, the prediction of those levels must be considered fundamental. Much attention must be paid on the neutral ligand, a possible non-radiative de-excitation channel. On the main molecule design, and besides the feasibility of the respective synthesis, the electrical nature of the ligands must be taking into account. Novel ligands, not necessary belonging to the β-diketonates family, must be explored. Some preliminary work shows the potentiality of tetrakis ligands, though some others are not impossible.

Also, a multi-nuclear complex, i.e. molecules with more than one RE ion, can be an interesting solution. Our actual work on a dimmer complex of europium with BTA ligands suggests a potential increase of the light emission by the presence of the two Eu^{3+} ions in the same molecule. This idea, though not completely now, requires special attention, because not all atomic arrangements are suitable to increase the whole efficiency. Three distinct situations can occur: (1) an independent RE ion emission; (2) a mixed emission; or (3) a cooperative emission. In the first case, the bond between the two RE ions is usually made using another atom in the middle, and no particular benefits are expected, comparatively to the emission of a complex with a single RE ion. In turn, this complex is very interesting that the two RE ions belongs of different nature, i.e. each of them can emit in different color. We report in Chapter 4 one example of this situation. The second case is perhaps the more problematic. Typically the two RE ions are directly bonded to each other but the distance between them are less than the required for an efficient blindage effect. The resulting emission is originated in mixed energy levels, with, *a priori*, no previsible results. The most interesting case is naturally the third one. The RE ions are also directly bonded, each of them emits *per se*, but in phase, i.e. cooperatively. The advantage of that molecular system needs to be clearly evaluated, and probably depends on the OLED architecture. Our first results are very promising.

Concerning the TM complexes, there are a lot of possibilities when novel ligands are synthesized. We must not forget that such complexes emit as a whole system and the mechanism is not addressable to the ligands or TM ion only. The imposed limits are, in a first observation, the resonance between the emitting levels and the MC levels. Though some much specific complexes can take advantage of that MC levels, the posterior application in OLEDs (when testes) exhibits very poor performance. So, such energy level resonance must be avoided. Also the ligand strength field must be taken into account. Considering that much work was done, actually the novel ligands design for TM complexes must specifically addressed further efficient application in OLEDs. As a matter of fact, much TM complexes are not successful employed in light emitting devices due to difficulties in the material processing. In a more particular TM complex design, actually one important field of work is the specific ligand fragment substitution, modifying the HOMO and/or LUMO levels, without completely redesign the ligand. This approach had shown very promising results. Besides such questions, the ligands' modifications in TM complexes must always comprises the improvement of the triplet harvesting concept, playing with the LC/MLCT "molecular" levels, and for color tuning.

In parallel, the LEC appears to be very promising in some technological applications. Clearly, ruthenium-based LECs are the most interesting as their delay time is, under some situations, acceptable. Unfortunately, at the moment, iridium does not show similar situation. For LECs, the minimization of the counterions appears to be an interesting solution.

A special note concerning the organic ligands: the final complex must be easily processed for device production, independently of the method. As expected, the intersection of these different desires can be a hard work. Each more, the inter-disciplinary work involving materials synthesis and device processing is needed.

As final consideration, the potential of technological application of OLEDs based on RE and TM complexes is expected to increase as more chemically and electrically stable complexes are synthesized. From quasi-monochromatic emission to very bright ones, covering all the main spectral colors, the actual interest is much more than simple curiosity.

Index